FINANCIAL AND ECONOMIC ANALYSIS FOR ENGINEERING AND TECHNOLOGY MANAGEMENT

FINANCIAL AND ECONOMIC ANALYSIS FOR ENGINEERING AND TECHNOLOGY MANAGEMENT

Second Edition

HENRY E. RIGGS
Professor Emeritus, Stanford University
Harvey Mudd College
Keck Graduate Institute

WILEY

JOHN WILEY & SONS, INC.

Library of Congress Cataloging in Publication Data:

Riggs, Henry E.
    Financial and economic analysis for engineering & technology
management / Henry E. Riggs.—2nd ed.
        p. cm.
Rev. ed. of: Financial and cost analysis for engineering and technology
management. c1994.
Includes index.
    ISBN 0-471-22717-X (Cloth)
  1. Accounting. I. Riggs, Henry E. Financial and cost analysis for
engineering and technology management.   II. Title.
    HF5635.R564   2004
    657—dc22

                                                    2003016098

Printed in the United States of America

10 9 8 7 6 5 4 3 2

CONTENTS

13 Analyzing Manufacturing Costs: Introduction to Cost Accounting / 294

14 Analyzing Other Operating Costs and Decisions / 348

15 Budgeting and Forecasting / 377

PREFACE

Financial and Economic Analysis is written for managers and aspiring managers—that is, decision makers, who must draw information from financial statements and accounting records of organizations with which or for whom they work. It is organized to serve as both a textbook and a professional book for practicing managers.

The purpose and contents of this book are substantially different from those of the plethora of other accounting and financial analysis textbooks on the market. First, the book assumes that you, the reader, have no previous background in accounting or finance—and, moreover, that this may be the last such book you ever study. Second, I assume you are not seeking to become an accountant (although some of you may get intrigued and decide to do so), but rather your goal is to be "financially literate"—able to read and understand financial statements and accounting reports in order to make better decisions. You want to be able to ask insightful questions about financial matters and to engage in thoughtful debate when accounting issues arise. You also need to know the vocabulary—even the jargon—of accounting. I hope you will develop both enthusiasm for, and healthy skepticism about, accounting reports. Third, in addition to the background in accounting that you will gain from the first 11 chapters, you will in the remaining chapters be introduced to the financial markets; receive a primer in corporate finance, including capital investment analysis; and gain some familiarity with managerial and cost accounting.

This book keeps to a minimum your involvement in bookkeeping. You need to know some of the key terms (like *debit*, *credit*, *ledgers*, and *chart of accounts*), but I assume you neither need nor want endless, tedious exercises in making accounting entries, developing journals, or drawing up financial statements. Accordingly, material in Chapters 6, 7, and 8 (relating to financial accounting processes) and Chapters 13 and 14 (relating to cost accounting) are substantially streamlined in this second edition as compared with the first.

The linkage between financial accounting and financial management is frequently obscure to students who take separate courses (or read separate books) in each. This edition seeks to integrate the two because you must integrate them in your role as a manager. Accordingly, in this edition, Chapters 10, 11, 12, and 15 are substantially new.

As with the first edition, this edition begins at "square one," asking you to consider what decisions you need to make and what financial information you require to improve the quality of those decisions. Unlike most accounting textbooks, it does not start by laying out a set of dogmatic rules. Instead, it leads you to understand that the key issue is valuation: valuation of what the company owes and owns. You will develop an understanding of why rules, or at least guidelines, are necessary and also why the rules we work with are imperfect. I also ask you to adopt different perspectives—a senior or mid-level manager in the company, a security analyst, a lender, an employee of the company, a tax collector, an investor—because different perspectives give rise to different questions and thus different informational requirements.

Following the introductory, context-setting chapter, Chapters 2 through 5 introduce financial accounting at the conceptual rather than the bookkeeping level, including the framework for accounting and alternative valuation methods. A chapter is devoted to each of the accounting system's primary "products": the balance sheet, the income statement, and the cash flow statement. Chapters 7 and 8 explore key valuation issues—from current assets, to fixed assets, to liabilities. Chapter 9 explains financial ratios that underpin financial statement analysis. As mentioned above, Chapters 10 and 11 explore the linkage between accounting and the financial markets and then invite readers to go deeper into financial statement analysis. Chapter 12 explores capital investment decisions (often called engineering economy). In Chapters 13 and 14, we shift gears to examine managerial accounting, with an emphasis on operating costs and the key cost accounting concepts. The book concludes with a chapter on budgeting and financial and cash forecasting.

At the end of each chapter are a summary and a list of key terms ("New Terms") introduced in the chapter. Also, for those of you who use the book as a text, each chapter concludes with exercises. Readers tell me they find these features particularly useful.

ACKNOWLEDGMENTS

The revisions incorporated in this edition owe much to the helpful feedback from the many students—from undergraduates at Stanford University and Harvey Mudd College, from graduates at Stanford and at Keck Graduate Institute, and from practicing managers in various executive courses—with whom I have worked, as well as from other faculty who have used the first edition. I am indebted to them all.

I'm eager also to acknowledge and thank Gayle Riggs, whose patience and encouragement are nearly boundless; my high school English teachers, who taught this engineer to enjoy writing; Delia Gurjala and Lynn Conception, who cheerfully labored through umpteen chapter revisions; Bob Argentieri, editor; and Susan Middleton, copyeditor extraordinaire.

CHAPTER 1

INTRODUCTION: WHY KEEP SCORE?

Accounting. You probably already have some conceptions about the subject of **accounting.** You may view accounting information as merely history, with little relevance for forward-looking management. You may think that accounting data are precise and represent indisputable facts about an operation; or you may feel that, since the accountants themselves often disagree, the results of their labors must represent little more than their individual opinions. You may equate accounting with your checkbook and monthly bank statement; or your savings account passbook; or your paycheck with its stub indicating amounts withheld from pay; or the record of receipts and expenses for your club, charity, or association. You may be accustomed to reading published financial statements of corporations, perhaps in connection with common stock investments you have made or contemplate making. Or, you may feel totally lost when confronted by these published financial statements, replete with large numbers and financial jargon that seem to speak only to the initiated.

Accountant. Again, the word may bring a variety of thoughts or images to mind. Your mental image may be of the Victorian clerk working for Ebenezer Scrooge, leaning over his desk, perched atop his tall stool. Alternatively, you may view the modern accountant as commanding a massive computer database with high-speed laser printers spitting forth financial facts at lightning speed.

Each of us is in contact with accounting information nearly every day of our lives. These frequent encounters—many of them baffling and some simply unpleasant—form our opinions on what accounting is and is not, what it can and cannot do for us or to us, and what accountants must be like.

For now, keep an open mind about both accounting and accountants. Those of you who are skeptics now may be persuaded that accounting information, properly analyzed and interpreted, can be of substantial benefit. Those who are now enthusiastic believers in accounting's value may, after further study, want to temper that enthusiasm. You may conclude that accounting information, like most other information, has its limitations.

USERS OF THIS BOOK

This book is addressed to present and future *users* of accounting information rather than to budding accountants. This user—you—may now be a business, engineering, or liberal arts major in college or graduate school; or a manager (or aspiring manager) in business, industry, government, or in one of the many other types of organizations, both public and private, that dominate our society. For every professional accountant, there are many others of us who need to understand basic accounting concepts because we must interpret accounting reports and make use of financial data both in our personal lives and in our jobs.

Some of you will proceed no further in your study of accounting or financial control than this book; for you, this text should provide both a satisfactory introduction and a conclusion. You will gain an understanding of the fundamental concepts, the framework of accounting, and the managerial uses of accounting information. Importantly, you will also come to appreciate the inherent limitations of accounting techniques and, therefore, of the resulting information.

For those of you who will continue to study accounting, this book provides basic knowledge upon which to build both greater technical skills and greater sophistication in financial analysis. Of course, this book will awaken in a few readers an interest in accounting as a career. If you are one of these, this discovery may come as a surprise, particularly if you were reluctant to crack open this book.

This book is quite unlike traditional accounting books. First, it assumes that you are or will be a manager, not a person preparing for a professional accounting career. Thus, it does not instruct you in bookkeeping. Second, the book seeks to develop within you a healthy skepticism regarding the precision and reliability of accounting data, as well as an appreciation of the benefits of accounting reports. Third, it omits discussion of a number of subjects of interest to professional accountants but of marginal usefulness to managers.

Fourth and most importantly, this book is nontraditional in its organization. It asks you first to consider the need for, and purposes of, financial information before studying a set of dogmatic accounting rules. Accounting can be taught, and often is, by a pedantic presentation of rules and procedures that you are asked to apply to a set of simplified and generally unrealistic exercises. The rules and procedures are slow to sink in if you have little appreciation of the condition or dilemma that gave rise to them. Moreover, if you don't wrestle with the dilemmas, you are likely to resist the rules as being unnecessarily arbitrary.

So this book will create in you a need to know by exploring financial data requirements, first in the area of general, or financial, accounting and then later in cost and managerial accounting. This flow is intended to make you aware that there is a need for financial information, that a body of knowledge has developed for compiling this information, and that some quite arbitrary rules and procedures are necessary to permit the accountant to communicate with his or her audiences.

This book's purpose is to make you a better user of accounting and financial information—gaining all such information that you can, while being mindful of its shortcomings and limitations.

SCOREKEEPING

Accounting is analogous to scorekeeping. Managers in business are interested in the financial score; managers of athletic teams are interested in the game score. Nonmanagement members of the business enterprise, like their athletic counterparts, the players, are also interested in the score. Besides the active participants, a host of other individuals are also interested in the score—in athletics, the fans, the sportswriters, and the concessionaires selling beer and hot dogs; in business, the shareholders, the bankers, the customers, and the government.

We are interested in the score for a variety of reasons, but most importantly because it helps us make better decisions. Managers—whether in business or athletics—change strategies and tactics depending on the score. The score can cause customers, employees, investors, and sports fans to renew or change their allegiances to a company or team.

DEFINITION OF ACCOUNTING

A useful definition of accounting, or financial scorekeeping, must both describe and circumscribe the tasks and responsibilities of accountants. It must tell us both what we can expect from our scorekeeping system and what kinds of information or insights are simply beyond its reasonable scope.

Accounting Is Historical

The score of an athletic contest gives the results of the game thus far, not an estimate of what may take place in the next period, inning, or minute; similarly, accounting information shows only what has gone on in the enterprise up to the date of the reports. Thus, an accountant is a historian.

Of course, history is relevant to the future. Plans for virtually all human endeavors—business, athletics, and international diplomacy, for example—are better formulated after a careful analysis of history. You should recognize, however, that accounting cannot provide a crystal-ball view of the future. Sales forecasts, production plans, and expense budgets—all prospective and all describing what we expect, hope, or plan to occur in the future—are not in the accounting records. On the other hand, the best financial forecasts, plans, and budgets build on solid knowledge of the past as revealed by the historical financial score.

Accounting Measures and Reports in Monetary Terms

Much about an enterprise cannot easily or usefully be expressed in **monetary terms.** These events are not less important than those that can be expressed monetarily, and they are no less relevant to the enterprise's future. Hiring a respected design engineer may be much more significant for the company than purchasing a new piece of laboratory equipment; however, we have a good deal less trouble valuing laboratory equipment in monetary terms than we do a newly hired engineer. Similarly, obtaining an appointment with the president of a major potential customer company may be more significant than shipping just another order to a present customer, but clearly the first event is more difficult to value in monetary terms (since an appointment is not a sale) than is the second (for which we can value both the sales transaction and the value of the goods shipped).

Accounting Relates to a Defined Entity

The **entity** may be a retail shop, an industrial concern, a church, the city government, the local fire department (a segment of the city government), the senior high school, a parish affiliated with an archdiocese, your family, or just you, an individual. But in each case, the entity must be precisely defined. Then, both the accountant and the user of the accounting information must not confuse that entity with other entities of close association or particular affinity.

For example, when you account for a partnership (say, a law firm), you account for the partnership's activities and not the personal activities of the partners. The partners probably have other incomes, and they surely have other expenses, but these are not incomes or expenses of the partnership. If the accounting entity is your family, you include the earnings and expenses of all family members, and the result is different than if the accounting entity is you alone. If you are accounting for your parish, you are concerned only with that organization—not the archdiocese and certainly not the worldwide Catholic Church. And you should make sure that you include all activities of this parish, including, for example, the membership committee and the social committee.

For accounting purposes, a large enterprise is frequently broken up into elements, each treated as an accounting entity. For example, normally a single division of a large, multinational industrial enterprise is an accounting entity; the division manager and his or her associates need to know the division's score separate from the score of the larger enterprise. Of course, the score must also be kept for the overall enterprise, but that can be readily accomplished by combining divisional accounting information.

Accounting Is an Action Process

What verbs define accounting actions? The first verb that comes to mind is *record*. We speak of accounting records, and therefore the accounting process must record what goes on in the entity. But, before the goings-on can be recorded, they must be *observed:* what of monetary significance is happening? Once observed, the goings-on must be translated into monetary terms so they can be recorded; they must be *measured* in monetary units, that is, valued.

Once the goings-on are recorded, however, the accountant's job is not completed. A sequential log of all that has transpired in a business might satisfy certain legal requirements for records, but it would be so long and unorganized as to be unreadable. Thus, the accountant *classifies* and then *summarizes* the data, so that data become information.

To recap, the action verbs in the definition of accounting, listed in the order in which the actions are taken, are **observe, measure, record, classify,** and **summarize.**

Now, to what are these actions directed? What kinds of goings-on need to be observed, measured, recorded, classified, and summarized? Surely, transactions to which the entity is a party: for example, making a sale, paying a bill, receiving a check from a customer, lending or borrowing money, paying the payroll, acquiring inventory, buying a machine tool. Can we concentrate solely on transactions between the entity and other, outside entities? Not quite. While such transactions trigger the great majority of accounting entries, we also need to take accounting action to recognize changed conditions internally, changes that have not—at least not yet—involved transactions with outside persons or organizations. Consider these several examples of occurrences or changed conditions:

- A piece of production equipment grows older and, thus, becomes less useful to the company.
- Certain inventory items become obsolete.
- The company's obligation to pay salaries or taxes arises before the date when they must actually be paid.

Stringing these elements together, then, we arrive at the following definition of accounting:

Accounting is the process of observing, measuring, recording, classifying, and summarizing the changes occurring in an entity, expressed in monetary terms, and interpreting the resulting information.

The final clause of this definition—"interpreting the resulting information"—is, of course, not solely the accountant's task; managers and all others interested in the financial score of the entity also interpret the resulting information. Nevertheless, this final clause is included in the definition to emphasize that part of the accountant's job is to extract meaning from the recorded history. Indeed, this book places major emphasis on interpretation.

LIMITATIONS OF ACCOUNTING INFORMATION

To many people the definition of accounting suggests that the accounting score tells the entity's full story. Such is most assuredly not the case. Accounting information is limited both in scope (or completeness) and in precision. Too frequently, readers of accounting reports assume they have in hand the truth about the organization's financial state. Even worse, they believe the reports present the *whole* truth. Unfortunately, some accountants, perhaps understandably, do little to dispel these erroneous impressions.

The first limitation arises because observing and recording is limited to those activities—transactions or changed conditions—that can be expressed in monetary terms. As mentioned earlier, much of importance cannot be so expressed. Second, we limit our attention to history, not to forecasting the future. Third, we focus on a particular entity, but that entity will be much affected by what goes on outside—for example, in the economy as a whole or within a competing organization.

Finally, the process of observing and measuring inevitably demands estimates, assumptions, and compromises. This situation is aggravated by the fact that the accountant is always working under time constraints, since the usefulness of accounting information is, in part, a function of its timeliness. For example, consider the following dilemmas. How precisely can you value obsolescing inventory? Can you be certain that no one has entered into an agreement that obligates the company without informing the accounting department? Can you be certain that all your customers will pay their bills? Is it not likely that reasonable people will disagree on the precise value to your company of a five-year-old piece of machinery?

In truth, accounting can only *estimate* the financial state of the enterprise. Don't expect the accounting documents to reflect all that is important about the enterprise—even its history—or to reflect that history in absolutely indisputable terms. Because accounting reports are often presented in very precise terms—down to the hundredth part of a dollar—they imply a higher level of precision than they deliver. It is possible to be precisely wrong; it is also true that vague accuracy can be highly useful!

AUDIENCES FOR ACCOUNTING INFORMATION

The accountant serves many constituencies; all those people and institutions are interested in an entity's financial score. Accounting reports address multiple audiences. Who are they, and what do they want to know?

Managers

Managers make the most extensive demands for financial scorekeeping. Charged with the responsibility for both day-to-day operations and long-range direction, managers are concerned with a broad array of issues and questions. They need rapid feedback to determine whether the company is operating according to plan. Are sales and expenses on target? If not, where are they out of line? Managers face a host of marketing decisions. Should a product be added? Should another be deleted? Should a particular order be accepted or declined? What price should the company bid on job X? Should it adjust prices on product line Y? Should it increase or decrease sales promotion expenditures?

In addition to using information to improve operating decisions, managers must also monitor the current financial health of the company. How much cash does the corporation have? How much do customers owe the company? Are they paying on time? How much does the company owe its suppliers? Is it paying them on time? How much inventory does it hold? Does the company have the ability to borrow additional money from the bank?

Furthermore, managers should stay alert to the messages that the company's financial statements are conveying to the following other audiences interested in the company's financial score.

Investors

The corporation's present shareholders are, of course, already investors, but any persons or institutions who might buy shares of the company's stock are potential investors, and both groups seek financial information. Is the company profitable? Is it sufficiently profitable to sustain or increase the dividend? What are the risks that the company will become bankrupt?

Security analysts and brokers who advise present and potential investors are an important audience. Indeed, the market prices for a company's securities are determined by the relatively few professional investors and analysts who become thoroughly knowledgeable of the company's financial score. Thus, this audience is both highly demanding and critically important.

Creditors

When considering a company's creditors, you probably think first of the bank. In addition, trade suppliers—companies selling on credit to the company—

are also creditors. All creditors are interested in being repaid on time and, in the case of formal loans, being paid interest; but they do not share, as investors do, in the future growth and profits of the company.

The sooner the creditor expects to be repaid—that is, the shorter the maturity of the credit—the greater the creditor's focus will be on immediate financial position (*liquidity*) and the less the concern with longer-term prospects. Conversely, if repayment is scheduled for the distant future—that is, if the loan has a long maturity—the lender is increasingly concerned with the company's long-term prospects, since the long-term lender may not be repaid if the borrower incurs a string of loss years and fails as a viable business.

Tax Collector

Various taxes are incurred by the corporation as a function of sales, profits, property owned, payroll, and occasionally other factors as well. In the United States, taxes on profit—income taxes—are exacted by the federal government, certain state governments, and a few municipal governments. Typically, property taxes are collected by the county or municipal government, while sales and payroll taxes are imposed by several levels of government—federal, state, county, and municipal.

Accounting records are used to calculate and subsequently verify tax liabilities. Government tax auditors routinely demand access to accounting records to verify the correctness of the company's tax payments.

Two preliminary comments regarding taxes are in order. First, in virtually all cases, corporate managers and individual taxpayers should and do strive to minimize current taxes of all kinds. This effort leads both to postponing the payment as long as possible and to taking advantage of every tax law provision to reduce taxes. Note the world of difference between avoiding unnecessary taxes and evading required taxes! Be aware, too, that the exact applicability of a tax law (particularly an income tax law) is often unclear. Thus, a company's tax liability may not be determinable with certainty. An adversarial relationship between taxpayers (corporate and individual) and tax collectors is almost inevitable.

Second, tax laws must not dictate accounting practice, particularly accounting for profits. Governments enact income tax laws to raise revenue and, in certain instances, to further national economic goals by providing incentives to taxpayers. Thus, the definition of *profit* in income tax laws is typically at variance to some degree with a definition that is useful to managers and investors. As a result, a company normally has to maintain in its accounting records certain information solely for the purpose of calculating its tax liability.

Others

Aside from these primary audiences, many others are interested in the financial score. Customers are interested in the company's financial viability, particularly if they are dependent on the company as a long-term supplier. Employees have similar interests; a growing, profitable, and financially strong company is an attractive employer. Labor unions, as representatives of the employees, have an avid interest in financial results, particularly just prior to contract negotiations, when the union is formulating demands based at least in part on what it feels the company can afford. Managers at competing companies are also eager readers of published financial statements.

Government regulatory bodies look to the company's accounting records for information. Companies are, of course, subject to varying amounts of regulation, depending primarily on their lines of business. Public utilities (electric power, natural gas distribution, telephone, and so forth), some transportation companies, and most broadcasting companies (radio and TV) are regulated by both federal and state agencies. All U.S. companies whose securities are actively traded in organized markets (for example, the New York Stock Exchange) are subject to financial reporting requirements stipulated by the federal Securities and Exchange Commission, and both federal and state securities agencies assess the financial information supplied in connection with the sale of newly issued securities. Antitrust legislation constrains the actions (particularly pricing actions) of even relatively small concerns. In their interactions with each of these regulatory bodies, managers rely to a great extent on financial information obtained from accounting records to advance the company's cause or to defend its position.

DIFFERENT MESSAGES FOR DIFFERENT AUDIENCES

This discussion should convince you that (1) different audiences are interested in quite different types of financial information, and (2) a company may legitimately wish to convey quite different messages to different audiences.

For example, while wanting to report a strong profit picture to shareholders, management seeks to minimize taxes calculated as a percentage of profits. The desired message to creditors may be that the company is in a strong financial position, entitled to more liberal credit terms, just when the desired message to the labor union is that the company cannot afford substantial wage or fringe benefit increases.

This book emphasizes frequently that accountants have considerable latitude in reporting both profit and financial position. Consider again the example of obsolescing inventory. If the accountant is relatively pessimistic about the value of this inventory, both profit and financial position will be reported less glowingly to both shareholders and the income tax collector than if the accountant takes a more optimistic view of its value. Of course, this and similar situations present opportunities for deception or, far worse, fraud. But no matter how honest and objective the accountant attempts to be, judgments are colored by the conscious or unconscious consideration of the message that the resulting financial reports convey to key audiences. After all, the accountant is subject to the same human frailties and is buffeted by the same human motivations that affect us all!

ACCOUNTING IN THE WORLD OF BUSINESS

The accounting function in business is a service function, not an end in itself. The accounting department does not create sales, fabricate products, or engineer new products. It keeps score and provides information and analyses. Within a business context, accounting is useful to the extent that it assists managers in achieving the company's objectives. An important (although not necessarily the prime) objective of business is to earn a profit.* The accountant keeps score with respect to that objective.

Most businesses, of course, articulate other objectives to accompany the profit objective. At the risk of provoking arguments about what should be business's objectives, we can say that managers pay attention to the following:

- Meeting budget
- Maximizing sales or market share
- Providing stimulating work and above-average compensation for employees
- Increasing economic power through sheer size
- Creating or maintaining technical leadership
- Developing enhanced reputation or prestige
- Perpetuating the enterprise for the benefit of employees, the community, the management, or the founding family
- Assuring that present management members retain their jobs

Accounting may not help much in keeping score on certain of these objectives. For example, try quantifying in monetary terms economic power or reputation or prestige. Remember, accounting reports tell only part of the company's total story.

The fundamental economic decisions that must be made by all economic enterprises involve the efficient allocation of available resources, since resources are inevitably scarce. These scarce productive resources are money, existing productive capacity, and labor, including technical expertise, management expertise, and human muscle. Many analytical techniques, some quite sophisticated and requiring extensive computing power, inform these resource allocation decisions. Much, though not all, of the data utilized in these mathematical models should be available from the accounting records—

---

*Actual behavior suggests that maximizing profits is not a widespread objective. Rather, managers seem to act as though they want their companies to be adequately (perhaps comfortably) profitable, but not necessarily maximally profitable.

for example, costs and revenues for certain products, for certain departments, or for certain geographic regions. Indeed, designers of accounting systems should have in mind data required by such analytical and planning tools as the following:

- Discounted cash flow analyses of alternative investment opportunities
- Simulation of segments of the company's operations
- Economic order quantity and other scheduling analyses
- Linear programming of the company's distribution activities

ACCOUNTING IN NONPROFIT AND GOVERNMENTAL ORGANIZATIONS

The primary focus of this book is on businesses—manufacturers, service organizations, and sales organization—that seek to make a profit. Of course, our society is replete with organizations that are not profit-seeking: government units, educational institutions, charities, certain hospitals, churches and synagogues, clubs, foundations, fraternal and service organizations, political parties, and consumer or farmer cooperatives. The accounting requirements, and therefore accounting systems, of these organizations are somewhat, but not fundamentally, different from those of profit-seeking companies. With different objectives, they require different sorts of data and reports to track progress. Yet all of them take in revenue and incur expenses. The well-managed among them operate on a plan, and the plan involves a budget in monetary terms. Their managers have the same needs as their profit-seeking counterparts to monitor actual financial performance with reference to the budget. Nevertheless, some accounting conventions are different for governmental, educational, or other types of nonprofit organizations.*

USEFUL NONACCOUNTING INFORMATION

While accounting information is limited to what can be stated in monetary terms, nonmonetary measures and records are vitally important to managers of both for-profit and nonprofit organizations. Thus, a caveat: You should seek out insightful *non*accounting measures of both performance and condition to supplement and amplify accounting information.

Consider again the analogy with athletic teams. Sports fans are concerned with other information about the team besides simply the score of the games played and the related statistics on individual team members. They are inter-

---

*In general, these organizations place a good deal more emphasis on the flow of cash, where profit-seeking businesses place more emphasis on the measurement of profit.

ested in the age and health of the athletes, and the depth of backup personnel for each critical position. They are interested in which athletes perform best under varying climatic, competitive, or time conditions.

Similarly, business managers are interested in such nonmonetary measures as the number of potential new customers contacted, the frequency of late deliveries, employee turnover, employee absenteeism, the market share vis-à-vis competitors, the educational background of technical and managerial personnel, the number of sales calls per salesperson per day, the percentage of contract proposals accepted, the percentage of production capacity utilized, and the quality yield. Indeed, most businesses define a handful of nonmonetary key indicators of performance that they monitor every bit as closely as accounting data. The most significant of these provide early warning signals—that is, foretell operating problems.

IMPORTANCE OF PERSONAL MOTIVATIONS

Managing is often defined as the process of planning, supervising, and controlling an organization in pursuit of its objectives. To repeat, this book focuses on accounting's role in providing data and analyses to assist managers, particularly in their efforts to plan and control.

Bear in mind however, that an organization is a collection of human beings. These people—middle- and lower-level managers, salespersons, scientists and engineers, clerks, machine operators, and everyone else in the organization—have their individual and collective objectives. Every student of human behavior knows that each of us is motivated to satisfy our own needs, whatever they may be. Accounting information may help us to satisfy certain of our needs (for example, you need to demonstrate that your department can operate on budget, and the accounting report verifies that it has), or it can threaten our need satisfaction (say, you committed to lowering overtime costs in your department, and the last accounting report shows no improvement). Inevitably, readers of accounting reports react based on their motivations; those reactions may be constructive to the organization's overall goals and objectives (for example, you are going to take corrective steps to increase sales of certain products because sales have been below target), or they may be destructive (for example, you will subcontract certain production activities at a sacrifice in total costs rather than incur additional overtime costs, since the high subcontract costs will not be charged to your department and management is on your back about overtime costs). Accounting reports can be used to threaten or coerce mid- and lower-level managers, or they can be used to provide feedback to managers to permit them to do a better job. More about these motivational issues in Chapter 15.

Accountants themselves are, of course, influenced in turn by how managers react to their reports. Too frequently, an adversarial relationship builds up between accounting managers and operating managers. If operating managers dispute the veracity of the accounting numbers, the accounting department becomes defensive. Operating managers may withhold unfavorable information from the accounting department. Accounting personnel may delight in highlighting areas of poor operating performance. By contrast, a healthy relationship exists when operating managers look to the accounting department for feedback and assistance, and the accounting department readily admits that accounting is not an exact science and that actual results may deviate from plan for perfectly valid reasons.

SUMMARY

Accounting is financial scorekeeping. More comprehensively, accounting is defined as the process of observing, measuring, recording, classifying, and summarizing the changes occurring in an entity, expressed in monetary terms, and interpreting the resulting information.

This book is primarily for managers, those seeking to gain accounting understanding to make them more proficient in using accounting and financial information. It is not intended for those who seek to learn basic bookkeeping skills.

Accounting is not a complete story—or history—of the enterprise. It ignores activities that cannot be measured and expressed in monetary terms. Furthermore, to a significant extent, accounting ignores events or conditions outside of the entity, even though they may importantly affect the future of the enterprise. Remember that nonmonetary, and therefore nonaccounting, measures of a company's activities and condition also can be both revealing and useful.

While accounting reports serve a myriad of audiences, the primary audience is the managers of the enterprise. Secondary, but very key, audiences include the present and potential investors in the company's securities, the company's creditors, and the various governmental taxing authorities. Other interested parties include customers, employees, suppliers, trade unions, government regulatory bodies, and competitors. The informational requirements of these audiences vary widely.

Accounting reports aid managers in tracking the progress toward one or more of the corporate objectives. Furthermore, accounting data are the key inputs for most analytical and planning techniques that assist decisions regarding the optimum allocation of the enterprise's scarce productive resources.

Accounting techniques for government, education, charity, and other nonprofit organizations are not fundamentally different from those for profit-seeking enterprises. However, informational requirements and certain accounting conventions do vary, since these nonprofit organizations pursue objectives different from those of private-sector businesses.

NEW TERMS

Accounting. The process of (1) observing, measuring, recording, classifying, and summarizing the changes occurring in an entity, expressed in monetary terms, and (2) interpreting the resulting information.

Classify. To categorize within the accounting system similar transactions, occurrences, or conditions.

Entity. The organizational unit for which the accounting is being performed.

Measure. To value in monetary terms the transaction, occurrence, or condition to be recorded in the entity's accounting system.

Monetary terms. The measure used in valuing all that is to be included in the accounting records

Observe. To determine what transactions, occurrences, or conditions have monetary implications to be recorded in the entity's accounting system.

Record. To evidence in the accounting system those transactions, occurrences, and changed conditions that have been observed and measured (valued).

Summarize. To combine and condense data in the accounting system in order to supply meaningful information to the various decision makers within and outside the entity.

EXERCISES

1. What are the five action verbs used to define the process of accounting?

2. Cite three common observable *transactions* that should be measured and recorded in the accounting records.

3. Cite three examples of observable changed *conditions* that should be measured and recorded in the accounting records.

4. Cite three common observable *events* in an industrial company that probably cannot be measured in monetary terms.

5. Name three sets of audiences interested in drawing information from an industrial company's financial statements.

6. Suggest three examples of nonmonetary (and therefore nonaccounting) data that would be useful to the operating managers in making decisions regarding the manufacture of artificial heart valves for humans.

7. Who are the primary users of accounting data and reports:
   a. Within the organization?
   b. External to the organization?

8. Indicate whether each of the following statements is true or false:

a. Probable future events have no bearing on today's accounting valuations.

b. Accounting records are only historical.

c. In the United States, income tax laws often define good accounting practices.

d. Most companies provide different financial statements to different audiences.

e. Accounting processes for nonprofit (not-for-profit) entities are, in general, the same as those for profit-seeking companies.

9. Describe the process by which you would go about measuring the value of the following:

   a. An inventory of men's shirts acquired 15 months ago by a department store

   b. A 10-year-old machine tool used by an automobile manufacturing company to fabricate parts

   c. A three-year-old automobile used by one of the company's salespersons

   d. A company's obligation to pay wages during vacation leaves earned by its employees

   e. The obligation to repay a $12,000 loan due in five years with interest payable annually at the rate of 8 percent

   f. A used truck recently purchased

   g. The exchange of a parcel of land owned by the company for a 15-year lease in a newly constructed facility

   h. The acceptance of a five-year interest-bearing note from a customer in settlement for the customer's past-due account

   i. The signing of a five-year employment agreement with the company's chief scientist

   j. The acceptance of an order for 120 units of product X requiring delivery at the rate of 10 units per month over the next year

   k. The filing by the company of a lawsuit seeking $200,000 in damages from a supplier that delivered inferior quality material

   l. The return to the company of defective goods previously supplied to the company's best customer

   m. The bankruptcy of the company's second largest customer

10. Why do investors buy shares of common stock of companies? Be specific.

11. You are a securities analyst for a major stock brokerage firm and have the responsibility for writing a report on the Schuster Corporation. You have just received Schuster's annual report containing the company's published financial statements. What specific questions can be answered by

a careful review of these statements? You now have an opportunity to interview the president of the Schuster Corporation for one hour. What are some of the questions you will ask the president? How do these questions differ from those to be answered by the financial statements?

12. You are a shareholder of the Schuster Corporation and have just received the latest annual report of the company, containing financial statements for the previous fiscal year. You have no access to key managers of the company. Where will you look for additional information about the company to decide whether to increase or decrease your share holdings in Schuster? How will you evaluate the reliability of this information?

13. What is meant by the phrase "accounting entity"? Give three examples of related yet separate accounting entities.

14. What different messages regarding its financial condition and profit performance might an industrial company wish to convey to (a) one of its major customers and (b) its shareholders?

CHAPTER 2

ACCOUNTING FRAMEWORK: THE CONCEPT OF VALUE

In Chapter 1, I defined the accountant's task as observing, measuring in monetary terms, recording, classifying, and summarizing the transactions and changed conditions in an entity. This chapter focuses on the first two action words in that definition—observe and measure—but in particular, on the second: measuring, or valuing, in monetary terms. The discussion of recording, classifying, and summarizing will be put off to subsequent chapters.

How does an accountant value what the enterprise owns, as well as its obligations? How is a monetary value assigned to the many and varied transactions to which the enterprise is a party? Some valuations are straightforward and relatively indisputable. Others present very real dilemmas. Before considering the three primary valuation methods, you need to understand the framework for incorporating values into accounting systems.

ACCOUNTING EQUATION

Every entity for which we account owns property and rights; it also has obligations that it must discharge. What the entity owns are called **assets;** what it owes—its obligations—are called **liabilities.**

Assets are defined as physical property, rights, and financial resources that hold the promise of providing ongoing future benefits. Examples are cash, securities, promises of customers to pay, inventory, production facilities such as machine tools and plant space, rights to use patents, and protection under a one-year insurance policy. Certain other important assets, such as customer orders and employment contracts with key personnel, are critically important but a good deal more difficult to value, as we shall see.

The company also has certain obligations—to its bank for money it has borrowed, to its vendors for inventory or services received, to its employees for wages and benefits earned but not yet paid, to its customers in connection with product warranty provisions, and to others as well. These obligations to outside persons* or organizations are referred to as liabilities. Most company liabilities are discharged by the payment of cash, but some are discharged by the performance of a service, such as warranty repair.

Thus, to repeat, *assets* are what the entity *owns,* and *liabilities* are what the entity *owes.* Both personally and as a manager, you feel one way about assets (you are better off to own them) and the opposite way about liabilities (you prefer to avoid them). Put another way, liabilities represent a call on the assets; to discharge some liabilities, you utilize the assets owned, most frequently cash. A healthy company owns more than it owes: the value of its assets exceeds the value of its liabilities. Moreover, the difference between what it owns and owes is a measure of the company's *worth.* This worth accrues to the benefit of a company's owners—in the case of a corporation, its shareholders, and in the case of a partnership, its partners. So, the greater this difference between assets and liabilities, the better off are the shareholders or partners collectively. This difference between assets and liabilities is **owners' equity,** and the fundamental **accounting equation** is

$$\text{Assets} - \text{liabilities} = \text{owners' equity.}$$

This statement must be true at all times. If the assets of the company increase with no change in its liabilities, owners' equity increases; that is, the owners collectively are in a better position. If, on the other hand, the liabilities of the corporation increase with no change in its assets, then owners' equity declines.

Using simple algebra, the equation can be restated as follows:

$$\text{Assets} = \text{liabilities} + \text{owners' equity}$$

This form is, in fact, the typical one in the United States. You may already be aware that most published financial statements in this country follow this format: assets are listed first (or on the left), while liabilities and owners' equity are shown below (or on the right), the two with equal totals.

Another way to describe the fundamental accounting equation is that the assets represent the entity's investments in physical property, intangible rights, and financial resources, while the liabilities and owners' equity represent the sources of funds used to make the investments. The left-hand side of the

---

*Are employees part of, or outside, the entity? While essential to the operation of the enterprise, they personally earn income and incur expenditures independent of the business entity and so each is a separate and distinct accounting entity.

equation shows what the company owns, and the right-hand side shows how the ownership of the assets was financed.

HOW IS OWNERS' EQUITY CREATED?

But how is owners' equity created? How can assets increase without an increase in liabilities, or how can liabilities decrease without a corresponding decrease in assets? Both events increase owners' equity. Two ways:

1. The owners can invest additional capital in the enterprise. In a sole proprietorship or partnership, the owners are the sole proprietor or the partners, and they simply agree to put more funds into the business. In a corporation, additional shares of stock are created and sold to investors.
2. The enterprise can earn a profit. Take the simple example of a company selling for $3 merchandise it bought for $2. The inventory asset decreases by $2, and the cash asset increases by $3. This $1 of profit adds to owners' equity.

On the flip side, it is easy to see the two primary ways that owners' equity is decreased:

1. The owners withdraw funds from the business (for example, by paying dividends to the owners of the corporation's common stock).
2. The company incurs a loss. If in the example above the merchandise purchased for $2 is sold for $1.50, the company has a loss of 50¢, which decreases owners' equity.

Therefore, the owners' equity of an enterprise is defined both as the difference between its assets and liabilities, and as

the sum of capital invested by the owners plus profits earned, less the sum of any losses incurred and any funds paid by the enterprise back to its owners.

Unfortunately, the meaning of owners' equity often gets confused. What is it *not?* First, owners' equity is not a pool of cash, not liquid funds available to be spent or repaid to shareholders. You must look to the company's assets, not its owners' equity, to determine whether cash exists for spending or for return to shareholders. If today the owners put additional capital into the company, the company's financial resources—assets in the form of cash—increase by exactly the same amount as owners' equity increases. If tomorrow the company uses this cash to purchase inventory, it trades one asset (cash) for another (inventory); thus, owners' equity remains unchanged but the company no longer has the liquid funds.

What else is owner's equity *not?* It is not the value that the financial securities markets assign to the ownership of the company. The market value of the entire corporate entity whose common shares* are traded publicly is equal to the current trading price of a single share of stock times the total number of shares owned by all investors. This *trading value* (generally referred to as its *capitalized value*) may be considerably at odds with—higher or lower than—the value arrived at by subtracting liabilities from assets. Thus, the value of owners' equity provides you no indication of the price at which you could buy or sell shares—whether a few or a very large number of shares. Trading (i.e., market) prices for common shares are a function of the investors' collective expectations about the future flow of benefits (primarily dividends) from ownership. Moreover, the stock trading price applies only to fairly small trades, 100 to 200 shares. If you sought to purchase all the shares of a corporation, your demand would greatly exceed the available supply of shares and so would cause the trading price to increase; conversely, if you sought to sell a very large number of shares, your supply might swamp demand and thus depress the trading price.

Therefore, for reasons having to do both with the mechanics of establishing the share trading prices and with the difficulties in valuing the assets and liabilities of the corporation, you can't reasonably expect that the two evaluations of owners' equity will agree; that is,

$$Assets - liabilities$$

typically won't be equal to

$$Trading\ price\ per\ share \times number\ of\ common\ shares.$$

VALUATION METHODS

The fundamental accounting equation requires that we value both assets (what the entity owns) and liabilities (what it owes). Valuing assets and liabilities is the fundamental accounting task. Disagreements about valuation are at the root of most controversies among accountants and between accountants and their audiences.

These controversies typically center on which of the following three valuation methods is most appropriate in a particular circumstance:

---

*Each share is evidence of partial ownership.

1. **Time-adjusted value** method: the value is viewed as a function of future benefits and costs arising from the item owned or from the obligation.
2. **Market value** method: the value is equal to the price at which the item owned could now be bought or sold or the obligation could be discharged.
3. **Cost value** method: the value is equal to the price paid for the item when it was originally acquired or the nominal amount of the obligation.

Keep in mind three blunt facts about these valuation methods: They do not typically result in the same values. No single method is inherently more correct than the other two. The cost value method is dominant in accounting practice today, although in some sense it is the least appealing intellectually. To better understand the advantages and shortcomings of each method, you need to understand all three.

Let's begin with an explanation of the time-adjusted value method. It is most easily visualized by valuing a promise to pay. Note that a promise to pay is an asset to the lender and a liability to the borrower.

VALUING A LOAN OR NOTE

How does the lender go about valuing the borrower's promise to repay the principal at maturity and, in the meantime, interest on the outstanding balance? As to both amount and timing, these two cash flow streams—interest and principal—are fixed by agreement between lender and borrower. Thus, the flows are very predictable, assuming the borrower is a good credit risk. Because money has a time value (that is, you would prefer to receive $1 now rather than a year from now, since you can make use of that dollar in the intervening year), those flows of cash that the lender is scheduled to receive in the distant future are of less value today than those flows that are due to be received sooner. Interest tables permit the lender to calculate the equivalent value today of cash flows occurring at various dates in the future, at any interest rate that the lender selects as appropriate. Such values are referred to as *present values of futures cash flows,* or *time-adjusted values*. Thus, a time-adjusted value is to today's equivalent value, at a specified interest rate, of a flow of cash (inflow or outflow) to occur at a fixed date in the future.

For example, assume that $1,000 has been lent at 5 percent: the borrower promises to make interest payments of $50 at the end of each of the next three years and a principal repayment of $1,000 at the end of the third year. What is the value to the lender of this borrower's promise to pay? The answer depends on the lender's **equivalency rate.** That rate, in turn, is a function of the lender's other opportunities to deploy money. If the lender can earn only

5 percent (at equivalent risk*), then 5 percent is the lender's equivalency rate, and the time-adjusted value of this series of flows is obviously $1,000. Put another way, the lender is indifferent between having $1,000 now or the promise from the creditworthy borrower to pay the following amounts: $50 at the end of 12 months; $50 at the end of 24 months; and $1,050 at the end of 36 months.

Now suppose the lender has opportunities to invest at 8 percent (again at equivalent risk). The lender's equivalency rate moves from 5 to 8 percent and the time-adjusted value of the promised payments is now less than $1,000. The lender would be willing to "sell" or "settle" this note for less than $1,000. How much less? Standard interest rate tables tell us that at 8 percent the time-adjusted value of the payment stream is about $923. Conversely, if a lower rate is used—because the lender's equivalency rate is lower—the time-adjusted value will be greater: at 3 percent it is about $1,057. The following simple table illustrates these three conditions:

Equivalent Value Now of the Cash Flows at Alternative Equivalency Rates†

		Equivalency Rates		
Cash Flow	Timing	3%	5%	8%
$ 50	End of 12 months	48.54	47.62	46.30
$ 50	End of 24 months	47.13	45.35	42.87
$1,050	End of 36 months	960.90	907.03	833.52
		$1,056.57	$1,000.00	$922.69

This illustration assumes a creditworthy borrower. Suppose the lender learns, soon after making the loan, that the borrower's financial position and earnings prospects are such that on-time repayment is by no means assured. Now the lender regards this loan as more risky. Lenders routinely charge higher interest rates on riskier loans; the higher interest compensates for the higher risk. Therefore, the lender concludes that the equivalency rate should be higher—say, 10 percent. Using a 10 percent equivalency rate, the lender values its asset, a Note Receivable, at about $876. The lender is willing to discount the note all the way to $876 to secure immediate payment of that amount. The lender would also be willing to sell this loan to another lender (or lending institution) for $876, a considerable discount from the $923 equivalency value at 8 percent.

Of course, borrowers use the same technique to value their own promises to pay interest and principal, that is, to value their loan liabilities. A borrower whose equivalency interest rate is 8 percent is willing to pay not more than

---

*We will consider differences in risk in a moment.
†For further discussion of these techniques, see Appendix 2A, at the end of this chapter.

$923 right now to be relieved of the obligation to make the scheduled payments of principal and interest. On the other hand, a borrower whose equivalency rate is 3 percent is better off paying any amount up to $1,057 now (a premium of up to $57 over the face value of the note), since a 3 percent equivalency rate indicates that the borrower does not have other opportunities to invest cash at a return higher than 3 percent. Note that the repayment of a debt is essentially a risk-free investment.

Exactly this mechanism drives prices in the bond market. Bonds are a form of borrowing by corporations and government units. Bonds are issued by the borrower and sold to investors (lenders), typically institutions; subsequently, bonds are traded in an organized market, just as common shares are. As prevailing interest rates increase, the trading prices of bonds decrease, and vice versa. Also, if the creditworthiness of the bond issuer (borrower) decreases, the lenders' (bond owners') equivalency rate increases to compensate for the increased risk, and the bond's price decreases.

VALUING COMMON STOCK SECURITIES

We move now to the more complex case of valuing a share of common stock. If the stock is publicly traded in organized markets (for example, the New York Stock Exchange), the value is determined by auction; it is equal to what the buyer is willing to pay and the seller is willing to take for the share.

But how do buyers and sellers judge value and thus decide whether they should buy or sell? Like the lender, the shareholder (or investor contemplating a purchase) estimates what benefits will flow from ownership of the share of stock; these benefits include dividends while the stock is owned, and the market price (less selling commissions) at the time the owner elects to sell the stock (ignoring income taxes). Thus, the stock value is the time-adjusted value of the future cash flow calculated at the shareholder's equivalency rate. That rate is influenced by the riskiness of the investment—that is, by the shareholder's assessment of the risk that the anticipated dividend payments and selling price will be realized—and by the returns available from competing investment opportunities.

As shareholders lower their interest rate equivalencies, they bid up the price of shares. Investors who are averse to high risks are attracted to companies with secure dividend payments; they calculate time-adjusted values at relatively low equivalency rates, thus placing a high value on the security.

Though this discussion may sound hypothetical, it is not. Time-adjusted values are calculated routinely. In a rational market, the price that future buyers will pay for a share of stock is a function of the expected cash returns. If all future buyers are treated as one, the future cash returns from ownership of the share of common stock are limited to the dividends paid plus the liquidation payment to shareholders when the company is liquidated. Thus, even for a company currently paying no dividends, the value of its common stock security is a function of its shareholders' expectations as to the amount and timing of all future cash returns.

CALCULATING TIME-ADJUSTED VALUES

This valuation method correctly assumes that money has a time value. You would rather receive a certain amount of money today than the same amount at a later date—say, a year from today—since you could earn a return on the money during the intervening year, this return coming in the form of interest earned or of "wants" satisfied by spending the money. Similarly, you would prefer to delay the payment of a specified amount as long as possible so that you could earn a return on the money during the delay period. Thus, values of cash flows (receipts or payments) are a function of their timing; timing differences have monetary value consequences. The time-adjusted valuation method values, as of today, future cash flows that will arise because of assets owned or obligations undertaken, utilizing an appropriate equivalency rate.

Suppose, for example, you are the manager of a retail store that owns a short-term security that earns interest at a specified rate, r percent per year. Now suppose that the owner of your building offers to reduce (or discount) the rent if you agree to pay the rent a year in advance of its due date. To make this an attractive deal, you demand a discount. You need to calculate how much to pay today in lieu of the $20,000 rental payment otherwise due a year from today.

If you do not accept this deal, the $20,000 will remain in the security earning at the rate r. Your equivalency rate is r. Therefore, the maximum amount you would be willing to pay today is the amount that would otherwise grow at the rate of r percent per year to $20,000 at the end of the year. Therefore, an equation for the maximum advance payment, P, that you would agree to is

$$P(1 + r) = \$20,000$$

$$P = \frac{\$20,000}{1 + r}.$$

If r is 8 percent, then

$$P = \frac{20,000}{1.08} = \$18,520.$$

Your decision? If the building owner today accepts $18,520 or less in lieu of $20,000 a year from today, you should pay the rent in advance. P is referred to as the present worth, or time-adjusted value, of the future rental payment at rate r. In other words, the time-adjusted value today of $20,000 a year from now is $18,520, assuming an equivalency rate of 8 percent.

Now suppose the building owner asks you to consider paying in advance the rentals due at the end of each of the next three years. Certainly, you will demand a greater discount on the second year's rent than on the first, and a still greater discount on the rent otherwise due at the end of year 3. Therefore, you need to know the time-adjusted value (or present value) P of a stream of three payments, each $20,000, occurring at one-year intervals, assuming an interest rate r. P is the sum of the individual time-adjusted values for each of the three years:

$$P = \frac{20,000}{1 + r} + \frac{20,000}{(1 + r)^2} + \frac{20,000}{(1 + r)^3}$$

If r is once again 8 percent, this equation will produce a value for P of $51,540. You are willing to pay an amount up to $51,540—but not more— to be relieved of the obligation to make the three annual rental payments of $20,000 each.

Assume now that the building owner, badly needing cash, will have to borrow money at very high interest rates if you don't prepay the rent. Under these conditions the owner might be willing to accept substantially less than $51,540. That is, the owner undoubtedly has an equivalency rate well above 8 percent and thus the time-adjusted value of these future cash flows is less for her than for you. Assuming r is 12 percent, P for the building owner— the time-adjusted value of the next three annual rental payments—is $48,040. You and the building owner should be able to strike a deal somewhere between $51,540 and $48,040.

Note that the higher the interest, or equivalency rate, r used in the calculation, the lower the time-adjusted value of future cash flows. Standard present value tables for calculating time-adjusted values are included in Appendix 2A, at the end of this chapter. Further explanation of how to use these tables is also given in the appendix.

VALUING PERSONAL AND COMPANY ASSETS

As mentioned earlier, the term *asset* encompasses all physical property, rights (such as patent and trademark rights), and any other resources that hold the promise of providing ongoing future benefits to the owner. Certainly, shares of common stock and debt instruments are assets to the investor. An automobile, household furniture, an account at the savings bank, or even a pantry full of canned foods—each is an asset to the person who owns it: each holds the promise of delivering future benefits. The car may become wrecked or the canned food in the pantry may spoil, but, at the moment, the owner has a reasonable expectation of future benefits flowing from ownership of each. Each asset has a value, and our task is to consider alternative methods for determining that value.

Can the time-adjusted value of an automobile be calculated? Yes, at least in concept. Ownership of the automobile gives rise to certain future cash outflows and inflows and provides transportation benefits. These transportation benefits can be valued by reference to the cost of alternative transportation methods, such as taxicab, train, rental car, or bicycle. Other benefits, more difficult to value in monetary terms, may include convenience, status, and time saved. If the owner's time-adjusted valuation of the car is less than the amount it would sell for on the secondhand market, the owner will probably sell it; conversely, if the time-adjusted value is higher than its secondhand price, the owner will retain the car.

The automobile example highlights the importance of the evaluator's viewpoint. You may feel that the convenience of automobile ownership is very important, while your neighbor, who seldom requires travel to out-of-the-way places or at odd hours, places little value on convenience. If so, you will place a higher time-adjusted value on an automobile than will your neighbor.

About now you probably think that a far easier and more practical way to value your automobile is simply to value it at its secondhand price; in this country, values of recent-vintage cars are published monthly. Or, if your car is almost brand new, you might argue that the value of the car is what you paid for it. Clearly, these are alternative valuation methods, and will be defined more fully shortly; they are known as *market value* and *cost value,* respectively. But, these methods of valuation are not inherently any more or less correct than the method based on the time-adjusted value of the future flow of benefits.

Consider now the difficulty of valuing another asset that you own, your household furniture. Again, the time-adjusted value of the future benefits derived from owning this furniture is one kind of evidence of its value, although not easily calculated. Secondhand prices are another, but used household furniture does not have the same organized secondhand market or readily ascertainable resale prices that used automobiles do. As in the case of your automobile, the price you originally paid for the asset represents still another indication of value.

What about the pantry full of canned goods? You probably filled the pantry either because you were able to get a very attractive price on the food or because you like the security of having a stock of food in the event of a tornado, hurricane, earthquake, terrorist attack, or food shortage. In any case, the benefits of owning this large stock of food stretch into the future, and you have some idea of the timing and magnitude of the financial benefits you derive from owning it. Thus, calculating a time-adjusted value for this stock of food is feasible; indeed, you probably went through an intuitive time-adjusted valuation when you purchased it. Of course, you could also value this hoard of food at its replacement price, or you could value it at the actual acquisition cost. Again, as with the automobile and the household furniture, all three valuation methods are applicable.

Let's look more closely at each of these valuation methods: the time-adjusted value method, the market value method, and the cost value method.

Time-Adjusted Value

The time-adjusted method requires us to forecast in monetary terms the future benefits we derive from ownership. For certain assets, such as loans, investments in common stock, paid-up insurance policies, and a customer's promise to pay, estimating future monetary benefits is straightforward and reliable. These benefits are either future cash inflows or the elimination of future cash outflows that would be required in the absence of the asset. Advance payment of one year's rent on a building creates an asset—the right to use the building for one year or, equivalently, the elimination of the requirement to make rental payments during the year.

This valuation method is particularly useful for assets that have a long life and a predictable benefit flow. What is the value of an investment (such as a lender's ownership of the borrower's promise to pay) returning $1,000 per year for 10 years? The value depends on the equivalency rate that the owner assigns to investments of this type. Its value is $10,000 only if one assumes the very unlikely situation that the owner has a zero equivalency rate, implying that the owner has nothing else productive to do with the money. The higher the equivalency interest rate assumed, the lower the value of the asset:

Equivalency Rate	Time-Adjusted Value
6%	$7,360
8%	6,710
10%	6,144
12%	5,650
15%	5,019

Consider this valuation method for other assets, for example, a machine tool owned by a manufacturing company and a cellar of fine wines owned by an individual. In both cases, asset ownership promises benefits extending into the future, but measuring these benefits is a real challenge.

The benefits of the machine tool depend on a host of factors, including the following:

1. The rate of obsolescence of the tool, which in turn depends on the rate of technological development by machine tool builders.

2. The future demand for the product or products produced on the machine tool; this demand in turn depends on the rate of change in the company's product lines, changes in the competitive climate, the strength of the economy, and changing consumer preferences.

The cellar of fine wines is even more troublesome to value. The benefits are the personal pleasure of consuming the wine and serving it to friends, but here value is also dependent on factors such as how well the wine ages, the volume and quality of wines to be produced in upcoming years, and future supply-demand imbalances that may affect wine prices generally.

So, this method of valuation, while appealing in concept, may be less useful in practice, at least for valuing such assets as machine tools and wine collections.

So far in this section the focus has been on valuing assets. Does the time-adjusted valuation method apply also to liabilities? Of course. In the example tabulated above, the 10 annual payments of $1,000 each can be valued using the borrower's equivalency interest rate.

What other obligations (liabilities) might be similarly valued? Consider a company's obligation to perform warranty service on the products it sells. If the company has some experience with the warranted product, it can estimate the timing and amount of cash outflows it will incur to perform the warranty service. Once the future cash flows have been estimated, the time-adjusted valuation method can be applied to arrive at a present-day equivalent value. Similarly, a company's obligation to make payments to employees who have retired (in other words, to honor their pensions) or to others on vacation or sick leave can be valued on a time-adjusted basis. Typically, the timing and amounts of these flows are predictable.

You may feel that these future obligations—warranties and employee leaves—need not be valued as liabilities. Why not postpone any accounting for warranties or leaves until the product is returned for repair or the employee commences the leave? The reason is quite simple: it is today's shipments of products that give rise to the warranty service obligations and today's employment of personnel that requires the company to make certain payments at the time of a promised leave. There is little doubt that the company will incur these expenditures, since inevitably a certain percentage of products will fail during the warranty period and employees do, indeed, take vacations or get sick. Activities today are creating these company obligations, and the timing and amount of these future expenditures are quite predictable. Thus, the obligations should be valued and recorded as liabilities.

Market Value

Another indicator of value of both assets and obligations is market value, or the price at which similar assets and obligations currently trade in the marketplace. If you can determine the price at which the asset can be purchased or sold, the price is its market value.

Recall the examples of the used automobile and actively traded common stocks and bonds, all of which have market prices. Loans are bought and sold among financial institutions and thus have market prices. On the other hand, common shares that are not traded in an organized market, and loans between individuals, do not have easily determinable market prices. Nevertheless, even these securities do have some market price, that is, a price at which they could be bought or sold.

Looking back over the other examples used in this chapter, we can make the following observations:

1. Household furniture surely has a market value. Newspaper classified advertisements evidence a relatively active, though largely unorganized, market for secondhand furniture. Persons knowledgeable about the used furniture market can reliably estimate market values.

2. The pantry full of canned foods can easily be valued at market prices by reference to today's prices at the food market.

3. Secondhand machine tools of a standard configuration have a very ready market at predictable prices, much like used automobiles. However, market prices of machine tools fluctuate widely, depending on the state of the economy and the delivery schedules for equivalent new machine tools. These fluctuations are troublesome; if we use this valuation method in accounting records, we will be forever changing recorded values in our attempt to reflect these fluctuations. Also, specialized machine tools—custom-made for a particular user—may have a very limited market and accordingly a low market value even though they are delivering important benefits to the user. Can we arrive at reliable market prices for such assets?

4. The cellar of fine wines can be valued by reference to today's prices for a bottle of the same wine from the same vineyard in the same year. In fact, vintage fine wines are sold and traded quite regularly.

What about valuing obligations? We saw earlier how the time-adjusted valuation method applies to warranty and employee leave obligations. Is there a market for warranty obligations? A manufacturer of warranted devices could contract with another organization, perhaps one with an extensive network of service centers, to undertake the warranty work. Such a contract fixes a market price for this obligation: the price the manufacturer will pay to be relieved of the obligation to make repairs during the warranty period. Contracts of this nature are entered into regularly in certain industries.

How about pension obligations to employees? Contracting with outside firms, generally insurance companies, to fulfill employee pension obligations is widespread—another example of paying a market price to be relieved of a future obligation. Many organizations also contract with insurance companies to pay wages to employees in the event of extended sickness.

Therefore, valuing in terms of current market price for both assets and liabilities is feasible; in many instances, the process is simple and the resulting values are defensible. This valuation method has some strong appeal: it does not require predictions about the future, as the time-adjusted value method does; it is rooted in the reality of today's marketplace; and it is understandable, explainable, and not complex.

However, the method has its shortcomings. Recall the example of the specialized machine tool for which the secondhand market is limited. Although the company would receive a low price for the asset in a sale, the specialized tool may have great value to the owning company, and, moreover, the company has no intention of selling it. So, just how relevant are current market prices to the task of valuing an asset that the owner does not intend to sell? The value of the specialized machine tool resides in it use, not in its resale.

Similarly, the household furniture, the pantry full of canned foods, and the wine cellar may have values to their particular owners that are higher or lower than the market prices of strictly comparable items today. The furniture may have a sentimental value, the emergency supply of food a security value, and the wine cellar a prestige value—important values to their owners, but not reflected in the market prices that others are willing to pay.

Thus, like the time-adjusted value method; the market value method of valuation, while appealing, has limitations. How about the cost value method?

Cost Value

A more complete name for this method is the *historical cost method,* since it calls for assets to be valued at the prices paid at the time they were acquired.

A decided advantage of the cost value method is that, typically, you can determine with both ease and accuracy what was actually paid for the asset. The acquisition prices are known for the stock or bond securities, the automobile, the household furniture, the pantry full of canned goods, the machine tools (whether standard or custom), and the cellar of fine wines.

Note, however, that these historical cost values bear no necessary relationship to the values arrived at by either the time-adjusted value or market value methods. If the automobile or household furniture was purchased long ago, its cost value may be considerably higher than its current market value. On the other hand, if the automobile is now considered a classic, or if the owner of the fine wine made a particularly astute purchase, current market prices may be considerably above the acquisition or historical cost value. Consider valuing a parcel of land, a building site. If the land was acquired some time ago, its original cost may be much below today's market value, unless, of course, the development of the community or roadway system has left the parcel in less desirable surroundings, in which case the reverse may be true.

If the decline in an asset's value is caused simply by the passage of time—for example, wear and, tear on, and obsolescence of buildings or machine tools or vehicles—we could value the asset at a declining percentage of its original cost, as the asset ages. You probably already know that this procedure is, in fact, widely followed: cost values in the accounting records are reduced each year by an amount called *depreciation*. We'll look at depreciation accounting in Chapter 8.

Even if we undertake a systematic depreciation of an asset's value over its life, the procedure takes care of only some causes of value decline. A number of other factors create wide differences between the market value and the historical cost value. The most important is inflation.* Inflation can cause an old asset recorded at historical cost to appear in accounting records to have substantially less value to the company than its replacement cost. This condition frequently affects public utilities (electric power and gas distribution). The facilities built by these companies many years ago may continue to be very productive, but inflation in land and construction costs causes them to be valued much below the cost value of newly constructed, equivalent facilities. This dilemma has been widely discussed by the accounting profession, but with essentially no resolution.

Can the cost method be used to value liabilities? Yes. By their nature, obligations are settled in the future; history has not yet caught up with the obligations. Liabilities such as bank loans, promises to pay vendors, and wage payments due to employees are typically settled in the near future by highly predictable cash outflows.

CHOOSING AMONG VALUATION METHODS

What are the advantages and disadvantages of the three valuation methods just discussed? By now, it should be clear that one can have different valuation methods fit different situations. The time-adjusted value method well suits the valuation of common stocks, notes, bonds, and other investment securities. The market value method seems workable in valuing automobiles, standard machine tools, a pantry full of canned goods, and perhaps even a cellar of fine wines. The cost value method seems widely applicable, but the longer the time between acquisition date and valuation date and the higher the rate of inflation, the less confidence there is in this method of valuation.

We need some criteria to compare these three valuation methods. The remainder of this section describes seven useful criteria.

---

*Or, more generally, changes in purchasing power, both inflation and deflation. The history of virtually all organized economies is one of persistent inflation at various rates, punctuated by relatively few periods of deflation.

Currently Relevant

Accounting information is used to make decisions, such as operating decisions by management, investment decisions by shareholders, and credit decisions by lenders. These decisions require currently relevant data: data that reflect today's situation and expectations. Almost by definition, the historical cost method suffers in fulfilling this criterion, and the more ancient the history, the more this method is likely to suffer.

Feasible

For a valuation method to be feasible, it must be possible to develop the data the method requires. If the future flow of an asset's benefits simply cannot be quantified, the time-adjusted value method is infeasible for that asset. An example is a patent on one feature of a potential new product. If the remaining design work is never completed, or if the resulting product proves to have no market or to be too expensive to produce, the patent in question may provide very modest, or even no, future benefits. Alternatively, if the product gets designed, is producible, and meets a market need, the flow of benefits may be long and large. The accountant simply can't know at the time of invention what the future holds for the patent.

Market values are readily ascertainable for some assets but not others. Some liabilities (obligations) can be transferred for a (market) price; others can't.

The cost value method is typically the most feasible—historical cost data are readily available.

Effective

While it may be feasible to develop certain data for time-adjusted valuation or market valuation, it is not always practical. The expense incurred to engage outside appraisers, other expertise, or computing power may be prohibitive. For example, while it is certainly possible to develop the current market value of a chemical processing plant or a trademark—you can advertise the asset for sale and solicit offers—the process is costly. A bona fide offer might never be forthcoming, if it became known that you had no intention of selling the asset.

You cannot afford to spend more to develop accounting information than the information is worth to you. As a result, the time-adjusted and market value methods are ineffective for valuing very many assets and liabilities that do not have a predictable flow of future benefits and readily available market prices.

Timely

The usefulness of accounting information declines rapidly with time. Management needs information to make today's operating decisions, to correct problems, and to seize opportunities. Investors need information to decide whether to buy, sell, or hold securities, and they want it as soon as possible. A month-end accounting report that takes three months to prepare is much less useful than a report that is available in one week following the end of the month. To get rapid valuation, therefore, accountants are willing to sacrifice some accuracy.

Free from Bias

By now, you undoubtedly see that an accountant has a good deal of latitude in deciding what should be recorded and at what value. The ideal valuation method is objective and little affected by an accountant's conscious or unconscious bias.

Repeatable

This criterion is closely related to the previous one. If the valuation method is objective, it should be repeatable. If an asset valuation done today and one performed next month or next year are consistent (but, of course, not necessarily identical), you can have more confidence in the resulting information. If valuation by another accountant leads to the same result, both the method and the information are more reliable.

Verifiable

Independent, professional accountants audit financial records and statements of major enterprises. (See further discussion of this in Chapter 6.) For these auditors to fulfill their role of confirming that the financial reports fairly represent the company's position, the valuations of assets and liabilities by the company's own accountants must be verifiable by the outside accountants. That is, evidence that is subject to independent verification must be used. Again, if the method used is objective, it will probably satisfy the final three criteria: it will be free from bias, repeatable, and verifiable.

CHALLENGE OF ALTERNATIVE VALUATION METHODS

Clearly, none of the three valuation methods—time-adjusted value, market value, or cost value—satisfies all seven criteria. The first criterion, that data be currently relevant, seems best satisfied by the time-adjusted value and market value methods, and least satisfied by the cost value method. However, the remaining criteria are well satisfied by the cost value method:

Uncovering historical cost data is clearly feasible.

The data can be arrived at in a timely manner.

Historical costs are relatively free from bias; arguments sometimes arise as to what constitutes cost, but these are minor compared with disagreements on the future flow of benefits.

Cost value determinations are repeatable; once determined, they don't change with time, and thus they are consistent and reliable.

Tangible evidence as to historic cost is readily available and can be independently verified.

Yet, the market value method also satisfies most, if not all, criteria in valuing certain assets such as many securities, automobiles and standard machine tools. Market values of these assets are certainly currently relevant; arriving at the data is feasible and cost-effective; the data are quickly available, objective, free from bias, repeatable, and verifiable.

The time-adjusted value method also satisfies substantially all of the criteria in certain instances, such as the valuation of a loan or an employee pension, where the data required are available, objective, verifiable, and so forth. Moreover, the time-adjusted value method provides information that is most relevant to today's decision. The problem is that the method is extraordinarily difficult to apply to many assets and liabilities. The data required are not readily or inexpensively available, biases are inevitable; as a result, different individuals, each doing the most conscientious job possible, arrive at different values.

The challenge that faces the accountant is to develop accounting information and reports that combine the relevancy and usefulness inherent in time-adjusted values with the efficiency and reliability inherent in cost values.

DOMINANCE OF THE COST VALUE METHOD

Today's reality is that current accounting practices and rules are built to a very large degree on the cost value method because historical cost data provide the best evidence of what the company owns (the value of its assets) and what the company owes (the value of its liabilities). While expedient and pragmatic, most agree that the method is not wholly adequate. It is precise but, in the view of many, not sufficiently accurate.

As a result, accountants are embracing, to an increasing degree, the other two valuation methods. Market (or replacement) values of assets are seen as relevant to the business and financial communities, particularly in times of rapid inflation when values escalate or in times of recession when market values plummet, often to values well below cost. The future flow of benefits and costs is now considered the most appropriate way to value certain liabilities. Thus, we are likely to see a continuing, although gradual, move away from strict adherence to historical cost valuations. As you proceed through this book, you should question the rules and conventions now in use, including the cost value method.

SUMMARY

This chapter focuses on the measuring part of the accounting task: valuing the various assets and obligations (liabilities) of the accounting entity, preparatory to recording the events and conditions of the enterprise.

The resulting valuations determine what the company owns (assets) and what it owes (liabilities); the difference between the two is the net worth attributable to the owners. Thus, the fundamental accounting equation can be written:

$$\text{Assets} = \text{liabilities} + \text{owners' equity}$$

The simplest view of owners' equity is that it is just the difference between assets owned and liabilities owed. But it is also equal to capital invested in the business by its owners, plus the net of profits earned less losses incurred in operating the business, less any amounts returned to the owners. Thus, the two fundamental sources of financing for the assets owned by the enterprise are (1) the funds obtained from the company's creditors, as represented by the company's liabilities, and (2) funds obtained from the company's owners, including earnings retained on behalf of those owners.

The three valuation methods are as follows:

1. The time-adjusted value method, which determines the equivalency value today of the future flow of benefits or costs that will arise as a result of owning the asset or being subject to the liability
2. The market value method, which equates value with the price at which the asset or liability, in its present state, could be bought or sold in the market today
3. The cost value method, which relies on historical cost—that is, the cost of the asset or liability at the time it was originally acquired or incurred—to represent its value

The relative advantages and disadvantages of each of the three valuation methods are judged by seven criteria. The ideal valuation method provides information that is (1) relevant to current decisions, (2) feasible to derive, (3) cost-effective (worth more than the cost to generate it), (4) timely, (5) free from bias, (6) repeatable, and (7) verifiable by independent persons. None of the three valuation methods is clearly the best in all circumstances. The time-adjusted value method provides the most currently useful data, but it is often difficult to implement. The cost value method best satisfies those criteria dealing with objectivity and verifiability of the data, but historical costs are frequently not relevant to today's decisions.

Today accepted valuation rules and techniques adhere closely to the cost value methodology; but increasingly, both accountants and financial audiences are pressing to correct some shortcomings of the cost value method. The time-adjusted value and market value methods are now accepted in more and more circumstances.

NEW TERMS

Accounting equation. Assets = liabilities + owners' equity.

Assets. All property and rights owned by the accounting entity.

Cost value. The value of an asset or liability based on the price at which it was originally purchased (asset) or will be discharged (liability).

Equivalency rate. The interest rate used in time-adjusted valuations to adjust future benefits and costs to their value today (present value).

Liabilities. The obligations of (amounts owed by) the accounting entity.

Market value. The value of an asset or liability based on the price at which it could be bought or sold today.

Owners' equity. The net value or worth of the owners' interest in the entity as expressed in the accounting records. Owners' equity is the sum of capital invested by the owners plus profits earned (less losses incurred), less any funds paid (e.g., dividends or repurchase of shares) to the entity's owners. Owners' equity also equals the difference between the total value of assets and the total value of liabilities.

Time-adjusted value. The value of an asset or liability based on the future stream of benefits and costs associated with it, with appropriate adjustments for their timing.

APPENDIX 2A: INTEREST TABLES FOR CALCULATING TIME-ADJUSTED VALUES (PRESENT WORTH)

Table 2A-1 contains factors for calculating the time-adjusted value of $1 to be received or paid at various dates in the future and at various interest rates. Table 2A-2 contains factors for calculating the time-adjusted value of a stream of payments or receipts, each of $1, occurring at annual intervals (at year-end) for the periods (n) indicated; these factors are also shown for various interest rates. We will also use these techniques and these interest tables when we discuss capital investment decisions in Chapter 12.

The use of these tables is illustrated with the examples discussed in Chapter 2:

TABLE 2A-1. Time-Adjusted (Present) Value of One Dollar at the End of *n* Years

Present Value of $1 at Year *n* discounted at rate (i):

Year (n)	1%	2%	3%	4%	5%	6%	7%	8%	9%	10%	12%	14%	15%
1	0.990	0.980	0.970	0.962	0.952	0.943	0.935	0.926	0.917	0.909	0.893	0.877	0.870
2	0.980	0.961	0.943	0.925	0.907	0.890	0.873	0.857	0.842	0.826	0.797	0.769	0.756
3	0.971	0.942	0.915	0.889	0.864	0.840	0.816	0.794	0.772	0.751	0.712	0.675	0.658
4	0.961	0.924	0.888	0.855	0.823	0.792	0.763	0.735	0.708	0.683	0.636	0.592	0.572
5	0.951	0.906	0.863	0.822	0.784	0.747	0.713	0.681	0.650	0.621	0.567	0.519	0.497
6	0.942	0.888	0.837	0.790	0.746	0.705	0.666	0.630	0.596	0.564	0.507	0.456	0.432
7	0.933	0.871	0.813	0.760	0.711	0.665	0.623	0.583	0.547	0.513	0.452	0.400	0.376
8	0.923	0.853	0.789	0.731	0.677	0.627	0.582	0.540	0.502	0.467	0.404	0.351	0.327
9	0.914	0.837	0.766	0.703	0.645	0.592	0.544	0.500	0.460	0.424	0.361	0.308	0.284
10	0.905	0.820	0.744	0.676	0.614	0.558	0.508	0.463	0.422	0.386	0.322	0.270	0.247
11	0.896	0.804	0.722	0.650	0.585	0.527	0.475	0.429	0.388	0.350	0.287	0.237	0.215
12	0.887	0.788	0.701	0.625	0.557	0.497	0.444	0.397	0.356	0.319	0.257	0.208	0.187
13	0.879	0.773	0.681	0.601	0.530	0.469	0.415	0.368	0.326	0.290	0.229	0.182	0.163
14	0.870	0.758	0.661	0.577	0.505	0.442	0.388	0.340	0.299	0.263	0.205	0.160	0.141
15	0.861	0.743	0.642	0.555	0.481	0.417	0.362	0.315	0.275	0.239	0.183	0.140	0.123
16	0.853	0.726	0.623	0.534	0.458	0.394	0.339	0.299	0.252	0.218	0.163	0.123	0.107
17	0.844	0.714	0.605	0.513	0.436	0.371	0.317	0.270	0.231	0.198	0.146	0.108	0.093
18	0.836	0.700	0.587	0.494	0.416	0.350	0.296	0.250	0.212	0.180	0.130	0.095	0.081
19	0.828	0.686	0.570	0.475	0.396	0.331	0.277	0.232	0.194	0.164	0.116	0.083	0.070
20	0.820	0.673	0.554	0.456	0.377	0.312	0.258	0.215	0.178	0.149	0.104	0.073	0.061
21	0.811	0.660	0.538	0.439	0.359	0.294	0.242	0.199	0.164	0.135	0.093	0.064	0.053
22	0.803	0.647	0.522	0.422	0.342	0.278	0.226	0.184	0.150	0.123	0.083	0.056	0.046
23	0.795	0.634	0.507	0.406	0.326	0.262	0.211	0.170	0.138	0.112	0.074	0.049	0.040
24	0.788	0.622	0.492	0.390	0.310	0.247	0.197	0.158	0.126	0.102	0.066	0.043	0.035
25	0.780	0.610	0.478	0.375	0.295	0.233	0.184	0.146	0.116	0.092	0.059	0.038	0.030

TABLE 2A-1. (*Continued*)

Year (n)	Present Value of $1 at Year n discounted at rate (i):												Year (n)
	16%	18%	20%	22%	24%	25%	26%	28%	30%	35%	40%	50%	
1	0.862	0.847	0.833	0.820	0.806	0.800	0.794	0.781	0.769	0.741	0714	0.667	1
2	0.743	0.718	0.694	0.672	0.650	0.640	0.630	0.610	0.592	0.549	0.510	0.444	2
3	0.641	0.609	0.579	0.551	0.524	0.512	0.500	0.477	0.455	0.406	0.364	0.296	3
4	0.552	0.516	0.482	0.451	0.423	0.410	0.397	0.373	0.350	0.301	0.260	0.198	4
5	0.476	0.437	0.402	0.370	0.341	0.328	0.315	0.291	0.269	0.223	0.186	0.132	5
6	0.410	0.370	0.333	0.303	0.275	0.262	0.250	0.227	0.207	0.165	0.133	0.088	6
7	0.354	0.314	0.279	0.249	0.222	0.210	0.198	0.178	0.159	0.122	0.095	0.059	7
8	0.305	0.266	0.233	0.204	0.179	0.168	0.157	0.139	0.123	0.091	0.068	0.039	8
9	0.263	0.225	0.194	0.167	0.144	0.134	0.125	0.108	0.094	0.067	0.048	0.026	9
10	0.227	0.191	0.162	0.137	0.116	0.107	0.099	0.085	0.073	0.050	0.035	0.017	10
11	0.195	0.162	0.135	0.112	0.094	0.086	0.079	0.066	0.056	0.037	0.025	0.012	11
12	0.168	0.137	0.112	0.092	0.076	0.069	0.062	0.052	0.043	0.027	0.018	0.008	12
13	0.145	0.116	0.093	0.075	0.061	0.055	0.050	0.040	0.033	0.020	0.013	0.005	13
14	0.125	0.099	0.078	0.062	0.049	0.044	0.039	0.032	0.025	0.015	0.009	0.003	14
15	0.108	0.084	0.065	0.051	0.040	0.035	0.031	0.025	0.020	0.011	0.006	0.002	15
16	0.093	0.071	0.054	0.042	0.032	0.028	0.025	0.019	0.015	0.008	0.005	0.002	16
17	0.080	0.060	0.045	0.034	0.026	0.023	0.020	0.015	0.012	0.006	0.003	0.001	17
18	0.069	0.051	0.038	0.028	0.021	0.018	0.016	0.012	0.009	0.005	0.002	0.001	18
19	0.060	0.043	0.031	0.023	0.017	0.014	0.012	0.009	0.007	0.003	0.002		19
20	0.051	0.037	0.026	0.019	0.014	0.012	0.010	0.007	0.005	0.002	0.001		20
21	0.044	0.031	0.022	0.015	0.011	0.009	0.008	0.006	0.004	0.002	0.001		21
22	0.038	0.026	0.018	0.013	0.009	0.007	0.006	0.004	0.003	0.001	0.001		22
23	0.033	0.022	0.015	0.010	0.007	0.006	0.005	0.003	0.002	0.001			23
24	0.028	0.019	0.013	0.008	0.006	0.005	0.004	0.003	0.002	0.001			24
25	0.024	0.016	0.010	0.007	0.005	0.004	0.003	0.002	0.001	0.001			25

TABLE 2A-2. Time-Adjusted (Present) Value of One Dollar per Year for n Years

Present Value of $1 per Year for n years discounted at rate (i):

Year (n)	1%	2%	3%	4%	5%	6%	7%	8%	9%	10%	12%	14%	15%	Year (n)
1	0.990	0.980	0.971	0.962	0.952	0.943	0.935	0.926	0.917	0.909	0.893	0.877	0.870	1
2	1.970	1.942	1.914	1.886	1.859	1.833	1.808	1.783	1.759	1.736	1.690	1.647	1.626	2
3	2.941	2.884	2.829	2.775	2.723	2.673	2.624	2.577	2.531	2.487	2.402	2.322	2.283	3
4	3.902	3.808	3.717	3.630	3.546	3.485	3.387	3.312	3.240	3.170	3.037	2.914	2.855	4
5	4.854	4.713	4.580	4.452	4.330	4.212	4.100	3.993	3.890	3.791	3.605	3.433	3.352	5
6	5.796	5.601	5.417	5.242	5.076	4.917	4.767	4.623	4.486	4.355	4.111	3.889	3.785	6
7	6.728	6.472	6.230	6.002	5.786	5.582	5.389	5.206	5.033	4.868	4.564	4.288	4.160	7
8	7.652	7.325	7.020	6.733	6.463	6.210	5.971	5.747	5.535	5.335	4.968	4.639	4.487	8
9	8.566	8.162	7.786	7.435	7.108	6.802	6.515	6.247	5.985	5.759	5.328	4.946	4.772	9
10	9.471	8.963	8.530	8.111	7.722	7.360	7.024	6.710	6.418	6.145	5.650	5.216	5.019	10
11	10.368	9.787	9.253	8.760	8.306	7.887	7.498	7.139	6.805	6.495	5.938	5.453	5.234	11
12	11.255	10.575	9.954	9.385	8.863	8.384	7.943	7.536	7.161	6.814	6.194	5.660	5.421	12
13	12.134	11.348	10.635	9.966	9.394	8.853	8.358	7.904	7.487	7.103	6.424	5.842	5.583	13
14	13.004	12.106	11.296	10.563	9.899	9.296	8.745	8.244	7.786	7.367	6.628	6.002	5.725	14
15	13.865	12.849	11.938	11.118	10.380	9.712	9.108	8.560	8.061	7.606	6.811	6.142	5.847	15
16	14.718	13.578	12.561	11.652	10.838	10.106	9.447	8.851	8.313	7.824	6.974	6.265	5.954	16
17	15.562	14.292	13.166	12.166	11.274	10.477	9.763	9.122	8.544	8.022	7.120	6.373	6.047	17
18	16.398	14.992	13.753	12.659	11.690	10.828	10.059	9.372	8.756	8.201	7.250	6.467	6.128	18
19	17.226	15.678	14.324	13.134	12.085	11.158	10.336	9.604	8.950	8.365	7.366	6.550	6.198	19
20	18.046	16.351	14.877	13.590	12.462	11.470	10.594	9.818	9.129	8.514	7.469	6.623	6.259	20
21	18.857	17.011	15.415	14.029	12.821	11.764	10.836	10.017	9.292	8.649	7.562	6.687	6.313	21
22	19.661	17.658	15.937	14.451	13.163	12.042	11.061	10.201	9.442	8.772	7.645	6.743	6.359	22
23	20.456	18.292	16.444	14.857	13.489	12.303	11.272	10.371	9.580	8.883	7.718	6.792	6.399	23
24	21.244	18.914	16.936	15.247	13.799	12.550	11.469	10.529	9.707	8.985	7.784	6.835	6.434	24
25	22.023	19.523	17.413	15.622	14.094	12.783	11.654	10.675	9.823	9.077	7.843	6.873	6.464	25

TABLE 2A-2. (*Continued*)

															Year
Year (n)	16%	18%	20%	22%	24%	25%	26%	28%	30%	35%	40%	50%				(n)

Year (n)	16%	18%	20%	22%	24%	25%	26%	28%	30%	35%	40%	50%	Year (n)
1	0.862	0.848	0.833	0.820	0.807	0.800	0.794	0.781	0.769	0.741	0.714	0.667	1
2	1.605	1.566	1.528	1.492	1.457	1.440	1.424	1.392	1.361	1.289	1.225	1.111	2
3	2.246	2.174	2.107	2.042	1.961	1.952	1.923	1.868	1.816	1.696	1.589	1.407	3
4	2.796	2.690	2.589	2.494	2.404	2.362	2.320	2.241	2.166	1.997	1.849	1.605	4
5	3.274	3.127	2.991	2.864	2.745	2.689	2.635	2.532	2.436	2.220	2.935	1.737	5
6	3.685	3.496	3.326	3.167	3.021	2.951	2.885	2.759	2.643	2.385	2.168	1.824	6
7	4.039	3.812	3.605	3.416	3.242	3.161	3.083	2.937	2.802	2.508	2.263	1.883	7
8	4.344	4.078	3.837	3.619	3.421	3.329	3.241	3.076	2.925	2.596	2.331	1.922	8
9	4.607	4.303	4.031	3.786	3.566	3.463	3.366	3.184	3.019	2.665	2.379	1.948	9
10	4.833	4.494	4.103	3.923	3.682	3.571	3.465	3.269	3.092	2.715	2.414	1.965	10
11	5.029	4.656	4.327	4.035	3.776	3.656	3.544	3.335	3.147	2.752	2.438	1.977	11
12	5.197	4.793	4.439	4.127	3.851	3.725	3.606	3.387	3.190	2.779	2.456	1.985	12
13	5.342	4.910	4.533	4.203	3.912	3.780	3.656	3.427	3.223	2.799	2.469	1.990	13
14	5.468	5.008	4.611	4.265	3.962	3.824	3.695	3.459	3.249	2.814	2.478	1.993	14
15	5.576	5.092	4.676	4.315	4.001	3.859	3.726	3.483	3.268	2.826	2.484	1.995	15
16	5.669	5.162	4.730	4.357	4.033	3.887	3.751	3.503	3.283	2.834	2.489	1.997	16
17	5.749	5.222	4.775	4.391	4.059	3.910	3.771	3.518	3.295	2.840	2.492	1.998	17
18	5.818	5.273	4.812	4.419	4.080	3.928	3.786	3.529	3.304	2.844	2.494	1.999	18
19	5.878	5.316	4.844	4.442	4.097	3.942	3.799	3.539	3.311	2.848	2.496	1.999	19
20	5.929	5.353	4.870	4.460	4.110	3.954	3.806	3.546	3.316	2.850	2.497	1.999	20
21	5.973	5.384	4.891	4.476	4.121	3.963	3.816	3.551	3.320	2.852	2.498	2.000	21
22	6.011	5.410	4.909	4.488	4.130	3.971	3.822	3.556	3.323	2.853	2.499	2.000	22
23	6.044	5.432	4.925	4.499	4.137	3.976	3.827	3.559	3.325	2.854	2.499	2.000	23
24	6.073	5.451	4.937	4.507	4.143	3.961	3.831	3.562	3.327	2.855	2.499	2.000	24
25	6.097	5.467	4.948	4.514	4.147	3.965	3.834	3.564	3.329	2.856	2.499	2.000	25

Present Value of $1 per Year for *n* years discounted at rate (i):

a. The time-adjusted value today of a cash flow of $20,000 occurring a year from today can be calculated from Table 2A-1; the factor in the "8%" column and the "year 1" row is 0.926. Therefore, $P = 20,000 \times 0.926 = \$18,520$.

b. The time-adjusted value today of a stream of three $20,000 cash flows, occurring at the end of each of the next three years, can be calculated from Table 2A-2; the factor in the "8%" column and the "year 3" row is 2.577. Therefore, P at 8 percent $= 20,000 \times 2.577 = \$51,540$. The factor in the "12%" column and the "year 3" row is 2.402. Therefore, P at 12 percent $= 20,000 \times 2.402 = \$48,040$.

For purposes of accounting valuations, we are typically concerned with making time adjustments of future cash inflows or outflows to value them as of today. However, these same tables can also provide answers to the following types of questions:

1. If you place $1,000 today in an investment earning 10 percent per year, what will be the value of that investment in five years? Since the factors in Table 2A-1 indicate the value today of $1,000 to be received at some date in the future, the *inverse* of the factors indicates the value in the future of an amount placed at interest today. The factor in the "10%" column and "year 5" row is 0.621; the future value after five years, of $1,000 placed at 10 percent interest today, is

$$1,000 \times \frac{1}{0.621} = \$1,610.$$

2. If you borrow $10,000 today at an interest rate of 12 percent, what equal annual payments (at year-end) will you be required to make (principal plus interest) so as to just repay the loan in eight years? (The typical home-mortgage borrowing agreement or installment purchase of an automobile or appliance follows this form.) Since the factors in Table 2A-2 indicate the value today of a stream of future payments or receipts, the inverse of the factors indicates the amounts, to be paid or received in the future, that are equivalent to a certain value today. In this example, the factor in the "12%" column and "year 8" row is 4.968; the stream of eight future annual year-end payments required to discharge a $10,000 loan, taken out today at 12 percent interest, is

$$10,000 \times \frac{1}{4.968} = \$2,013.$$

3. If you place $10,000 today in an investment earning 12 percent, how much can you withdraw at the end of each of the next eight years so

as to just deplete the investment at the end of the eighth year? (This is the same problem as 2 above, now phrased from the point of view of an investor rather than a borrower.) If you remove $2,013 per year for each of the eight years, the investment account will be zero at the end of eight years.

4. If the seller of a residential building site agrees to accept either a lump-sum payment of $25,000 now, or a payment schedule of $5,000 now and $3,300 per year at the end of each of the next 10 years, the seller is, in effect, offering to lend $20,000 (the difference between $25,000 and $5,000) to the buyer in return for the 10 annual payments of $3,300. What is the interest rate inherent in this borrowing? Table 2A-2 (the "year 10" row) provides factors for adjusting the 10 annual payments to today's value:

$$\text{at } 10\%: \quad 6.145$$
$$\text{at } 12\%: \quad 5.650$$

Thus:

$$P(@10\%) = 6.145 \times \$3,300 = \$20,279$$
$$P(@12\%) = 5.560 \times \$3,300 = \$18,348$$

Since the amount of the loan ($20,000) lies between these two values of P, the inherent interest rate on the loan must also lie between 10 and 12 percent. The interpolated rate is 10.1 percent. Thus, the tables can be used to estimate returns (interest rates) when both the time-adjusted (present) values and future values are known.

EXERCISES

1. State the fundamental accounting equation.

2. Which of the three valuation methods discussed in this chapter is typically the most feasible?

3. How would you calculate your personal "owners' equity" (or net worth)? Under what conditions would it be negative?

4. How are liabilities and owners' equity similar? How are they different?

5. Consider a five-year-old office building owned by the Parsons Company and utilized as its Jacksonville sales office. What factors might cause the building's time-adjusted value to Parsons to be above its market value?

6. Is it possible for a company or an individual to have a negative owners' equity and not be bankrupt? Explain.

7. Give examples of three assets whose valuation might be affected by anticipated future events or conditions. Describe the future events or conditions.

8. Why does the owners' equity value shown in a company's financial statements seldom equal the product of the current market price per common share and the number of outstanding shares of the company's common stock?

9. Why do market prices of outstanding bonds tend to decrease when each of the following occurs?
   a. General prevailing interest rates increase.
   b. The perceived credit risk of the borrower (the entity that issued the bonds) increases.

10. Rate the difficulty of valuing each of the following assets and liabilities (on a scale of 1 to 3, where 3 is most difficult) by the time-adjusted value method in comparison with the other two methods:
    a. A company's obligations under an employee pension plan
    b. An inventory of autos owned by a used-car dealer
    c. The lease payment obligations owed over five years on a computer:
       i. As valued by the lessor
       ii. As valued by the lessee
    d. A company's cash deposit in a bank
    e. Three years of insurance protection paid for in advance
    f. Rent paid monthly on a company's regional sales office

11. Refer to exercise 10(c). Under what conditions would the value of the computer to the lessor and lessee be different?

12. Assume you have $5,000 on deposit at your bank. Is this an asset or a liability to you? To your bank?

13. What are the principal assets and liabilities, and who are the probable owners, of the following:
    a. A commercial bank
    b. A law firm that practices general commercial law
    c. A private athletic club

14. Are current market prices of assets owned by a manufacturing company generally equivalent to the cost to replace those assets? Explain your answer.

15. Under what set of circumstances might a company have substantial owners' equity and yet not be able to pay its trade vendors in a timely manner?

16. What does it mean to say that valuations of assets and liabilities should be "free from bias"?

17. Why is the valuation of assets and liabilities by the time-adjusted method sometimes simply infeasible?

18. The following obligations apply to a particular company. Which of them should be valued in the company's accounting records, and how would you go about valuing them?

a. The company's promise to deliver 10 units of instrument A by the 15th of next month to customer M

b. As part of the contract described in part a, the company agrees to pay customer M a $1,000 penalty for every day the delivery is late

c. A two-year employment contract with a well-respected scientist

d. The employees' expectation that they will receive a handsome end-of-year bonus

e. The five-year lease on the company's premises, where 4.5 years of the lease term remain, and the lease calls for monthly payments of $20,000

f. A contract with a major supplier that requires the company to accept delivery of, and pay for, 14,000 metric tons of material over the next year, where the fixed price on the contract being 20 percent below the price the company would now have to pay to other suppliers for equivalent material

g. A three-year contract with a consulting firm to assist in improving scheduling in all of the company's plants

h. Warranty provisions on a recently introduced product, where recent evidence suggests that about 25 percent of all units supplied to customers since the product introduction will require warranty repair

i. The company's promise to customer N to rebate $2,000 if the customer is not fully satisfied with the quality of the products shipped to the customer this month

19. The following *things* or *rights* are owned by a particular company. Should they be valued in the company's accounting records, and, if so, how would you go about valuing them?

a. A trade name that is widely known and trusted by customers

b. A two-year employment contract with a well-respected scientist

c. A contract requiring that a major supplier deliver 14,000 metric tons of material to the company at a fixed price over the next year, where the fixed price is 20 percent below the price the company would now pay to other suppliers for equivalent materials

d. A small parcel of land that is not large enough to serve as a building site

e. Two hundred dozen pencils, representing a nine-month supply, for use by office personnel

f. A simple conveyor system that was built by the plant maintenance person during her spare time from scrap material she found around the plant

g. Obsolete material in inventory that was originally purchased for $3,000 and could now be sold to a salvage dealer for about $400

h. The five-year lease on the company's premises, where 4.5 years of the lease term remain, and the lease calls for monthly rent of $20,000

i. Electric power distribution equipment that was installed at a cost of $50,000 in the company's leased premises, but which, by the terms of the lease, cannot be removed if and when the company vacates the premises

j. An option to purchase the leased premises at the end of the current lease term at a predetermined price

k. A list of names of about 200 customers with whom the company has done business for 10 years or more

l. A five-year site license for several software products

20. Why is the cost value method the most widely used of the three valuation methods discussed in this chapter?

21. Using the tables in Appendix 2A, and assuming an equivalency rate of 10 percent, derive the time-adjusted values today of the following:

a. The promise of $3,000 to be received three years from now

b. The following cash payments: $1,000 to be received in one year, $2,000 to be received in two years, and $3,000 to be received in three years

c. Eight year-end payments of $20,000 to be received at the end of each of the next eight years

d. Three $1,000 payments to be made at the ends of years 6, 7, and 8

22. Using the tables in Appendix 2A, estimate the interest rate you would earn if you loaned a friend $5,000 today and the friend promised in return to pay you $1,000 at the end of each of the next seven years.

23. Using the tables in Appendix 2A, and assuming you place $1,000 in an investment that promises an 8 percent compound rate of return, what will be the value of this investment after 20 years? After five years?

24. Using the tables in Appendix 2A, and assuming you can earn 9 percent on a particular retirement investment, how much will your investment be worth in 10 years if you invest $1,500 at the end of each year?

25. Refer to exercise 24: If you want to have accumulated $30,000 in this retirement investment at the time you retire, for how many more years will you have to work?

CHAPTER 3

FINANCIAL POSITION: THE BALANCE SHEET

Following our discussion of the task of valuing assets and liabilities, we turn now to the accountant's tasks of recording, classifying, and summarizing. To gain useful information from accounting records, to avoid a jumble of financial data, and to facilitate the compiling of accounting reports, accountants record values in a systematic and organized way.

KEY ACCOUNTING REPORTS

A wide variety of accounting reports are useful to managers, owners, creditors, and others, but two are fundamental to all systems: the **balance sheet** (discussed in this chapter) and the **income statement** (discussed in the next).

The balance sheet details the firm's financial position at a particular date. Its form is based on the accounting equation,

$$\text{Assets} = \text{liabilities} + \text{owners' equity.}$$

A balance sheet tells how much the business owns and owes as of a particular date. (A reminder: the value of owners' equity is the difference between assets and liabilities.) Comparing a company's balance sheet at two different dates reveals the change in financial position, including the change in total owners' equity. Recall that changes in owners' equity are occasioned by one or both of (1) the owners investing or withdrawing funds, and (2) the company earning a **profit** or **loss.** Thus, if owners' equity has increased and the owners neither invested nor withdrew funds, the company was profitable for the period between the two balance sheet dates.

Note that you can derive a period's total profit or loss by comparing balance sheets at the beginning and end of that period. However, the other accounting report, the income statement, is the more convenient and comprehensive source of information about a business's **earnings** (profits) or losses. It provides detail about the company's operations for the period, not simply the bottom-line profit or loss figure, and thus helps you analyze how the company is performing, critical information for the statements' various audiences. That is, the income statement provides substantial backup data for one section of the balance sheet, namely, owners' equity.

FEATURES OF THE BALANCE SHEET

The balance sheet is sometimes more formally called a **statement of financial position.** It shows what the company (or individual or other entity) owns and owes as of a single, particular date. Every balance sheet is dated and provides a snapshot of the entity's financial condition at that date. By the next day, the stream of events and transactions will have changed that financial position, however slightly. Of course, most businesses don't draw up a balance sheet each day; monthly or even quarterly (three-month-interval) balance sheets provide management, owners, and creditors with sufficiently timely information.

Each major section of the balance sheet—assets, liabilities, and owners' equity—can display as much detail, or as little, as you think useful. Surely, you want more than simply the aggregate value of all the assets owned by the company; values by asset category are useful: How much cash does the company hold? How much **inventory?** What is the total value of productive plant and equipment? Similarly, you want detail about the liabilities: How much is owed to trade creditors (vendors)? to the bank? to taxing authorities? Finally, with respect to owners' equity, you want to separate the amount of owners' equity attributable to funds invested by the owners from the amount derived from operating profits or losses.

But you could provide a lot more detail. Perhaps in your manufacturing company you want separate inventory categories for raw materials, work in process, and finished goods. Perhaps also, the balance sheet should distinguish between cash in the bank and cash invested in very liquid securities, or between loan principal repayments (liabilities) due soon to lenders and amounts that won't be due for some time. Of course, more detail requires more accounting work, and thus the usefulness of detail must be balanced with the cost of operating a more elaborate accounting system. Many companies err on the side of collecting and disseminating more detail than is truly useful.

Table 3-1 provides a sample balance sheet of the Metcalfe Company. We need now to define some of the nomenclature used on this typical balance sheet.

Definition of Current and Noncurrent

Virtually all accounting systems, including the Metcalfe Company's system, draw an important distinction between current and noncurrent (or long-term) assets and liabilities. **Current assets** are those assets that will, in the normal course of business, be converted into cash or used up within the next 12 months. Thus, amounts that customers owe and are likely to pay within the next year (typically labeled **accounts receivable** and abbreviated **A/R**) are considered current. If Metcalfe permits a customer to pay off its account by monthly payments over the next three years, only payments to be received during the next year are considered current; the remainder are noncurrent, or long-term. Inventory typically qualifies as a current asset, as it will be sold in the next several months. The primary current assets for both manufacturing

TABLE 3-1. The Metcalfe Company, Inc.: Balance Sheet at June 30, 2004

Assets		
Current assets		
Cash	$12,466	
Accounts receivable	93,050	
Inventory	46,540	
Prepaids	3,900	
Total current assets		$155,956
Investments		36,500
Property and equipment		113,650
Intangibles		13,350
Total assets		$319,456
Liabilities and Owners' Equity		
Current liabilities		
Accounts payable	$26,750	
Salaries and employee benefits payable	12,402	
Taxes payable	9,274	
Notes payable within one year	25,000	
Total current liabilities		$ 73,426
Long-term debt		75,000
Owners' equity		
Invested capital	75,000	
Retained earnings	96,030	
Total owners' equity		171,030
Total liabilities and owners' equity		$319,456

and merchandising firms are cash, accounts receivable, and inventory. Companies that sell for cash only (no credit sales) have no accounts receivable; and service companies often have little or no inventory.

The definition of **current liabilities** parallels that of current assets: those liabilities that must be discharged within the next 12 months. Most liabilities are discharged by cash payments, but some may require performing services or shipping products. Amounts owed to vendors (typically labeled **accounts payable** and abbreviated **A/P**) are current since they are usually due within 90 days or less. If Metcalfe has a bank loan requiring monthly repayments of principal over five years (generally referred to as a **term loan**), that portion of the principal due to be repaid within the next year is a current liability, and the remainder is a long-term (noncurrent) liability.

What is the magic of the one-year time frame? None, in particular. This definition of current and noncurrent is simply one of many widely accepted accounting conventions. Obviously, the job of reading and interpreting financial reports issued by a variety of companies is facilitated by the uniform adoption of such conventions.

Do you really care if your company's assets are predominantly current rather than noncurrent? Or if your liabilities are due within the next 12 months or thereafter? The distinction between current and noncurrent does not affect your owners' equity, which is still the difference between assets and liabilities. Since your creditors care about these distinctions, you had better as well. If your company has a high value of current assets and few liabilities due within the next 12 months, the probability that your company will be able to meet its liabilities on schedule is a good deal greater than for another company that has high current liabilities and low current assets. Your company is *liquid,* and the second company *illiquid.* A liquid company has less risk of running out of cash and being unable to meet the string of obligations inherent in every business operation: meeting the payroll, acquiring supplies, paying the utilities (on schedule).

Therefore, the relationship between current assets and current liabilities is really more important than the absolute level of either. A measure of **liquidity** is the ratio of current assets to current liabilities, defined as the *current ratio,* and discussed further in Chapter 9. The difference between current assets and current liabilities is **working capital.**

Companies with low, or negative, working capital may encounter difficulty meeting their day-to-day commitments and frequently must scramble for cash to pay their bills. A company with a strong working capital position knows that, even if a temporary business slowdown occurs, or the collection of accounts receivable lags for a month or two, the company will probably still be able to pay its bills. Of course, the composition of the working capital is also key: suppliers' bills and the payroll must be paid with cash, not with inventory! Thus, the relationship between the composition of current assets—particularly the amounts of cash and of accounts receivable to be collected soon—and current liabilities is critical.

Table 3-1 indicates that the Metcalfe Company's current assets total $155,956 on June 30, 2004, more than twice the total current liabilities of $73,426. Its working capital is $82,530—the difference between current assets and current liabilities.

In summary, a careful review of current assets and liabilities is revealing: for an employee anxious about getting paid on time, for a banker considering making a short-term loan, for a supplier contemplating providing credit terms. For these decisions, information regarding current assets and liabilities is more useful than information about company profitability or the value of owners' equity. A company may be profitable and growing, and yet be unable to meet its short-term liabilities because it simply runs out of cash.

Statement Format

The typical balance sheet lists assets first, followed by liabilities and owners' equity. Again, this convention represents simply a convenience, not a law of either nature or the government. Some companies present their balance sheets in the format "assets − liabilities = owners' equity." Still others use the format "working capital plus noncurrent assets equal to noncurrent liabilities plus owners' equity." Each of these formats is simply a variation on the equation assets = liabilities + owners' equity. While good arguments favor each of these variations in format, this book adheres to the traditional U.S. format illustrated in Table 3-1.

The subcategories of assets and liabilities are listed in order of liquidity, with the most liquid assets and the most immediate liabilities listed first. Cash, the ultimate in liquidity, is listed first among Metcalfe's assets. Accounts receivable are more liquid than inventory, which when sold creates an account receivable. **Prepaids,** (short for **prepaid expenses,** also called **deferred expenses**) typically not large in amount, represent advance payments for services or rights not yet received; for example, rent or insurance premiums paid in advance are prepaid expenses. More about prepaids later.

Long-term (noncurrent) assets of manufacturing or merchandising companies consist largely of land and building facilities (offices, manufacturing plants, warehouses, and salesrooms) and of equipment (machine tools, display cases, automobiles and trucks, and office machines). Investments in shares of other companies, or in loans (with maturities longer than one year) to employees or customers, are examples of other noncurrent assets. Finally, a company may own patents, trademarks, or other intangible rights having a multiyear life. For the Metcalfe Company, these are lumped together as **intangibles.**

You might reasonably argue that certain noncurrent assets are, in one sense, more liquid than some current assets. For example, a pickup truck (a noncurrent asset) may have a ready secondhand market; the company could realize cash from its sale without delay. The truck is classified as noncurrent because, assuming the business continues normal operations, the truck will, in fact, not be sold. That is, the truck is owned because it's useful and used; unlike inventory, it is not owned in order to be sold.

Of course, not all inventory is immediately salable. Nevertheless, in-process and raw material inventories are classified as current because the manufacturing company expects to complete and sell the inventory within the next 12 months.

The liability side of the balance sheet is arranged in a parallel manner. The labels used are, for the most part, self-explanatory. Recall that accounts payable are amounts owing to suppliers as the result of purchasing supplies, inventory, and services for day-to-day business operations. Within the owners' equity section of the balance sheet, a distinction is made between (1) capital contributed by the company's owners, and (2) the company's accumulated earnings net of dividends paid.

Once Again: What Is Owners' Equity?

In the owners' equity section of Metcalfe's balance sheet (Table 3-1), two items—

> Invested capital
> Retained earnings

—require some explanation. For a corporation, **invested capital** is the sum of all monies the company has received from shareholders as new shares of common stock were from time to time sold by the company (and perhaps at quite different prices per share). In a partnership, invested capital is the sum of all investments by the partners. **Retained earnings** is the cumulative total of all earnings reinvested in the business (that is, not paid out to shareholders) since the corporation's formation. Thus,

> Retained earnings = cumulative profits earned
>   − cumulative losses incurred
>   − cumulative dividends paid to the company's shareholders.

Recall that owners' equity is also simply the difference between total assets and total liabilities. That's not just a coincidence; accounting procedures result in making the two definitions of owners' equity equivalent.

You may be tempted to think of owners' equity as a pool of cash available to management or the owners. It is not. It has been used, along with the liabilities, to purchase the assets on the balance sheet. Moreover, owners' equity rarely equals exactly the aggregate market value of the corporation's shares of common stock (that is, the number of shares of common stock times market price per share). If shareholder A decides, two years after purchasing new shares from Metcalfe, to sell her shares to investor B, the price per share will not be determined by the owners' equity section of the balance sheet.

The market price per share may be higher or lower than the balance sheet value per share, depending on how eager A is to sell and B is to buy.

Note that the transaction between shareholder A and investor B has no effect on Metcalfe's owners' equity. A and B are different accounting entities from Metcalfe, and Metcalfe was not a party to the transaction. The sale from A to B may have been at a higher or lower price than shareholder A paid for the shares originally; this difference is relevant to A's balance sheet but not to Metcalfe's. The capital invested in the Metcalfe Company is whatever A originally paid in cash for the shares.

Incidentally, owners' equity is often referred to as **net worth,** or, for corporations, **shareholders' equity.** These three terms are used interchangeably in this book.

MORE ACCOUNTING DEFINITIONS AND CONVENTIONS

Before turning from the balance sheet to the income statement, we need to focus on some additional accounting definitions and the conventions that surround them.

Double-Entry Bookkeeping

You may have heard the phrase **double-entry bookkeeping.** It describes the accepted method for recording accounting data. It is neither mysterious nor particularly complicated. Like the format of the balance sheet, it derives from the accounting equation.

Recall that, when assets increase with no increase in liabilities, owners' equity must also increase; that is, the owners now own more without owing more, and are therefore better off. To record this event—an increase in assets and an increase in owners' equity—requires two entries, hence the name *double entry.* Another event may trigger both an increase in an asset and an increase in a liability: for example, suppose the company borrows money. Here both the increase in cash and the increase in loans payable must be recorded. Notice that owners' equity is unaffected—the company owns the additional cash (an asset), but it owes the same amount to the lender (a liability), and therefore the entity is neither better nor worse off. Owners' equity is unaffected because the two entries are to Cash (an asset) and to Loans Payable (a liability). Put another way, the increase in the asset Cash was financed by an increase in a liability; the double entry thus preserves the fundamental accounting equation: what the company owns continues to equal what it owes.

Another example may help: assume that a customer pays an amount he owes to the company. One asset has been swapped for another; cash increases and accounts receivable decrease. We need the following double entry:

An increase in Cash by the amount received
A decrease in Accounts Receivable by the amount received

Here, neither the liabilities nor the owners' equity section of the balance sheet is affected, but two entries are nevertheless required to record completely the swap in assets.

What is single-entry accounting? Your bank check stubs represent a common single-entry system. You record a deposit as a single entry and a withdrawal as a single entry. Your checkbook can still be balanced making only these single entries. Implicitly, however, when you make a deposit, you are increasing your net worth, and when you make a withdrawal, you are decreasing it. This simple accounting system, probably quite adequate for your purposes, involves the single asset cash; when your cash increases, your worth, or equity, also increases. A more complex accounting system than your checkbook, one designed to provide extensive financial information, requires a double-entry system.

Accounts, Ledgers, and T-Accounts

Accountants classify entries by putting like entries in the same **account.** Each subcategory of asset, liability, or owners' equity is represented by a separate account in the accounting records. Some typical account names are:

Cash on Hand
Cash on Deposit, First Bank
Cash on Deposit, Fidelity Savings and Loan
Accounts Receivable—Trade
Accounts Receivable—Employees
Raw Material Inventory
Finished Goods Inventory
Supplies Inventory
Inventory on Consignment at Customers

All of these accounts are current assets. Published balance sheets typically combine the first three accounts on this list simply as "cash," the next two as "accounts receivable." Although combined on published financial statements, separate accounts provide useful additional detail. The treasurer of the company, responsible for managing the company's cash, needs to know how much cash is located where. Similarly, the manager of inventories needs detail on the value of inventory of various kinds; yet, on published financial statements, combining the last four accounts on the list into a single account, Inventory, is typically satisfactory.

A listing of all accounts available to receive entries within a particular accounting system is the system's **chart of accounts,** a kind of road map of the accounting system. It shows the nature and extent of classification, or categorization, in the accounting records. Simple accounting systems may have only 20 or 30 accounts; as the size and complexity of both the organization and the accounting system increase, it is not unusual for the chart of accounts to grow to hundreds of separate accounts.

The **general ledger** is the set of accounting documents that details the current status of each of the accounts. Thus, all entries ultimately find their way to the general ledger. The balance sheet and the income statement are constructed from the balances in the general ledger.

A shorthand way of referring to general ledger accounts is **T-accounts.** Their name describes their appearance:

Account Name

We'll use T-accounts to illustrate accounting entries.

Debits and Credits

The accounting equation,

$$\text{Assets} = \text{liabilities} + \text{owners' equity,}$$

implies that assets are recorded on the left and liabilities on the right. Accordingly, asset accounts show balances on the left-hand side of their T-accounts, while liabilities and owners' equity accounts show right-hand balances in their respective T-accounts. As a result, the sum of all the left-hand balances equals the sum of all the right-hand balances. The double-entry convention assures this equality.

The terms *left-hand balance* and *right-hand balance* are neither convenient nor elegant. We use instead the names **debit** balance for left-hand balances and **credit** balance for right-hand balances. To repeat, asset accounts typically have debit balances, and liability and owners' equity accounts have credit balances; the sum of all debit balances always equals the sum of all credit balances.

All professions seem to promulgate their own particular conventions, definitions, and jargon. The accounting profession is no exception. You may ask, "Why can't I put assets on the right, and call the balances in asset accounts Charlie?" You can, as long as you remember the rules of your system and as long as no one else has to work with or seek information from it. As a practical matter, conventions greatly facilitate communication; please accept these conventions and definitions, and express your creativity in other ways!

In T-account format, the Metcalfe Company's abbreviated general ledger for June 30, 2004, appears in Table 3-2.

How are transactions or events recorded in the T-accounts? If a company borrows $2,000 on a short-term note, the transaction is recorded in the company's T-accounts as follows:

Cash		Notes Payable	
Balance* 2,000			Balance 2,000

An addition is made both to the debit balance of the asset and to the credit balance of the liability. A debit entry is made to Cash (an increase) and a credit entry to Notes Payable (also an increase). The equality demanded by the accounting equation is maintained. Obviously the double-entry concept requires equal debit and credit entries for a complete recording of every transaction in the general ledger.

Note that this transaction is recorded by the bank making the loan in essentially the opposite way. To the bank, the note is a receivable—an asset—and the company's cash deposit is the bank's liability. The debit entry is to the Note Receivable account, and the credit entry is to the Customer Deposit (liability) account. More on this in a moment.

Take another example: A customer pays $150 on her account. The company receiving the payment records the transaction as follows:

Cash		Accounts Receivable	
Balance 150		Balance	150

An increase in an asset (a debit entry to Cash) is matched by a decrease in another asset (a credit entry to Accounts Receivable). Since a debit entry matches a credit entry, the accounting equality holds.

Here is still another example: The company pays $850 to a vendor. Since the company previously recorded this obligation, $850 must be currently in the accounts payable balance. The company records this new transaction as follows:

Cash		Accounts Payable	
Balance	850	850	Balance

---

*The term *balance* indicates that the account had a debit (or credit) balance before the particular transaction was recorded.

TABLE 3-2. The Metcalfe Company, Inc.: Abbreviated General Ledger at June 30, 2004

Cash	Accounts Receivable	Inventory	Prepaids	Long-term Assets
12,466	93,050	46,540	3,900	153,500

Accounts Payable	Wages and Employee Benefits Payable	Taxes Payable	Notes Payable Within One Year
26,750	12,402	9,274	25,000

Long-term Liabilities	Owners' Equity
75,000	171,030

Both assets and liabilities are decreased, the debit and credit entries are equal, and no change in owners' equity occurs. The decrease in the asset Cash is a credit entry, and the decrease in the liability Accounts Payable is a debit entry. We can generalize the following:

- Asset accounts typically have debit balances.
- Liability and owners' equity accounts typically have credit balances.
- An increase in an asset is created by a debit entry.
- A decrease in an asset is created by a credit entry.
- An increase in a liability or owners' equity account is created by a credit entry.
- A decrease in a liability or owners' equity account is created by a debit entry.

Accordingly, we define debit and credit entries as follows:

A debit entry *increases an asset account or decreases a liability or owners' equity account.*

A credit entry *increases a liability or owners' equity account or decreases an asset account.*

By convention, we do not use negative entries. That is, a decrease in assets is not represented by a negative debit entry, but rather by a credit entry. Of course, a credit entry is the opposite of a debit entry, just as a negative debit entry would be—if such were used.

While asset accounts typically have debit balances, could they have credit balances? What would a credit balance mean? Certain asset accounts can temporarily have credit balances—that is, their typical debit balances can be forced by excessive credit entries to a negative (credit) position. A credit balance will occur in the account Cash on Deposit, First Bank when Metcalfe overdraws its bank account. Conceivably, an account receivable could incur a temporary credit balance if customers overpaid their accounts; this condition signals that the company owes refunds. Similarly, liability accounts can have temporary debit balances. Finally, if a company earns a cumulative loss or pays dividends in excess of cumulative profits, the Retained Earnings account in owners' equity will carry a debit balance.

Are you puzzled by the names, *debit* and *credit,* chosen for both balances and entries? Past habits may cause you to think of debits as bad and credits as good: you have always been eager to earn credits, and you may have experienced the bank debiting your checking account for service charges. Yet now assets—which you probably think of as good—have debit balances and liabilities—which you try to avoid—have credit balances.

Can we reconcile the popular connotations of these terms with their accounting definitions? Think of the terms in the context of the owners' equity accounts. These typically carry credit balances; an increase in owners' equity (a good event) is recorded by a credit entry, and a decrease in owners' equity is recorded as a debit entry. You may be relieved to learn (in the next chapter) that a sale triggers a credit, and the incurrence of an expense triggers a debt—quite consistent with the habit of thinking of credits as good and debits as bad. Incidentally, when the bank tells you that it is crediting (increasing) your account or debiting (decreasing) it, the bank, from its viewpoint, is being consistent with our definitions. Your checking account is the bank's liability (since it is obligated to return your money to you on request). An increase in your account is an increase in the bank's liability, a credit entry. Similarly, when you withdraw money from the bank, you credit your asset account, Cash, but the bank debits (decreases) its liability account, Customer Deposits.

If these perspectives on the terms debit and credit don't help, simply block their habitual connotations from your mind.

SUMMARY

While a wide assortment of accounting reports is useful to various audiences, the two key reports are the balance sheet and the income statement. The balance sheet, a statement of financial condition, provides a snapshot of the entity, as of a particular date; in particular, the balance sheet shows the assets owned, the liabilities owed, and the balance of owners' equity. Comparing two balance sheets, one at the beginning of an accounting period and the other at the end, reveals the total change in the company's retained earnings during the period. This change equals the profit for the accounting period (assuming no dividends were paid to the owners). However, this single profit figure tells little about the company's operations, and thus a separate report, the income statement, to be discussed in the next chapter, is developed.

The typical format for the balance sheet follows the accounting equation, assets = liabilities + owners' equity. The company's assets and liabilities are each listed on the balance sheet in order of decreasing liquidity, and a careful distinction is made between their current and noncurrent components. The difference between these two amounts, the working capital of the company, indicates the company's overall liquidity, that is, its ability to meet its near-term obligations.

The double-entry system of accounting is widely used and derives directly from the fundamental accounting equation. A full accounting entry must involve equal debit and credit entries in order to preserve the equation. Typically, asset accounts carry debit balances, and both liabilities and owners' equity accounts carry credit balances. A debit entry increases an asset or decreases a liability or owners' equity account; a credit entry increases a liability or owners' equity account or decreases an asset account.

NEW TERMS

Account. The fundamental element of the accounting system that permits categorization and combination of like transactions and of like assets and liabilities. All accounts appear in the company's general ledger.

Accounts payable (A/P). The liability account showing the amounts the entity owes to suppliers.

Accounts receivable (A/R). The asset account showing the amounts that customers owe the entity.

Balance sheet. A statement of condition of an enterprise as of a particular date, expressed in the form of the accounting equation: assets = liabilities + owners' equity. An alternative name is *statement of financial position.*

Chart of accounts. The simple listing of all accounts used in the general ledger.

Credit; credit entry; credit balance. Liability, owners' equity, and revenue (sales) accounts typically carry credit balances. A credit entry increases a credit balance or decreases a debit balance.

Current assets. The assets of the enterprise that will be converted to cash within the next 12 months.

Current liabilities. The obligations of the enterprise that will be discharged within the next 12 months.

Debit; debit entry; debit balance. Asset and expense accounts typically carry debit balances. A debit entry increases a debit balance or decreases a credit balance.

Deferred expenses. An alternative name for *prepaids.*

Double-entry bookkeeping. The type of accounting system requiring that debit entries be balanced with equal amounts of credit entries in order to preserve the accounting equation.

Earnings. An alternative name for *profit.*

General ledger. The set of accounting records detailing the current status of all of the accounts.

Income statement. A statement of performance of an enterprise for an accounting period, indicating the profit or loss earned by the entity during the period. Alternative names are *operating statement, profit-and-loss (P&L) statement,* and *earnings statement.*

Intangibles. The asset valuation of certain rights or other intangible property such as patents, trademarks, and licenses.

Inventory. The asset valuation for materials or merchandise owned by the enterprise and available for resale or for use in manufacturing operations.

Invested capital. The owners' equity account showing the cumulative value of all investment in the enterprise by its owners. In a partnership, this account is often called *partners' capital,* and in a corporation it is called *common stock* or *capital stock.*

Liquidity. An indication of an enterprise's ability to meet its near-term obligations. Liquid assets include cash, marketable securities, certain accounts receivable, and any other assets that will be, or can be, converted to cash in the near term.

Loss. The amount by which entries that decrease owners' equity exceed those that increase owners' equity; the opposite of profit.

Net worth. An alternative name for owners' equity.

Prepaids (prepaid expenses). Current asset accounts showing amounts paid in advance of that accounting period when the corresponding expenses will be recognized. Examples are prepaid rent and prepaid insurance. An alternative name is *deferred expenses.*

Profit. The amount by which entries that increase owners' equity exceed those that decrease owners' equity. An alternative name is *earnings.*

Retained earnings. The owners' equity account showing cumulative earnings retained by the corporation. Net income increases retained earnings, and dividends paid reduce retained earnings.

Shareholders' equity. An alternative name for a corporation's *owners' equity.*

Statement of financial position. Alternative name for the *balance sheet.*

T-account. A shorthand notation for a general ledger account.

Term loan. A loan (liability to the borrower) having a maturity greater than one year.

Working capital. The difference between current assets and current liabilities. Working capital is negative when current liabilities exceed current assets.

EXERCISES

1. Why are accounts receivable considered more liquid than inventories?

2. What is an intangible asset? Give an example.

3. Suppose that you construct a balance sheet for your company (similar to Metcalfe's balance sheet shown in Table 3-1), and you find that the total value of the assets does not equal the sum of the liabilities and owners' equity. What does this situation indicate to you?

4. Is it possible for a company's working capital to be negative? Explain.

5. What are the two ways to calculate total owners' equity?

6. Consider your personal balance sheet. What do you own or owe in the way of the following?
   a. Fixed assets
   b. Current assets

c. Current liabilities

d. Long-term liabilities

Is your "net worth" positive or negative?

7. In certain countries, the typical balance sheet is presented so that the left-hand side of the accounting equation reads (instead of simply "assets")

Working capital + noncurrent assets.

How would the right-hand side of the equation read?

8. Classify the following into assets (current or long-term), liabilities (current or noncurrent), and owners' equity:

a. A borrowing of $10,000, payable as a lump sum in three years

b. A computer leased from the manufacturer for two years

c. Prepaid rent for the next two years

d. A borrowing of $15,000 that is payable in monthly installments over the next three years

e. Money owed to a supplier for raw materials and due in 30 days

f. Retained earnings of $27,000

g. Buildings and land valued at $90,000

h. Overpayment of income taxes from last year, which will be applied to the taxes due in the next fiscal year

9. Indicate whether each of the following is an asset or a liability of Company Y:

a. The down payment of $25,000, by Y, on a custom software program that will be delivered and installed at Y next year by the vendor

b. $23,500 of vacation leave wages "earned" by Y's employees, for vacations to be taken by the employees next year

c. Rights to a patent for which Y paid $12,000 last year

d. Y's warranty obligations on products sold this year, where Y estimates that making warranty repairs on these products will cost $20,000

e. $4,000 of dividends declared this month by Y's board of directors, where payments of these dividends to shareholders is to be made next month (that is, in the next accounting period)

10. For each asset or liability balance described in exercise 9, show in T-account format the full accounting entry (debit and credit entries) that might have led to the creation of that balance. Be sure to name each account that you use.

11. Describe a transaction that would involve a credit entry to each of the following accounts:

    a. Accounts Receivable

    b. Accounts Payable

    c. Machinery and Equipment (a long-term asset account)

    d. Capital Investment (an owners' equity account)

    e. Cash

    f. Inventory

    g. Salary and Wages Payable

12. For each transaction described in your answer to exercise 11, in what balance sheet account might the corresponding debit entry be made (in order to preserve the equality demanded by the fundamental accounting equation)?

13. Refer again to exercise 11: describe a transaction that would involve a debit entry to each of the seven accounts (parts a through g).

14. For each transaction described in your answer to exercise 13, in what balance sheet account might the corresponding credit entry be made?

15. What two balance sheet accounts would be affected by each of the following events or transactions, and would the entry to the account be a debit or a credit?

    a. The purchase of a one-year general liability insurance policy

    b. The repayment of $50,000 of a 90-day borrowing from the bank

    c. The sale of additional, newly issued common stock by the company

    d. The sale of outstanding shares of the company's common stock, by the company's president to the company's treasurer

    e. The purchase of a laptop computer for use by salesperson X

    f. The return of defective inventory to the supplier (assuming that receipt of this inventory was previously recorded)

    g. The investment of some of the company's cash in a short-term U.S. Treasury bond

    h. The payment, in cash, of dividends to the company's shareholders

16. In T-account format, record each of the following transactions. Indicate the name of each account that you use and whether it is an asset, liability, or owners' equity account:

    a. $4,000 paid in cash to a vendor for materials received (and properly accounted for) in a previous accounting period

    b. $800 received from a customer for services rendered (and properly accounted for) in a previous period

    c. $10,000 borrowed from Union Bank (due in 90 days)

    d. The receipt of merchandise (to be resold) valued at $4,500, with this amount to be paid in the next accounting period

e. The repurchase, by the company for $12,000 cash, of 1,000 shares of its outstanding common stock from an employee who recently resigned

f. The receipt of a custom software package, for which the company made an advance down payment of $5,000 last year and on which it owes an additional $20,000, this amount to be paid next year

g. The issuance of 1200 shares of the company's common stock to Marian Oaks in full settlement of a $100,000 long-term loan she made to the company three years ago

17. The following three events are somewhat unusual and therefore are challenging to value and to record properly. Indicate what accounts you think will be affected, if any, and whether the entry will be a debit or credit. Also describe how you would arrive at the values required for the entries.

a. The Greene Company acquires from a local inventor a patent that the company will use in the coming years, in exchange for 500 shares of Greene's common stock issued to the inventor.

b. The Mendocino Corporation leases a specialized machine tool from a lease financing company for five years, the estimated useful life of the machine tool. Mendocino estimates that the machine tool will be valueless at the end of five years. The lease requires 60 monthly payments of $2,750.

c. Ford Biotech files a lawsuit against one of its suppliers for $150,000 for breach of contract. The supplier has offered to settle for $25,000, but Ford has decided to take the matter to court.

18. The Earle Company, a manufacturing company, has working capital of $300,000. Following is a list of the balances in all current assets and current liabilities accounts except Cash. Determine the balance in the Cash account.

Accounts Payable	$190,000
Accounts Receivable	320,000
Inventories	110,000
Loans Payable	65,000
Prepaid Expenses	85,000
Salaries and Wages Payable	105,000

19. Prepare a balance sheet in conventional format for the Chase Corporation as of August 31, 2005, using the following data:

Accounts Payable	$ 80,000
Accounts Receivable	110,000
Accrued Vacation Pay	16,000
Cash	20,000
Dividends Payable	10,000

Estimated Taxes Payable	20,000
Interest Payable	6,000
Inventories	172,000
Invested Capital	200,000
Investment in Marketable Securities	10,000
Investment in Ramsey Corp. (20% ownership)	20,000
Land	24,000
Loan to Ramsey Corp.	10,000
Notes Payable	60,000
Patents and Trademarks	30,000
Plant and Equipment	72,000
Prepaid Insurance Premiums	4,000
Retained Earnings	80,000

20. The working capital of the Parsons Company as of May 31, 2004, is $70,000. Current assets are $120,000, total assets are $250,000, and the company owes no long-term liabilities. What is the balance of owners' equity on May 31, 2004?

CHAPTER 4

FINANCIAL PERFORMANCE: THE INCOME STATEMENT

The income statement is a companion to the balance sheet. In simplest terms, it elaborates on—provides detailed information about—changes in the Retained Earnings account within the owners' equity section of the balance sheet. Recall that sales and certain other events and transactions increase owners' equity, while expenses and other events decrease owners' equity. If, between one balance sheet date and the next, the cumulative effect of the increasing transactions exceeds those of the decreasing transactions, then the company has earned a profit for the period. The balance sheet account Retained Earnings has increased by the amount of this profit, less any dividends paid to shareholders.

Why do we need an income statement? The balance sheet shows only the net effect on owners' equity of the very many transactions that serve to both increase and decrease owners' equity. Alone, the balance sheet simply does not provide sufficient information. It tells nothing about the magnitude and classification of either the increasing or decreasing transactions. Audiences of the financial statements want and need more detailed information.

Whereas the balance sheet is a statement of *condition* of an enterprise as of a particular date, the income statement is a statement of *performance,* detailing how the company performed in terms of revenues and expenses during the period covered by the statement. Thus, while a balance sheet is a snapshot as of a particular date, the income statement is a "moving picture" of what happened during the period. Just as a balance sheet must carry a specific date to be meaningful, an income statement must specify the exact period for which it details the changes in owners' equity.

The income statement is also commonly referred to by other names: **profit-and-loss (P&L) statement, operating statement,** or **earnings statement.**

This book uses these terms interchangeably, but with a preference for the term *income statement.*

DEFINITION OF REVENUE, SALES, AND EXPENSES

Revenue increases owners' equity. Typically, the most important revenue source is sales of goods or services to customers. **Expenses** decrease owners' equity. If revenues exceed expenses for a particular period, the company earns a profit; conversely, if expenses exceed revenues, the company incurs a loss for the period.

While sales transactions with customers are the key source of revenue, the company may also earn interest on its savings, on its investments, or on its loans to employees or customers. The company may rent out to others excess plant or office space, or it may sell an old physical asset for more than the value currently reflected in the company's general ledger. All these activities also create revenue.

Expenses are categorized to help management judge the performance of the various segments, departments or activities, or divisions of the business. A primary expense category is **cost of goods sold (COGS),** or **cost of sales (COS):** the expenses that can be traced (matched) directly to the goods or services provided to customers. For a merchandising company using the cost value method, COGS equals the amount the company spent to acquire the particular inventory items sold during this period to customers.

Determining cost of goods sold for manufacturing or service companies is more difficult than for merchandising companies. Manufacturers need to include in COGS wages and related expenses of those personnel directly involved in creating the particular product being sold. Valuing COGS for manufactured products is a complex process, referred to as *cost accounting,* to which we will return in Chapter 13. For now, our illustrations are drawn primarily from merchandising companies.

Of course, both merchandising and manufacturing firms incur many other expenses not traceable directly to the products or services involved in the sales transactions. Typically, these include selling expenses and administrative expenses. Depending on how the business is financed, it may incur significant interest expenses. Tax obligations are still another form of expense.

Note, however, that dividend payments to shareholders, although an appropriate and necessary form of return on invested capital, are traditionally not shown as an expense of the company. Although dividend payments decrease the company's Retained Earnings account, they do not represent a diminution in the collective financial position of the shareholders; cash formerly held by the corporation for the benefit of the shareholders is now paid out to them as dividends.

STATEMENT FORMAT

Table 4-1 presents the Metcalfe Company's income statement in typical format. Note that the net income, or net profit, of the firm was $20,612 for the six-month period January 1 through June 30, 2004. Table 3-1 shows that the retained earnings for Metcalfe Company at June 30, 2004 were $96,030. Although we do not have Metcalfe's balance sheet for December 31, 2003,* we know that, if no dividends were paid during the six-month period, the retained earnings at December 31, 2003 must have been $75,418—$96,030 at June 30 less the $20,612 earned during these past six months.

The income statement tells much more, however, than simply the single profit figure. It begins with information on the sales transactions. Since Metcalfe is a merchandising company, it encounters sales returns and is thus required to make certain adjustments for, say, quality, delivery problems, and incorrect merchandise. Because the total of such returns and allowances is significant for this company—about 2.6 percent of gross sales for the first half of 2004—Metcalfe's accountants have provided this information in the accounting records and on the income statement. For other companies where returns are minimal, this refinement on the income statement may not be warranted.

TABLE 4-1. The Metcalfe Company, Inc.: Income Statement for January 1, 2004–June 30, 2004

Sales, gross	$582,050	
less: Returns and allowances	15,376	
Net sales		$566,674
Cost of goods sold		393,416
Gross profit		173,258
Operating expenses		
Selling and promotion	90,026	
General and administration	44,550	
Total operating expenses		134,576
Operating profit		38,682
Other income and expense		
Interest income	(1,816)	
less: Interest expense	4,586	
Net interest expense		2,770
Income before taxes		35,912
Taxes on income		15,300
Net income		$ 20,612

*The balance sheet at the close of December 31, 2003, must be the same as the balance sheet at the start of business on January 1, 2004.

Next, the income statement provides information on the cost of goods sold and **gross profit.** Gross profit, sometimes referred to as **gross margin,** is simply the difference between sales revenue and the corresponding cost of goods sold. Remember that the COGS account carries the cost of merchandise sold during the period, not the cost of merchandise acquired during the period. In that sense, the COGS account matches the Sales account. Thus, gross profit represents the cumulative total difference between the purchase cost and the sales price of all merchandise sold during the period—obviously useful information to management.

A great deal more detail could be provided in the revenue and expense sections of the accounting records, if the cost of generating this detail is worth the effort. For example, information on sales by major product category, by department, by region, or even by individual salesperson might be useful; indeed, most information processing systems permit multiple categorizations of sales. A parallel set of details could be developed for cost of goods sold, so that the gross profit could be calculated by product line, by department, or by region. This level of detail could be very helpful in making decisions about whether to add or delete products, how to allocate the promotional budget among products or outlets, or whether salespersons should be dropped or added.

RECORDING SALES TRANSACTIONS

What entries are required to record a sales transaction? Again, we must be true to the double-entry convention and to the accounting equation. A $250 merchandise sale for cash is recorded as follows:

Cash		Sales
Balance		250
250		

The debit entry is to the asset account Cash, and the credit entry is to the Sales account. The assets of the company have increased with no increase in liabilities, so owners' equity has increased. If the sale had been on account (on credit)—that is, if the customer did not pay cash but promised to pay in the near future—the debit would have been to Accounts Receivable, but the credit would still have been to Sales.

The entries required to record the cost of goods sold are not quite so obvious. Typically a company like Metcalfe sells merchandise from its inventory; when that happens, the sale creates a decrease in Inventory, an asset account. In order to match the cost of merchandise sold with the sales revenue, the following entries are made:

Inventory		Costs of Goods Sold	
Balance	150	150	

The debit entry is to the Cost of Goods Sold account, an expense account; the credit entry reduces Inventory, an asset account. This double entry decreases owners' equity, just as the double entry to record the sale increases owners' equity. (The debit balance in the Inventory account accumulated as Metcalfe purchased merchandise from its suppliers.)

Here, the sales price of the merchandise exceeded its purchase cost—a desirable condition! That is, the credit to the Sales account was greater than the debit to the COGS account. The owners of the business are better off to the extent of the difference, which is the gross profit.

You may see a possible—though undesirable—shortcut through these entries: simply record a credit to a Gross Profit account for the difference between the sales value and the cost value of the merchandise. The accounting entries would then be as follows:

Cash		Inventory		Gross Profit	
Balance 250		Balance	150		100

Here, two credit entries balance a single debit entry, but since the sum of the two credit entries equals the single debit entry, the accounting equation is preserved: the net difference in assets (differences between accounts receivable increase and inventory decrease) is balanced by the increase in gross profit. Why is this shortcut undesirable? Because useful information is lost. The shortcut precludes determining either total sales or total cost of goods sold for the company. Table 4-1 shows that net sales for the six months were $566,674 and the gross profit was $173,258, or 30.6 percent of sales. That information is both more complete and more useful than the information on gross profit alone.

To summarize, then, we record the $250 cash sale of merchandise that was valued in inventory at $150 as follows:

Cash		Inventory		Sales		Cost of Goods Sold	
Balance 250		Balance	150		250	150	

Note that two double entries are required for this transaction: one to record the sale and one to record the matched cost of goods sold. Note also that no account entitled Gross Profit is used here. The gross profit shown on the income statement (Table 4-1) was derived in preparing the statement.

Moreover, no debit or credit entries were made directly to an owners' equity account. You may want to think of the revenue and expense accounts as being, in effect, subaccounts within owners' equity.

OPERATING EXPENSES

Operating expenses for the Metcalfe Company are shown in Table 4-1 as "selling and promotion" expenses and "general and administrative" expenses. Within each of these expense categories, Metcalfe undoubtedly has a number of individual expense accounts, each appearing in Metcalfe's chart of accounts. Within the "selling expense" section of the chart of accounts, the following individual expense accounts might appear (these account names are meant only to be illustrative):

Salaries of Sales Personnel
Salaries Bonuses and Commissions—Sales Personnel
Office Salaries Expense
Rent Expense
Depreciation Expense
Telecommunications Expense
Automotive Expense
Other Travel Expense
Postage Expense
Supplies Expense
Advertising Expense
Brochure Expense
Miscellaneous Expense

Again, more or less detail can be included. For example, rather than combining hotel expenses and air travel expenses together as Other Travel Expense, separate accounting of them might be useful; conversely, advertising and brochure expenses might be combined into a single account called Promotional Expenses.

The entries to record these expenses are straightforward. Assume the company pays a $1000 bonus in cash to salesperson Chen:

Cash		Commission Expense	
Balance	1,000	1,000	

The decrease in the asset Cash (a credit entry) is balanced by the debit entry to an expense account.

If the company purchases promotional brochures on open account for $1,750, the accounting would look like this:

Accounts Payable		Brochure Expense	
	Balance 1,750	1,750	

No change in assets has occurred, but the increase in liabilities (a credit entry) is balanced by a debit entry to an expense account.

Table 4-1 shows that Metcalfe earned an **operating profit** of $38,682 for the six-month period. Like gross profit, operating profit is derived when constructing the income statement and does not appear as an account in the general ledger. Operating profit is calculated before considering other revenues and expenses that did not arise directly from merchandising operations. Thus, the interest revenue from Metcalfe's investments (see the "assets" section of Table 3-1) is not part of operating profit, since holding investments is not Metcalfe's primary line of business. Similarly, interest obligations on the company's long-term and short-term debts (see the "liabilities" section of Table 3-1) are nonoperating expenses, although clearly, they are very real expenses to the company. While these other incomes and expenses are not included in the calculation of operating profit, the U.S. tax laws require that they be included in determining taxable income. Thus, the next item on the income statement is "income (or profit) before taxes." "Taxes on income," a function of the profit earned, is the last expense item shown on the income statement. The net income—the *bottom line*—represents the net improvement in the shareholders' position for the period, before any dividends are paid to those shareholders.

Good arguments can be made for presenting income statements in other formats. However, Table 4-1 is currently the most widely accepted format for merchandising, and manufacturing companies follow a very similar format. On the other hand, income statements for other types of businesses sometimes may look quite different. For example, the primary revenue and expense categories for an insurance company or commercial bank are vastly different from those of a manufacturing or merchandising firm; similarly, a bank's or insurance company's income statement is designed to highlight information relevant to its particular business.

We have now touched on each of the major account categories that appear in Metcalfe's chart of accounts: asset, liability, owners' equity, revenue (sales), and expense. To repeat, the revenue and expense accounts can be thought of as subsets of the owners' equity category; the net difference between the two subsets adds to (or, in the case of negative earnings—that is, a loss—subtracts from) Metcalfe's retained earnings. Table 4-2 presents a simplified chart of accounts for Metcalfe. The account numbers appearing along side the account names facilitate information processing in Metcalfe's accounting department.

Remember, the more detail on revenues and expenses that a company wants, the longer its chart of accounts will be—up to hundreds or even thousands of general ledger accounts. When developing a chart of accounts, the accountant should avoid recording so much detail that those using the financial statements get lost in the trees and never see the forest.

Finally, notice that the statements appearing in both Tables 3-1 and 4-1 show amounts only to the nearest whole dollar; that is, the cents have been omitted for presentation purposes. This practice is typical, although the accounting records themselves carry figures to the hundredth part of a dollar. Indeed, large companies may, for statement purposes, round off to the nearest thousand dollars, or even to the nearest million dollars.

ACCOUNTING PERIOD

Table 4-1 is Metcalfe's income statement for the six-month period from January 1, 2004 through June 30, 2004. That is, the **accounting period** to which this statement applies is six months long. An accounting period can be any length. The most typical is the calendar month. While most publicly owned companies publish financial statements for shareholders every three months (that is, each quarter), for internal use they generally produce financial statements monthly. Then, by simple combination, they can develop income statements for longer periods, such as the six-month statement shown in Table 4-1. Income statements for accounting periods shorter than one year are generally referred to as **interim statements.** Most companies focus particular attention on their annual income statements and year-end balance sheets.

For convenience or because of business seasonality, many companies define their financial, or *fiscal,* year differently than a calendar year. Retailers often end their years on January 31, so that the busy holiday season is fully reflected, while other companies may select September 30, June 30, or another date. Accounting periods not tied to calendar months are sometimes used. The year may be divided into 13 four-week periods, with each four-week segment treated as an accounting period, or into four 13-week periods. This definition of the accounting period assures that each period has the same number of business days, ignoring holidays, and thus operating results are more comparable period to period and year to year. Periods defined in this way are particularly useful for retailing firms and some service firms.

A one-year period permits a full seasonal cycle to be included in each accounting year, but some years may be adversely affected by economic cycles (recessions), while others are favorably affected (prosperity). Thus, one might argue for accounting periods that are longer than one year so as to encompass a full economic cycle. Because the frequency, duration, and se-

TABLE 4-2. The Metcalfe Company: Chart of Accounts

Number	Name
	Assets
1001	Cash on Hand
1051	Cash on Deposit, First Bank
1110	Accounts Receivable—Trade
1115	Accounts Receivable—Other
1210	Inventory—Department X
1220	Inventory—Department Y
1301	Prepaids
1510	Land
1520	Warehouse and Store Facilities
1525	Depreciation—Warehouse and Store Facilities
1530	Fixtures and Equipment
1535	Depreciation—Fixtures and Equipment
1540	Transportation Equipment
1545	Depreciation—Transportation Equipment
1701	Investments—Shares in Unrelated Companies
1711	Investments—Municipal Bonds
1901	Intangibles—Trademark
	Liabilities
2010	Accounts Payable—Trade
2020	Accounts Payable—Other
2110	Salaries Payable
2120	Commissions and Bonuses Payable
2210	Payroll Taxes Payable
2220	Sales Tax Payable
2230	Property Taxes Payable
2410	Short-term Notes Payable—Bank
2420	Short-term Notes Payable—Other
2701	Long-term Debt
	Owners' Equity
3001	Invested Capital
3100	Retained Earnings
	Revenues
5010	Sales—Department X
5015	Sales Returns and Allowances—Department X
5020	Sales—Department Y—Department Y
5025	Sales Returns and Allowances—Department Y
5100	Interest Income
5200	Dividend Income

TABLE 4-2. (*Continued*)

Number	Name
	Expenses
6000	Cost of Goods Sold Expenses
6010	Cost of Good Sold—Department X
6020	Cost of Goods Sold—Department Y
7000	Selling Expenses
7110	Salaries of Sales Personnel
7130	Sales Bonuses & Commissions
7150	Office Salaries Expense
7200	Rent Expense
7300	Depreciation Expense
7410	Telecommunications Expense
7420	Automotive Expense
7430	Other Travel Expense
7510	Postage Expense
7520	Supplies Expense
7600	Promotional Expense
7900	Miscellaneous Expense
8000	General and Administrative Expenses
8110	Salaries—Executive
8150	Salaries—Clerical
8200	Rent Expense
8250	Insurance Expense
8300	Depreciation Expense
8410	Telephone and Telegraph Expense
8430	Travel and Entertainment Expense
8510	Postage Expense
8700	Professional Fees Expense
8900	Miscellaneous Expense
9000	Other Expenses
9110	Interest Expense
9210	Other Nonoperating Expenses

verity of economic cycles remain quite unpredictable, and because tax laws require companies to file annual tax returns, 15-month, 18-month, or longer accounting periods are generally impractical.

A more convincing argument can be made for accounting periods that are shorter than a month. Much can happen in a month, and managers often need very rapid feedback on operations. This feedback can often be achieved using partial financial statements on perhaps even a daily basis. Daily reports might pertain to cash position, sales, sales returns, and a host of other nonaccounting data such as equipment downtime, number of overtime hours, production yields, and new orders received.

WHEN IS INCOME EARNED AND WHEN ARE EXPENSES INCURRED?

The Metcalfe Company is a retailer, a merchandising operation. When customers come to one of Metcalfe's stores, they select merchandise, purchase it for cash or on credit, and take the merchandise away. Metcalfe does not typically receive orders from its customers in advance of delivery time, and direct selling activity is limited to time when the customer is in the store. When should Metcalfe account for the sale and corresponding inventory reduction? Is all gross profit earned at the moment the customer buys and carries away the merchandise?

The cost valuation method does, indeed, require recognizing the full amount of the gross profit as being earned when the sale to the customer is consummated—no part before, and none later. Metcalfe's managers may have done a clever job of selecting merchandise or negotiating price, and may have spent thousands of dollars over several weeks on newspaper and radio advertising to promote the merchandise. Nevertheless, 100 percent of the gross profit is assumed to be earned at the moment of the sale.

The time-adjusted value method views the situation quite differently. Effective promotion and advertising for certain merchandise inventory increase the chance that it will be sold at attractive prices, and they decrease the risk that it will have to be disposed of at distress prices; as a result, the merchandise increases in value even before the sale. Similarly, if the managers obtain particularly attractive prices, perhaps by placing large orders or by making early commitments, the time-adjusted value method will record a portion of the total realizable gain long before the final sale. The market value method may also cause earlier recognition of a portion of the gain. Moreover, both methods imply that Metcalfe should record something less than the full gain as the customer leaves the store, since Metcalfe still carries the risk that the customer will return the merchandise or keep it but not pay for it.

For a manufacturer of large equipment, the differences among the time-adjusted value, the market value, and the cost value are even more pronounced than those for a merchandising company. The manufacturer of large equipment receives orders from customers well in advance of delivery date, and the equipment ordered is typically manufactured over a period of weeks or months. The cost value method insists that the critical moment in the buyer-seller transaction is the moment of shipment, when ownership changes hands; at that moment, the sale is recorded and the gross margin earned. The time-adjusted and market value methods recognize that receipt of the order itself is a valuable event—the company (and its owners) are almost surely better off with the order than without it. In addition, both methods recognize that gross profit is earned progressively during those weeks or months when the equipment is being manufactured.

You could reasonably argue that income is earned in small steps, in the course of development, sales, manufacturing, order processing, shipping, and after-sale servicing. However, reaching agreement on just how much of the income is earned at each step would be very difficult. The engineering design, embodied in drawings, has value; the inventory increases in value as manufacturing occurs; the selling process adds value; and after-sales services may be critical to keeping the product sold. But to measure each increment of value objectively, consistently, and without bias would be exceedingly difficult. As a result, accountants revert to the cost value method as being objective, verifiable, efficient, and timely.

In summary, the cost value method concentrates the recognition of earnings from a sales transaction at that moment when the goods are delivered, or services provided, in response to a firm order from the customer. Such a convention, although unquestionably arbitrary, simplifies the accounting task.

Similarly, accountants generally choose a moment in time to recognize expenses. Take, for example, the expense of servicing a large air-filtering system at the Gonzalez Company. When an outside contractor does this work, several events occur: Gonzalez issues a purchase order to the contractor, the service is performed, the contractor renders an invoice, and Gonzalez pays the invoice. Just as the outside contractor recognizes revenue at the time the service is performed, so also Gonzalez recognizes the expense at the same time.

When cash inflows or outflows are separated in time from the corresponding recognition of the revenue or expense, the use of *accrual accounting* is required.

ACCRUAL CONCEPT

Consider again the simple accounting method you use for your personal bank checking account. You equate your worth with your cash balance. As your cash balance increases, you become better off. You earn wages as you work, but your net worth increases only when your paycheck increases your bank balance. Similarly, when you purchase $30 worth of food, you consider that you have incurred $30 of expense, even though you take home an inventory of food to consume over the coming days. You are operating a simple **cash-basis** accounting system.

Such a system works well for individuals and for certain businesses, particularly professional services, such as doctors' or lawyers' offices. These businesses have essentially no inventory, only minimal physical assets, and one dominating expense category: salaries and wages.

The great majority of business enterprises, however, must use an accounting system based on the more complex concept of: the **accrual basis:**

1. Revenue is recognized as being received when goods are delivered or services rendered, regardless of when the customer orders or pays for them. The customer may pay in advance, simultaneously with delivery, or 30 or more days later.

2. Expenses are recognized as having occurred when the goods or services are received, regardless of when the cash outflow occurs. For example, the rent expense for a company's rented facilities is incurred in the period the facilities are used, whether the rent was actually paid during the previous period or will be paid in a future period; similarly, the management salaries expense is incurred when the work is performed by the managers, regardless when the salaries are actually paid in cash.

Thus, accrual accounting, in contrast to your checkbook accounting, permits us to adjust for the difference in timing between revenue and expense flows and cash flows.

Remember that proper profit reporting (and asset and liability valuation) requires that revenues and expenses be matched to the accounting period. Therefore, revenue for the accounting period includes only those sales involving delivery of merchandise or services during that period. Similarly, expenditures incurred in the previous period but paid for in this period are considered expenses of the last period, not this one, while expenditures properly assigned to a future period are not expenses of this period even if the associated cash outflow occurred in this period.

Examples of the Accrual Concept

A few examples will help clarify the accrual concept. The simplest example is a credit sale. The critical transaction—delivery of merchandise—occurs during this period, although cash from the customer will be received in a subsequent transaction which may occur in a subsequent period. As illustrated earlier, the appropriate entries are these:

Debit to Accounts Receivable
Credit to Sales

The entries associated with transferring the merchandise to the customer also must be recorded in this period. The cost of goods sold transaction in this period is recorded as follows:

Debit to Cost of Goods Sold
Credit to Inventory

Let's assume the merchandise was acquired in a previous period. When the merchandise was received, the transaction was recorded as follows:

Debit to Inventory
Credit to Accounts Payable (or to Cash)

Another example: A customer makes a $100 down payment on an order that will be delivered in a future accounting period. The $100 received is not revenue to the company in this accounting period; the critical sales transaction has not yet occurred (according to the cost value method). Rather, the receipt of the $100 creates a liability for the company: it must either make the delivery in a future accounting period or return the customer's deposit. Here are the appropriate entries:

Debit $100 to Cash
Credit $100 to Customer Advance (a liability account)

Note here that the increase in the asset is exactly matched by the increase in the liability; no gross margin has been earned and, thus, owners' equity is unaffected.

What entries will be required in the future period when the merchandise is finally delivered, that is, when the sale is completed? Assume that the total sale is $1,000 and that the customer pays the remaining $900 in cash at the time of delivery. The appropriate entries at that time will be as follows:

Cash		Customer Advances		Sales	
Balance			100	Balance		1,000
900					

Here, two debit entries balance the single credit entry to the Sales account. (This example ignores the cost of goods sold.)

Now an example involving expenses, not revenue: Today the company purchases and pays for a one-year comprehensive insurance policy providing protection commencing with the next accounting period. Recording the $1,200 annual premium payment requires these entries:

Debit $1,200 to Prepaid Expenses (an asset account)
Credit $1,200 to Cash

One asset was swapped for another; no expense was incurred, nor did either liabilities or owners' equity change. Some cash was given up in return for another asset, the right to future insurance protection.

What will be the appropriate entry next month when 1/12 of the insurance benefit will have expired? At that time, the asset Prepaid Expenses will have declined in value to $1,100; an insurance expense of $100 will need to be included in next month's expenses:

Insurance Expense		Prepaid Expenses (an asset)	
100		Balance	100

Obviously, at the end of 12 months, when the insurance coverage expires, the prepaid expense account will have been reduced to zero; each of the months will have been charged with $100 of insurance expense.

A final example: By agreement with the owner of the building it occupies, a company may delay monthly rental payments until 15 days following month-end. The company, if using a monthly accounting period, needs to recognize that this month it received the benefit of using the property, even though it does not have to pay rent until the next accounting period. The use this month creates a liability, an obligation to the property owners. The appropriate accounting entries this month are as follows:

Debit the monthly rent to Rental Expense
Credit the monthly rent to Rent Payable (or to Accounts Payable)

Next month, when the rent for this month is finally paid, the entries will be:

Debit to Rent Payable (or to Accounts Payable)
Credit to Cash

Notice that next month's entries—a decrease in an asset and a decrease in a liability—do not affect owners' equity. The effect on owners' equity—the expense—is recorded this month.

Review again these four examples:

1. A sale this period, with the inflow of cash occurring in a subsequent period.
2. A cash inflow this period, with the corresponding sale recognized in a subsequent period.
3. A cash outflow prior to the accounting period in which the expense will be recognized.
4. An expense recognized this period, although cash outflow occurs in a subsequent period.

Table 4-3 summarizes the accounting entries to recognize revenues and expenses under alternative assumptions as to the flow of cash. Adherence to the accrual concept is essential for accurate profit reporting, and, consequently, for accurate valuation of assets and liabilities. However, applying this concept creates much of the disagreement among accountants and between

TABLE 4-3. Accrual Entries Under Alternative Cash Flow Conditions

	Transaction Neutralizes Prepayment	Cash Flow Simultaneous With Transactions	Transaction Now with Future Cash Payment
Revenue: merchandise or service delivered this period	Debit to Customer Advances (liability)	Debit to Cash	Debit to Accounts Receivable (asset)
	Credit to Sales	Credit to Sales	Credit to Sales
Expense: incurred this period (service or benefits received this period)	Debit to Expense Credits to Prepaid expense (asset)	Debit to Expense Credit to Cash	Debit to Expense Credit to Accounts Payable (liability)

accountants and their audiences; in practice, it is often far from clear just when a sale is consummated or an expense incurred. Chapter 6 looks at some rules to assist in resolving these disagreements.

EXAMPLE: ACCOUNTING FOR A FULL PERIOD

A simple example will illuminate the relationship between the balance sheet and the income statement: accounting for a small business for a full accounting period, one month. The example also illustrates the use of T-accounts (representing general ledger accounts) and debit and credit entries.

We are accounting for a flower stand for the month of April. The flower stand sells all its flowers for cash and, thus, has no accounts receivable. It buys new flowers each day and, thus, carries no inventory over from one day to the next. The March 31 balance sheet for the flower stand looks like this:

ASSETS

Cash	$ 250
Stand (fixed asset)	1,250
Total assets	$1,500

LIABILITIES

Accounts payable	$ 300
Owners' equity	1,200
Total liabilities and owners' equity	$1,500

TABLE 4-4. Flower Stand: T-Accounts

Assets

Cash		Stand (Fixed Asset)	
250[a]	100[b]	**1,250**[a]	50[f]
1,700[c]	600[d]		
	850[e]		

Liabilities and Owners' Equity

Accounts Payable		Owners' Equity	
850	**300**[a]		**1,200**[a]
	900[g]		

Revenue

Sales	
	1,700[c]

Expenses

Cost of Goods Sold Expense		Rent Expense		Wage Expense		Decline in Value of Fixed Asset (expense)	
900[g]		100[b]		600[d]		50[f]	

[a] Balance at March 31.

[b] A $100 payment was made to the shopping center for rent. The debit to the expense account Rent Expense balances a credit to Cash.

[c] Sales for cash totaled $1,700. (Although the operator of the stand made daily bank deposits, the entry is shown as if only a single deposit were made at month-end.) The credit to Sales balances a debit to Cash.

[d] Wages paid in cash to the stand's attendants totaled $600. The debit to the expense account Wages Expense balances a credit to Cash.

[e] Cash paid out to the flower suppliers aggregates to $850. Recall that flowers were purchased on credit; payments to suppliers during the month were less than credit purchases, and thus the Accounts Payable balance is higher at month-end. This cash outflow does not affect the income statement. The credit to Cash balances a debit to Accounts Payable.

[f] The flower stand operator estimates that the value of the fixed asset (the physical stand) declined by $50 during the month. No transaction is involved here; rather, a review of the assets owned reveals that one of those assets should be valued lower at the end of the month than at the beginning. The credit entry to the Stand (Fixed Asset) account balances a debit to an expense account (Decline in Value of Fixed Asset).

[g] Flowers were purchased fresh each day, and the total of all purchases (on credit) was $900. Since the stand had no beginning or ending inventory, these purchases represent the cost of goods sold. The debit to Cost of Goods Sold (expense) balances a credit to Accounts Payable.

TABLE 4-5. Flower Stand: Financial Statements

INCOME STATEMENT FOR APRIL

Sales, gross		$1,700
Cost of goods sold		900
Gross margin		800
Operating expenses		
Rent	$ 100	
Wages	600	
Decline in value of fixed assets	50	750
Profit		$ 50

BALANCE SHEET AT APRIL 30

Assets

Cash		$ 400
Flower stand structure		1,200
Total assets		$1,600

Liabilities and Owners' Equity

Accounts payable		$ 350
Owners' equity as of March 31	$1,200	
Profit for April	50	
Total owners' equity (at April 30)		1,250
Total liabilities and owners' equity		$1,600

Table 4-4 shows the T-accounts for the flower stand. The balances at March 31 in the four accounts listed on the balance sheet above are shown in bold-face in the table. Take a few minutes to work your way through the entries.

The accounting process for April is complete, so we can construct an April income statement and a balance sheet as of April 30, utilizing the balances that appear in the T-accounts (general ledger accounts). These statements are shown in Table 4-5.

Note that the flower stand made a $50 profit in April. This profit amount does not appear in any T-account but is derived on the income statement as the difference between sales and expenses. The profit amount also appears in the "owners' equity" section of the balance sheet as the balancing item—here is the linkage between the income statement and the balance sheet.

Note also that total assets increased during the month by $100, and this $100 increase is balanced by a $50 increase in liabilities (accounts payable) and a $50 increase in owners' equity (profit for the month). Is this good or bad? Neither. The balance sheet shows the flower stand's condition at month-end—it is quite solvent—but says little about its performance. Performance (a $50 profit) is assessed by the income statement.

SUMMARY

Although comparing balance sheets at the beginning and end of the accounting period can derive summary profit information, this single profit figure provides little insight into the company's operation. The income statement is designed to provide more extensive information. While the balance sheet is a snapshot of financial condition as of a date, the income statement is a statement of performance for a particular accounting period.

On the income statement, expenses directly attributable to sales made during the period (that is, cost of goods sold) are matched against those sales to permit derivation of the gross profit. The operating expenses (such as selling and administrative) for the period are then subtracted from the gross profit to derive the operating profit. The bottom line of the income statement—net income after nonoperating income and expenses and after income tax expense—increases the Retained Earnings account of owners' equity.

Recall that the cost value method of valuation dominates actual accounting practice in this country, in preference to the time-adjusted value or market value methods. So also does accrual accounting in preference to cash-basis accounting. The accrual concept requires that revenues and expenses be matched to the accounting period when the revenues are earned and the expenses are incurred, regardless of when cash is received or paid out. The cost value method assumes that income is earned only when a sale is realized: at the single point in time when the goods or services are delivered to the customer. Thus, the delivery transaction with an external party, the customer, gives rise to revenue and earnings. Expense recognition follows parallel rules.

The categories of general ledger accounts are assets, liabilities, owners' equity, revenue, and expense. Each category has from several to hundreds of individual accounts, depending on the size and nature of the business and on management's preferences. A full listing of all accounts utilized in the company's accounting system is the chart of accounts, and the set of records containing this account-by-account information is the general ledger.

NEW TERMS

Accounting period. The duration of time for which the income statement reflects performance.

Accrual basis. Revenues and expenses are recorded in the period when earned or incurred regardless of when the corresponding cash inflow or outflow occurs.

Cash basis. The accounting concept that, in contrast to the accrual basis, requires that revenues and expenses be recognized in the accounting period when the corresponding cash flow occurs.

Cost of goods sold (COGS). The expense account containing costs directly identifiable with (and therefore matched to) the sales for the accounting period. An alternative name is *cost of sales.*

Cost of sales (COS). An alternative name for *cost of goods sold.*

Earnings statement. An alternative name for *income statement.*

Expense. A decrease in owners' equity as a result of operations.

Gross margin. An alternative name for *gross profit.* Gross margin may be given as gross profit/net sales, expressed as a percentage.

Gross profit. An amount equal to the difference between sales and cost of goods sold.

Interim statements. Income statements for accounting periods shorter than one year, and balance sheets at dates other than fiscal year-end.

Operating profit. The amount of profit (or earnings) derived from normal operations of the enterprise; the amount before other (nonoperating) income and expense, and taxes on income, are recognized.

Operating statement. An alternative name for *income statement.*

Profit and loss (P&L) statement. An alternative name for *income statement.*

Revenue. An increase in owners' equity as a result of operations.

EXERCISES

1. Explain the difference between cash accounting and accrual accounting.

2. What would your personal income statement look like? What categories of income and expense would you include?

3. Generally, operating profit is accounted for later using the cash-basis accounting method than the accrual method. However, there are exceptions:

    a. Identify a sales transaction for which cash-basis accounting would recognize revenue earlier than would accrual-basis accounting.

    b. Identify an expenditure transaction for which accrual-basis accounting would recognize the expense later than would cash-basis accounting.

4. Why are accounts listed on a company's chart of accounts typically identified both by a number and a descriptive name?

5. What is the difference between operating profit and net income?

6. Why is interest income not considered part of operating revenues, and why is interest expense not considered part of operating expenses?

7. If a retailing firm acquires more merchandise for resale than it actually sells, how is the retailer's Cost of Good Sold account affected? How is

the difference in value between merchandise acquired and merchandise sold accounted for?

8. Why might a company chose a 13-week account period rather than a calendar quarter? (They are not quite identical.)

9. An investment in a large computer does not have an infinite life. Should the cost of this computer be recognized as an expense only at the end of its useful life? Explain your answer.

10. Why do Gross Margin, Operating Margin and Net Income accounts not appear in a chart of accounts?

11. Give two examples of transactions that lead to a company utilizing a Prepaid Assets account.

12. Only one entry is made during each accounting period to the Retained Earnings account. What is that entry, and when is it made?

13. Explain the probable use of an Accrued Warranty Expense liability account. What events trigger debit and credit entries to this account?

14. Indicate whether each of the following statements is true or false:
   a. When a company's Cash account balance is less than its Accounts Payable account balance, bankruptcy is likely to follow.
   b. When a company receives a down payment from a customer for merchandise to be delivered during a subsequent accounting period, the company incurs a liability.
   c. All companies are required to calculate their income tax liability as of December 31 each year.
   d. Payment to a vendor of an outstanding account payable has no effect on the operating profit of either the payer or the recipient of the payment.
   e. If company X guarantees each of its salespersons a minimum year-end bonus of $3,000, recognition of this expense should occur only at year-end when the bonuses are actually paid.
   f. Accrued Interest Expense Payable is a liability account, not an expense account.
   g. It is possible, but undesirable, to operate an accounting system solely with a balance sheet (in other words, without an income statement).
   h. When Leland Corporation determines that one of its customers is unlikely to pay its outstanding receivable (because of its financial difficulties), Leland should make no accounting entry until it has exhausted all avenues for securing payment.
   i. The repayment of a loan reduces the borrowing company's retained earnings.

j. In cash (as contrasted with accrual) accounting, a company does not record amounts due from customers in its accounting records.

15. Fill in "debit" or "credit" to record properly each of the following transactions in the accounts indicated:

    a. Payment of $2,000 for an insurance policy with coverage commencing next month: _____ to Prepaid Expenses, and _____ to Cash

    b. Receipt of $4,000 from customer M for merchandise bought by M last month: _____ to Accounts Receivable, and _____ to Cash

    c. Recognize that $400 of inventory is obsolete and must be scrapped: _____ to Inventory, and _____ to Other Expense

    d. Recognize $400 of interest expense attributable to this accounting period that will be paid two months from now: _____ to Interest Expense, and _____ to Interest Payable

    e. Customer P just returned merchandise that she bought for $200 last month. She has not yet paid for the merchandise. The cost of goods sold for the merchandise was recorded last month at $140. The returned merchandise is resalable: _____ to Sales Return, _____ to Accounts Receivable, _____ to Inventory, and _____ to Cost of Good Sold

    f. Payment of $150 in cash for repair of an air-conditioning unit: _____ to Cash, and _____ to Repair Expense

    g. Receipt of $300 in cash for merchandise to be delivered next month: _____ to Cash, and _____ to Due to Customer (a liability account)

    h. Payment of $600 for service to be performed by Sawyer & Company during the next accounting period: _____ to Cash, and _____ to Prepaid Assets

    i. Company T pays $800 in dividends to its shareholders: _____ to Cash, and _____ to Retained Earnings

    j. Receipt of a five-year note (evidence of borrowing) from a customer in settlement of the customer's outstanding receivable: _____ to Accounts Receivable, and _____ to Notes Receivable.

16. To record properly each of the following transactions, indicate the account name in which the debit and credit entries should be recorded:

    a. Company B receives $5,000 from a vendor in full settlement of a lawsuit (for breach of contract) filed by B against the vendor: debit to _____, and credit to _____

    b. Company B receives $2,000 as a down payment for service to be rendered to the customer in the next accounting period: debit to _____, and credit to _____

    c. Company B receives $1,000 for service rendered to the customer in the previous accounting period: debit to _____, and credit to _____

d. Company B pays $500 for material to be received from a vendor in the next accounting period: debit to _____, and credit to _____

e. Company B pays $1,700 for material received from a vendor during the previous accounting period: debit to _____, and credit to _____

f. Company B recognizes that $180 of prepaid insurance has expired (that is, a portion of the coverage has expired): debit to _____, and credit to _____

g. Company B returns $900 of merchandise to a supplier, alleging that it is defective; B has not yet paid for the merchandise: debit to _____, and credit to _____

h. Company B trades in a used truck for a new truck; B paid $6,000 in addition to the trade-in: debit to _____, and credit to _____

i. Company B trades material, which it acquired in the last accounting period for $300, with a vendor in full settlement of the $500 owed to the vendor: debit to _____, debit to _____, credit to _____, and credit to _____

j. Company B issues 1,000 shares of its comany stock to lender Y in full settlement of a $20,000 long-term borrowing: debit to _____, and credit to _____

17. Use the following information to construct an income statement in conventional format. The company's income tax rate is 30 percent.

Administrative expenses	$ 400
Cost of good sold	2,500
Gross sales	4,400
Interest expense	200
Rent income	400
Sales returns	100
Selling and promotion expenses	200

18. Record each of the following transactions or recognitions in T-account format, label the accounts used, and indicate for each account whether it is an asset (A), liability (L), revenue (R), expense (E), or owners' equity (OE) account. Make only those entries required to record an event or transaction; do not include balances that may exist in the account as a result of earlier entries. Make sure that the sum of your debit entries equals the sum of your credit entries.

a. Company Q sells $4,000 of merchandise to a customer; this merchandise is valued in inventory at $3,200.

b. Company Q purchases, on credit, merchandise normally valued at $8,000; because of the size of the order, Q's vendor provides a 5 percent quantity discount.

c. Company Q pays $600 of accrued interest payable on its bank loan.

d. Company Q receives $300 due in rent for this month from a tenant who rents an office suite in Q's headquarters building.

e. Company Q borrows an additional $10,000 from its bank.

f. Company Q pays $1,000 in dividends to its president, the sole owner of the corporation.

g. Company Q scraps merchandise valued at $500 because it is obsolete.

h. At year-end, company Q makes profit-sharing payments to its employees totaling $1,700; these profit-sharing obligations were previously accrued to the extent of $1,500.

i. Company Q receives $2,000 from Ms. D in partial payment of the $5,000 that she owes for services rendered in previous accounting periods.

j. Company Q pays $1,500 to company L in partial payment of the $3,500 it owes for services it received from L in previous accounting periods.

k. Company Q pays $5,300 to Centerville Bank: $5,000 repayment of principal on a loan, and $300 interest for the current month.

l. Company Q pays $3,600 in rent on office space: $1,800 for this month, and the same amount for next month.

19. Following the instructions in exercise 18 above, record the following:

a. For $5,000 paid in cash, company R acquires 100 shares of its common stock from its sales manager, who recently retired.

b. For $10,000 in cash, company R sells 200 shares of its common stock to its president (and majority shareholder).

c. Company R executes a contract for insurance coverage to commence in the next accounting period, the $1,000 premium for which is also due in the next accounting period.

d. Company R pays $600 for three display advertisements that will appear next week and at two-week intervals thereafter.

e. Company R receives $670 of merchandise in partial fulfillment of an order placed with vendor T three months ago.

f. Company R sells, on credit for $800, merchandise carried in its inventory at a value of $650.

g. Company R receives (and installs) an auxiliary generator valued at $5,000, on which it previously made a $1,000 down payment.

h. Company R recognizes that the value of its office equipment has declined in value by $700 during this accounting period.

i. Company R recognizes that it likely has to make a $400 compensatory payment to customer N for faulty merchandise that N has had to scrap.

j. Company R recognizes that, during this accounting period, its employees have accrued vacation leave valued at $2,000.

20. The Gold Delivery Company specializes in delivering products from local manufacturers to retail shops. At the end of November 2003, the company's general ledger (in which the accounts are listed alphabetically rather than by classification) carried the following balances (not separated as to debit and credit balances):

Accounts Payable	$10,200
Accounts Receivable	23,000
Capital Stock	25,000
Cash	8,000
Delivery Equipment	22,000
Delivery Expenses (gas, oil)	1,300
Delivery Revenues	12,000
Insurance Expenses	500
Loans Payable	10,500
Notes Payable	3,500
Office Equipment	3,700
Prepaid Insurance	2,000
Rent Expense	1,500
Retained Earnings (at October 31)	8,000
Wages Expense	7,200

Construct, in conventional format, both an income statement for the month of November 2003 and a balance sheet at month-end, using the data above. Derive Gold Delivery's profit for the month, and make sure the balance sheet balances.

21. The Montgomery Shoe Shine Stand in Union Station, employing three full-time and four part-time persons, has the following balance sheet for February 1, 2006:

Assets	
Cash	$300
Supplies inventory	100
Fixed assets (stand)	2,000
Total Assets	$2,400

Liabilities and Owner's Equity	
Accounts payable	$ 80
Wages payable	500
Owners' equity	1,820
Total Liabilities and Owners' Equity	$2,400

The following transactions occur during the month of February:

• Total cash revenue for shoe shines was $1,830. The stand does not shine shoes on credit.

- Total tips given to the employees by customers was $450.
- The stand purchased $300 of supplies and had an inventory of $150 as of February 28.
- The stand paid out $1,300 in cash for wages during the month and had wages payable at the end of the month of $450.
- The stand paid a quarterly rental fee to Union Station of $300 for the months of February, March, and April 2006.
- The owner of the stand concluded that the value of the stand had declined by $100 during February.

   a. Create T-accounts as required. Record the opening balances as of February 1 and then make the entries to record the transactions listed.

   b. Construct a balance sheet as of February 28, and an income statement for the month of February 2006.

   c. Then consider the following questions:

   1. What is the significance of the increase in assets during the month?

   2. How did the Montgomery Shoe Shine Stand perform during the month? Would your answer be different if you knew that the Stand's owner was not among those to whom wages were paid during the month?

   3. What is your assessment of the Stand's financial position at month end?

CHAPTER 5

CASH FLOW STATEMENT

The last two chapters focused on what most of us think of as the two principal statements produced by a company's financial accounting system: the income statement and the balance sheet. A third statement, called the **cash flow statement,** has gained popularity in the last decade or two, and is a requirement in all annual reports of publicly owned companies. The purpose of the statement is both simple and critical: to explain the sources of cash that the company has used to invest in new assets of the business or to repay its obligations.

The cash flow statement, like the income statement, relates to a specific time period; unlike the income statement, however, it focuses on the effect of events on cash inflow and outflow, rather than on profit. Cash is the lifeblood of business. Cash insolvency—running out of cash—is a crisis, and often the fatal blow for a weak business. While cash insolvency is most frequently precipitated by persistently unprofitable operations, a profitable, growing business can also run out of cash simply because it pays out more cash than it receives. Such a cash crisis can lead to business failure in spite of strong signs of vigor and growth, particularly those shown in the income statement. Conversely, a sick, declining business may withstand years of unprofitable operations if, through a variety of actions, it maintains an adequate cash flow.

WHY A CASH FLOW STATEMENT?

Why is a third statement necessary? Since the income statement and the balance sheet contain all of the balances from the general ledger, why are they not sufficient for both internal management and external audiences?

Chapter 4 distinguished between cash-basis accounting and accrual-basis accounting, and emphasized the importance of accrual accounting for companies that sell goods and purchase supplies on credit, and own substantial assets. The accrual method of accounting provides essential information on operating performance and financial condition. At the same time, it obscures information regarding cash flow. Therefore, we can think of the cash flow statement as a recasting, or translation, of the income statement to a cash-basis view of company performance.

If cash-basis statements are so useful, why do we go to the trouble of developing accrual statements? The point is that both are useful. If you want to assess current profitability of the business, the accrual-basis income statement is essential. If you want to assess what the company owns and owes, the accrual-based balance sheet provides just that information. To repeat, what these accrual-based statements do not provide is a good view of the flow of cash in and out of the business.

The cash flow statement summarizes the primary sources of cash (typically, transactions with customers and various external financing sources) and the primary uses of funds (primarily, investments in working capital and fixed assets, debt repayment, and dividend payments). This statement helps evaluate such questions as the following: How important to the company is external financing? What portion of the company's cash needs (i.e., uses) is derived from internal rather than external sources? Are the company's needs dominated by investments in working capital or fixed assets, or payments to creditors or shareholders? Did the company experience unusual cash needs during the year? Were extraordinary cash sources tapped during the year?

SOURCES AND USES OF CASH

Cash Flow Provided by Operations

Customers provide cash. Little of that cash, however "sticks" with the company, since most of it must quickly be paid out to employees (as salaries) and to trade suppliers (for goods and services). How much does stick? As a first approximation, the cash that remains with the company is the amount of profit earned, the difference between revenue and expense. But this is only an approximation. Not all revenue brings immediate cash receipts, since some cash payments may not be due for several months while others may have been received as down payments. Likewise, not all expenses consume cash; for example, depreciation and amortization (to be discussed in more detail in Chapter 8) are allocations of portions of past cash expenditures to the current accounting period.

Think of the ways in which net income differs from—by either overstating or understating—the cash that is developed from operating the business:

How Net Income Understates Cash. Expenses on the income statement that do not require the payout of cash. Depreciation of fixed assets and amortization of intangible assets—**noncash expenses**—are the major category. Frequently, too, income tax cash payments in a year are less than the income tax expenses shown on the income statement, because according to provisions in the tax laws, some of these taxes need not be paid (and thus are deferred) until future periods. Another example: Suppose the company stretches out the time it takes to pay vendors. Even if its total purchases from vendors were exactly the same this year as last, the balance of its accounts payable would increase; the company will not have paid out as much cash as is indicated by its total expenses calculated on an accrual basis. In this case, the increase in accounts payable (A/P) represents a source of cash. Moreover, you would expect A/P to increase as the company grows, because year-to-year total purchases from vendors increase. This, too, represents a source of cash. Of course, if the opposite occurs—if the A/P balance decreases—the reduction in A/P represents a use of cash.

How Net Income Overstates Cash. Similarly, suppose a company's total sales to customers are the same for two years running. If, in the second year, its customers take longer to pay, then the company's accounts receivable balance will increase. Accordingly, the revenue (or income) as shown on the income statement overstates the amount of cash received from customers. Also, as the company grows, its A/R and inventory balances are likely to grow. These additional investments represent uses of cash.

Net income adjusted in these ways is considered the *cash flow provided by operations:*

Net Income
Plus: Noncash expenses
        Increases in current liabilities, and decreases in current assets
        Differences between income tax expenses and income tax payments
Less: Increases in current assets, and decreases in current liabilities

Cash Flow Used for Investing

What are the primary uses of cash? Investments. Examples are investments in property, plant, equipment, and software; in securities of other companies (including, in some cases, the acquisition of all outstanding shares and, thus, of the entire company); in intellectual property (patents and know-how) purchased from other companies. Suppose a company sells a truck, a piece of real estate, or securities in an unrelated company. These "disinvestments" represent sources of cash. All together, these are called *cash flow used for investing.*

Cash Flow from Financing

Where else does a company obtain its cash? Our discussion in Chapter 3 of the balance sheet and the fundamental accounting equation emphasized that the right side of the accounting equation defines where financial resources are obtained to invest in the assets shown on the left side. Two important sources are creditors and investors in newly issued common stock of the company. Thus, if the bank lends your company additional money, the resulting bank loan is a source of investable funds in the period when the credit is extended. Another source is the sale of new shares of the company's securities, which is represented by an increase in the Invested Capital account, part of owners' equity. Again, the opposite—the company's repayment of loans* and its repurchasing of its own outstanding common shares—constitute uses of cash. Another use of cash is paying dividends on common stock; recall that dividends are not included in expenses. These flows are referred to as *cash flow from financing*.

In summary, then, a use of cash is created by an increase in an asset account or a decrease in a liability or owners' equity account, beginning with net income increasing the Retained Earnings account in owners' equity. It is also obvious, but worth repeating, that for each period, sources must equal uses. Otherwise, the fundamental accounting equation—assets equal liabilities plus owners' equity—will not be satisfied. The cash flow statement consists of three parts:

- Cash flow provided by operations
- Cash flow used for investing
- Cash flow from financing

The net of these three parts must equal the increase or decrease in the company's cash balance (including cash equivalents) between the beginning of the accounting period and the end.

EXAMPLE: THE METCALFE COMPANY

To illustrate the form of the cash flow statement, and the procedures used to develop it, we return to the Metcalfe Company example used in Chapters 3 and 4.

To repeat, a reduction during an accounting period in asset balances and an increase in liabilities and owners' equity balances represent sources of

---

*A loan repaid by issuing to the creditor new shares of common stock is equivalent to selling stock to the creditor for cash (a cash source) and immediately using the cash to pay off the loan (a cash use).

cash; the opposite represents a use of cash. We can begin, then, by simply comparing Metcalfe's balance sheets for June 30, 2003, and June 30, 2004, the beginning and end of the accounting period. (Metcalfe uses a June 30 fiscal year-end.) Information from both balance sheets appears in Table 5-1. The last column in the table shows the simple arithmetic differences in account balances between these two dates (where cash uses—increases in asset balances, and decreases in liabilities and owners' equity balances—are shown in parentheses).

A quick look at Table 5-1 tells us that assets have increased by $22,223, as have, of course, the sum of liabilities and owners' equity. But not every category has increased. The cash flow statement in Table 5-2 allows us to understand in more detail where Metcalfe obtained cash in 2004 and where it invested the cash. Metcalfe is growing modestly and thus we would expect its assets to grow. It is not so clear where Metcalfe obtained the cash to finance the growth.

TABLE 5-1. The Metcalfe Company: Balance Sheets for June 30, 2003, and June 30, 2004

	2004	2003	Difference
Assets			
Current assets			
Cash	$ 12,466	$ 13,613	$ 1,147
Accounts receivable	93,050	85,772	(7,278)
Inventory	46,540	40,337	(6,203)
Prepaids	3,900	4,200	300
Total current assets	155,956	143,922	(12,034)
Investments	36,500	40,000	3,500
Property and equipment	113,650	101,711	(11,939)
Intangibles	13,350	11,600	(1,750)
Total Assets	$319,456	$297,233	($22,223)
Liabilities and Owner's Equity			
Current liabilities			
Accounts payable	$ 26,750	$ 21,119	$ 5,631
Salaries and employee benefits payable	12,402	11,061	1,341
Taxes payable	9,274	10,663	(1,389)
Notes payable within one year	25,000	30,000	(5,000)
Total current liabilities	73,426	72,843	583
Long-term debt	75,000	65,000	10,000
Owners' equity			
Invested capital	75,000	71,800	3,200
Retained earnings	96,030	87,590	8,440
Total owners' equity	171,030	159,390	11,640
Total Liabilities and Owners' Equity	$319,456	$297,233	($22,223)

TABLE 5-2. The Metcalfe Company: Development of the Cash Flow Statement, Year Ended June 30, 2004

	From "Difference" Column in Table 5-1	Adjustments	Cash Flow Statement
Cash flow provided by operations:			
Profits	$8,440[a]	$12,172 + 15,300	$35,912[b]
Add noncash expenses (revenues):			
Depreciation and amortization		28,941 + 3,200	32,141
(Gain) loss on sale of assets		(4,570)	(4,570)
Net change in working capital:			
Accounts receivable	(7,278)		(7,278)
Inventory	(6,203)		(6,203)
Prepaids	300		300
Accounts payable	5,631		5,631
Salaries and employee benefits payable	1,341		1,341
Taxes payable	(1,389)	1,389	0
Notes payable within one year	(5,000)		(5,000)
Income taxes paid	—	(16,689)	(16,689)
Subtotal	(4,158)	39,743	35,585
Cash flow used for investing:			
Investments	3,500[c]		3,500
Property and equipment:			
Acquisition	(11,939)	(28,941) + (2,130)	(43,010)
Disposition		6,700	6,700
Intangibles	(1,750)	(3,200)	(4,950)
Subtotal	(10,189)	(27,571)	(37,760)
Cash flow from financing:			
Long-term debt	10,000		10,000
Invested capital	3,200		3,200
Dividends paid	—	(12,172)	(12,172)
Subtotal	13,200	(12,172)	1,028
Net (increase) decrease in cash	$1,147	0	$ 1,147

[a] Increase in retained earnings.
[b] After adjustments, this amount is income before taxes, as shown in Table 4-1.
[c] Sale of investments, and thus a source of cash.

We begin the construction of Metcalfe's 2004 cash flow statement by placing the figures from the "difference" column in Table 5-1 into the first number column of the cash flow statement template shown in Table 5-2. Note that cash is the balancing item. Since cash decreased during the year, the cash account represents a $1,147 source—just equal to the arithmetic sum of the subtotals in each of the three categories:

1. Cash flow provided by operations
2. Cash flow used for investing
3. Cash flow from financing

In order to create a more useful and insightful cash flow statement (the rightmost column on Table 5-2), the "Difference" column, which is necessarily rough, requires same adjustment (the middle column). Some of the information for these adjustments comes from the income statement in Chapter 4 (Table 4-1). But notice we have left some blanks. If we had a full set of notes to Metcalfe's financial statements, they would provide still more information, for refining the "Difference" column.

Profit and Income Taxes Paid

Table 4-1 shows Metcalfe's net income in 2004 as $20,612, and yet the change in retained earnings is only $8,440. What accounts for the difference? In 2004, Metcalfe paid to shareholders dividends of

$$20,612 - 8,440 = \$12,172.$$

Thus, we should show as a cash source from operations the full $20,612 and show as a financing use the $12,172 paid as dividends. These are shown in the "Adjustments" column of Table 5-2: a source balanced by a use.

Moreover, we can take a separate look at income taxes paid. The income statement in Table 4-1 shows tax expenses in 2004 of $15,300. But the balance in the Taxes Payable account (see Table 5-1) actually was reduced by $1,389, a use of cash. Accordingly, Metcalfe must have paid

$$15,300 + 1,389 = \$16,689$$

in cash for income taxes during the year. If we show these tax payments as a separate line item on the cash flow statement, the "Profits" entry on the statement becomes income before taxes rather than net income. These adjustments are also shown in Table 5-2's second column.

Noncash Expenses

The notes to Metcalfe's financial statements would tell us (although the income statement in Table 4-1 does not give this level of detail) the noncash expenses included in the cost of goods sold and the operating expenses for Metcalfe in 2004. Noncash expenses are dominated, in most companies, by depreciation expenses (for property plant and equipment) and amortization expenses (for intangible assets.)

Metcalfe's notes to the financial statements tell us that depreciation expenses for 2004 were $28,941, and the amortization of the intangibles was a $3,200 expense. If Metcalfe had purchased no new property and equipment in 2004, the value of its fixed assets would have decreased by $28,941, the amount of the depreciation expense. Instead, they increased by $11,939, as shown on Table 5-1. Thus, the following must have transpired in 2004:

Property and equipment, 6/30/04	$113,650
Property and equipment, 6/30/03	(101,711)
	11,939
plus: depreciation expense, 2004	28,941
Additions to property and equipment, 2004	$ 40,880

One more adjustment in the "property and equipment" entry should be considered. Metcalfe may have sold fixed assets as well as purchased them. For most companies, the sale of fixed assets is not a major item and, if the assets are fully depreciated (that is, if accumulated depreciation equals the original cost), the sale generates cash but does not affect the "net" balance sheet value. Metcalfe's notes tell us that the company sold, for a total of $6,700 in cash, various fixed assets that had a book value (original cost less depreciation accumulated to date) of $2,130. Incidentally, the difference between the price received for these disposed assets and their book value—$6,700 − $2,130 = $4,570—is taxable income to Metcalfe but is "nonoperating revenue." Thus, purchases of new property and equipment must have amounted to an additional $2,130 to make up for the book value of the assets sold, for a total of $40,880 + $2,130 = $43,010. The net result of all this is:

Increase in nonoperating revenue	($4,570)
Increase in acquisition of property and equipment	(2,130)
Disposition of property and equipment	6,700
	$ 0

All of these property and equipment adjustments are shown in the middle column of Table 5-2.

Now let's return to the noncash amortization of intangibles. These adjustments parallel the adjustments for depreciation, and thus are also shown in Table 5-2:

Intangible assets, 6/30/04	$13,350
Intangible assets, 6/30/03	(11,600)
	1,750
Plus: amortization expense	3,200
Acquisition of additional intangibles, 2004	$ 4,950

While these are the most common noncash expenses and revenues, there are others, all of which are discussed in more detail later in this book:

- An increase in reserves for the valuation of assets; for example, reserves for bad debts (uncollectible accounts receivable) or inventory obsolescence
- Gains or losses associated with the translation of foreign currency into local currency; for example, Euros into U.S. dollars

INTERPRETATION

Take another look at Table 5-2. What can we observe from the third column, which represents the fully adjusted cash flow statement? Metcalfe's cash inflow provided by operations just about equals its cash outflow for investing, with those investments concentrated in property and equipment. During this period, cash flows provided by operations is nearly twice the net income of $20,612 (see Table 4-1), primarily because of high noncash expenses. Cash flows from financing were modestly negative, but note that dividends exceeded new long-term debt. Is Metcalfe paying higher dividends than it can sustain over the future? Note the following:

- Dividends are about 60% of net income, hardly an unreasonable level.
- Dividends are only about 30% of total cash inflows from operations.
- While long-term debt increased by $10,000, notes payable declined by $5,000, and thus total new debt—long-term and short-term—increased by only $5,000.

Property and equipment investments, at nearly $41,000 (see the calculation in the previous "Noncash Expenses" section), were high in 2004. Is this unusual—due, say, to an expansion of manufacturing capacity? How fast is Metcalfe growing? As Chapter 9 will discuss, we need to answer those questions before we can conclude whether Metcalfe is generating enough cash to finance its growth.

Two other points are worthy of comment. First, the "bottom line" of the cash flow statement—net (increase) decrease in cash—is often not of much interest, as is the case here. Working balances of cash jump around day to day. The $1,147 decrease in the cash balance is less than 10 percent of Metcalfe's June 30, 2004, cash balance (see Table 5-1). It is really the subtotals of the three sections of the cash flow statement that contain the interesting information.

Finally, you might wonder about the small ($3,200) increase in invested capital. This increase probably resulted from the exercise of employee stock options, discussed further in Chapters 10 and 11, and small share purchases (typically at a modest discount from market price) by employees pursuant to a shareholder-approved employee stock purchase plan. Not infrequently, companies repurchase their shares of common stock, particularly when they have excess cash and believe the market price of their stock is low. Had this been the case at Metcalfe, we would have made adjustments to the cash flow statement to show separately this repurchase as an investing use of cash, on the one hand, and the sale of additional common stock as a financing, on the other hand, rather than simply the net of the outflow and the inflow.

OTHER EXAMPLES

Cash flow statements help particularly in understanding the financial condition and performance of companies that are expanding or contracting rapidly, or are undergoing sharp changes in their financial structures. Here are two simplified examples.

Table 5-3 shows the cash flow statement for Consolidated Industries, Inc., a medium-sized company undergoing substantial structural changes. First, note that earnings are quite modest in relationship both to investments in property, plant and equipment ($221 million) and to dividends paid (twice 2005 earnings). The sale of fixed assets is the largest source of cash, but the uses of cash are dominated by investment in new property, plant, and equipment; apparently, the company is redeploying its assets away from some activities in favor of others. The source of cash from working capital reduction is explained either by improved asset management (accounts receivables and inventories) or, more likely, by a decline in the company's business activity in 2005. (Note that accounts payable also declined.)

Why did Consolidated Industries have to incur $60 million more in long-term debt? About one-third of this borrowing was used to repay short-term notes payable ($22 million), but the balance funded substantial investments in property, plant, and equipment; these investments in production facilities could not be funded by internal sources (earnings plus noncash expenses) of $68 million and the sale of other fixed assets. Obviously, two conditions revealed in this cash flow statement cannot persist for Consolidated Industries if it is to remain a viable company: dividend payments in excess of earnings, and increased long-term debt in a business that is apparently not growing.

Table 5-4, the cash flow statement for Comptronics, Inc., presents quite a different picture. A much smaller company than Consolidated Industries, Comptronics is apparently growing. It pays no common stock dividends and has this year succeeded in selling $20 million of additional common stock.

TABLE 5-3. Cash Flow Statement, Consolidated Industries, Inc., for 2005

	$ Millions
Cash flow provided by operations	
Net earnings	$25
Adjustments for noncash expenses	
Depreciation and amortization	33
Other, net	10
Net (increase) decrease in:	
Accounts and notes receivable	53
Inventories	33
Net increase (decrease) in:	
Accounts payable and accrued liabilities	(42)
Subtotal	112
Cash flow used for investing	
Investments in property, plant, and equipment	(221)
Sale of property, plant, and equipment	102
Subtotal	(119)
Cash flow from financing	
Increase in long-term debt	60
Increase (decrease) in notes payable	(22)
Dividends	(50)
Subtotal	(12)
Increase (decrease) in cash and cash equivalents	($19)

The proceeds from the stock sale were used to repay $10 million of long-term debt, and the balance, combined with internally generated cash, funded investments in new plant and equipment and in increased working capital necessitated by the company's growth. The cash inflow from operations (including noncash expenses) plus the increase in accounts payable just balanced the increases in accounts receivable and inventories. This cash flow statement reveals the kind of dynamic financing and investing typical of smaller, growth companies.

SUMMARY

The cash flow statement is a third financial statement, supplementing—and providing different information from—the income statement and the balance sheet. The cash flow statement focuses on the inflows and outflows of cash and is comprised of three sections:

- Cash flow provided by operations: net income adjusted for noncash expenses and changes in working capital (current assets and current liabilities)

TABLE 5-4. Comptronics, Inc.: Cash Flow Statement for 2002

	$ Millions
Cash flows from operating activities:	
Net earnings	$ 6
Noncash expenses:	
Depreciation and amortization	3
Other, net	1
(Increase) decrease in:	
Accounts receivable	(11)
Inventories	(5)
Increase (decrease) in:	
Accounts payable and accrued liabilities	6
Subtotal	0
Cash flows for investment activities:	
Investment in property, plant, and equipment	(9)
Other, net	2
Subtotal	(7)
Cash flows from financing activities:	
Proceeds from sale of common stock	20
Repayment of long-term debt	(10)
Subtotal	10
Increase (decrease) in cash and cash equivalents	$ 3

- Cash flow used for investing: investments in long-term assets, adjusted for depreciation and amortization and for disposition of such assets
- Cash flow from financing: sales and repurchases of common stock, increases and decreases in long-term borrowing, and payment of dividends

The bottom line of the statement, net change in cash balances, is less useful than is the relationships among and within the three sections. Both internal and external audiences are keenly interested in cash flows, particularly as a company undergoes rapid growth, decline, or change in fundamental business strategy.

NEW TERMS

Cash flow statement (statement of cash flows). The financial statement that details for a specified period the primary sources of cash (typically, operations and additional financing) and uses of cash (typically investing in assets, repayment of debt and payment of dividends). The cash flow statement supplements the two primary statements, the income statement and the balance sheet.

Noncash expense. Expenses included on the income statement, and thus in the determination of profit, that do not require the payout of cash.

EXERCISES

1. What information does the cash flow statement of a company show that is not readily apparent from the income statement and balance sheet?

2. What are the primary sources and uses of cash in each of these sections of the cash flow statement?
   a. Cash flow from financing
   b. Cash flow provided by operations
   c. Cash flow used for investing

3. Indicate whether each of the following is a source of cash or use of cash:
   a. Depreciation expense
   b. The sale of production equipment at the end of its useful life
   c. Dividends paid to shareholders
   d. An increase in accounts payable
   e. An increase in inventory
   f. The sale of securities held for investment
   g. The repayment of borrowed capital
   h. The amortization of intangibles
   i. An increase in the Deferred Taxes account

4. Explain in your own words why the depreciation expense is considered to be a source of cash. What is the mechanism by which an *increase* in this expense also leads to an *increase* in cash inflow?

5. If company T's balance sheets at two successive year-ends show a $400,000 increase in net fixed assets, and during the year the company incurred depreciation expenses of $200,000, what was company T's gross investment in new fixed assets during the year?

6. In the same year, why would a company both pay dividends to shareholders and also sell newly issued common stock? Why are both transactions considered part of the "cash flow from financing" section of the cash flow statement?

7. Why might a company have a higher expense for income taxes for a particular year than the amount of cash income taxes that it is required to pay for the year?

8. Some small rapidly growing companies finance part of their growth by "stretching" the payment of their accounts payable—for example, paying

vendors 70 days rather than 30 days after receipt of invoice. Is this sensible? Discuss the pros and cons of this approach.

9. Name three "financing" activities that represent uses rather than sources of cash.

10. Under what condition(s) is a company likely to have negative cash flows from operating activities?

11. Explain in simple terms why an increase in accounts receivable balances between two dates represents a use of cash for the period, and why a decrease in inventory balances between the two dates represents a source of cash.

12. As a company expands, its working capital typically increases, thereby using cash. What primary sources of cash offset this use by the growing company?

13. Here is the Ohrstrom Company's cash flow statement for 2006:

Cash flow from operations:		
Net income	$	624
Depreciation		511
Decrease (increase) in working capital		(398)
		733
Cash flow from investing:		
(Investment) in fixed assets		(1,355)
Disposition of fixed assets		83
		(1,272)
Cash flow from financing:		
Sale of common stock		1,000
Reduction in long-term debt		(300)
Dividends		(220)
		480
Net change in cash and cash equivalents	($	59)

Ohrstrom's 2005 year-end balance sheet revealed the following summary balances:

Working capital	$5,840
Fixed assets, net	7,971
	$13,811
Long-term liabilities	$ 4,000
Shareholders' equity	9,811
Total	$13,811

a. What is the value of Ohrstrom's net fixed assets on December 31, 2006?

b. What is the value of Ohrstrom's shareholders' equity on December 31, 2006?

c. What evidence from the cash flow statement indicates that Ohrstrom's sales volume grew in 2006 compared with 2005?

d. Did Ohrstrom's total long-term debt increase or decrease as a percentage of the company's shareholders' equity during the year?

e. Do you think Ohrstrom's executives should be concerned about the reduction in cash and cash equivalents in 2006? Explain your answer.

14. The Fitzgerald Partnership balance sheets for September 30, 2003, and September 30, 2004, are shown below in summary format:

	September 30	
	2004	2003
Current assets	$ 762,000	$ 812,000
Fixed assets, net	986,000	986,000
	$1,748,000	$1,798,000
Current liabilities	$ 532,000	$ 468,000
Long-term debt	180,000	360,000
Partnership capital	1,036,000	970,000
	$1,748,000	$1,798,000

Construct a cash flow statement in as much detail as you can for the year ended September 30, 2004, assuming the following: no new investment in cash was made by the partners; half of the partnership's net income was paid out to the partners; $10,000 of fixed assets were sold, and $135,000 of new fixed assets were acquired.

15. The Schuster Corporation's balance sheets at the end of the two most recent fiscal years are shown below:

	Fiscal Years Ended October 31, ($000)	
	2006	2007
Assets		
Cash	$1,732	$1,915
Accounts receivable	7,219	8,613
Inventory	4,086	5,118
Other current assets	781	836
Total current assets	13,818	16,482
Fixed assets, at cost	16,083	16,991
Less: accumulated depreciation	(7,616)	(9,085)
Net fixed assets	8,467	7,906
Total assets	$22,285	$24,388

	Fiscal Years Ended October 31, ($000)	
	2006	2007
Liabilities and Owners' Equity		
Accounts payable	$3,923	$4,772
Other current liabilities	4,086	4,931
Total current liabilities	8,009	9,703
Long-term debt	5,500	5,000
Owners' equity		
Capital stock	2,861	3,015
Retained earnings	5,915	6,670
Total liabilities and owners' equity	$22,285	$24,388

For fiscal year 2007, Schuster recorded $2,300,000 in depreciation expense and paid dividends totaling $1,100,000. Construct a cash flow statement for Schuster for 2007 in as much detail as possible.

CHAPTER 6

PRINCIPLES, RULES, AND MECHANICS OF FINANCIAL ACCOUNTING

By now, you recognize that, because different individuals observe and measure the same event or condition in different ways, differences of opinion can and do arise about how best to record in monetary terms that event or condition. Moreover, an accountant may be influenced, properly or improperly, to observe and measure so as to improve the apparent financial performance or condition of the company; more on this issue in Chapter 11.

You need to understand the principles and some of the rules that are widely followed in accounting practice today. We will also look at the U.S. rule-making bodies that promulgate specific constraints and rules that guide day-to-day accounting practice. This chapter also introduces you to the mechanics of "recording, classifying, and summarizing," accounting entries—recall Chapter 1's definition of accounting.

A BRIEF REVIEW

As you consider accounting principles and rules, bear in mind the following:

- The valuation of assets and liabilities determines the valuation of owners' equity, as owners' equity is simply the difference between assets and liabilities.
- Because the income statement is an amplification or elaboration of changes in owners' equity, the valuation of assets and liabilities ultimately determines the magnitude and timing of revenues and expenses on the income statement.

• The two basic accounting products—the balance sheet and the income statement—form one integral whole: the balance sheet is a snapshot of assets owned and liabilities owed on a certain date; the income statement details changes in assets and liabilities (and thus owners' equity) for an accounting period.

Recall that there are three bases for valuing assets and liabilities:

1. *The time-adjusted value method:* the present value of the future stream of positive benefits (for assets) and negative benefits (for liabilities), discounted at an interest rate appropriate to the particular entity.
2. *The market value method:* the current market value of each asset and liability; for assets this often approximates replacement cost.
3. *The cost value method:* the original cost of the asset or liability. This method recognizes certain adjustments to original cost as fixed assets age, inventory becomes obsolete, and so forth.

Recall, also, the advantages and disadvantages of these three methods. The time-adjusted value method is the most intellectually satisfying basis for valuation, recognizing estimates of the future in today's valuation, but it is also the most difficult and least reliable method. The cost value method, on the other hand, minimizes the need to assess the future; this very real advantage greatly mitigates its disadvantages and is the reason for its dominant use.

Accounting struggles and debates over the years have led to the formulation of principles and specific rules to resolve many, but certainly not all, accounting valuation questions. The principles and rules, based largely, but not entirely, on the cost value method evolve in a series of small steps as the accounting profession attempts to resolve recurring dilemmas.

Remember that the principles and rules incorporate compromises between conflicting accounting objectives. You may not always agree with the rules, but you should at least understand them. As you interpret financial information, be alert to the limitations they impose.

PRINCIPLES

The basic principles derive directly from the accounting definition in Chapter 1 and the distinction drawn in Chapter 4 between the accrual and cash-basis accounting systems.

Expression in Monetary Terms

Only those events and transactions that can be expressed in monetary terms can be recorded. Necessarily, then, the accounting story is incomplete, since many important business events cannot be reduced to monetary terms; among these are the hiring and leaving of personnel; the discovery of a new process, product, or technique; and the resolution of a key management disagreement.

Entity

The entity being accounted for—a business, a nonprofit organization, a person, a family, or a government unit—must be carefully delineated, and only events affecting that specific entity should be measured and recorded in its accounting system. For a business, accounting is restricted to events affecting that business; the effects of the same events on individuals and outside organizations—for example, employees, stockholders, customers, labor unions, and suppliers—are ignored. For a governmental unit, accounting is confined to events affecting that unit; the effects of the same events on citizens served or individuals employed are ignored.

Going-Concern Assumption

Unless you know for a fact to the contrary, your valuations should assume that the entity will continue to operate for an indefinite future period, providing essentially the same services as it does today. This **going-concern assumption** requires that the valuations of assets and liabilities not reflect the amount that could be realized if the entity sold its plants, equipment, or other producing assets, since we assume the entity has no intention of selling them. Instead, the focus is on the value of those assets in the context of their present use.

Of course, should the entity decide to liquidate or sell its assets (or a significant portion of them), the going-concern assumption would no longer be valid; those assets and liabilities must then be revalued under this new assumption, and the resulting values would likely be considerably different from those under the going-concern assumption. If currently very productive, the assets might be valued substantially higher, but much lower if they were generally unproductive and were to be abandoned or sold at a distress price. In recent years, many large corporations have discontinued certain lines of business. Such a decision negates the going-concern assumption and triggers a reevaluation of the assets and liabilities related to that line of business, often resulting in a substantial decrease in owners' equity.

Conservatism

When faced with reasonable doubt as to whether an asset or revenue should be stated at one value or another *lower* value, the accountant should choose the conservative lower value; correspondingly, when faced with reasonable doubt as to whether a liability or expense should be stated at one value or another *higher value,* the accountant should choose the conservative higher value. That is, opt for the choice that results in lower owners' equity. The key phrase is "reasonable doubt."

This **conservatism** principle does not direct the accountant to understate owners' equity purposely, although some understatement may result from its application. Rather, it demands reasonable caution: provide for probable losses, and do not record revenues before they are clearly earned.

Heated arguments between accountants and business executives arise over the application of this principle. The executive is typically optimistic; the accountant must lean in the direction of pessimism. The surprises that occur in business—unexpected or unanticipated events—most frequently, although by no means always, have the effect of reducing owners' equity rather than increasing it. Thus, in practice, this conservatism principle results in greater realism in financial statements.

Realization

The **realization** principle follows from the accrual concept introduced in Chapter 4. We noted that all revenue is earned at the single moment when the particular goods or services are delivered or furnished. Intuitively, you may feel that not all of the revenue, and thus not all of the profit, from a sales transaction is earned at one moment in time. You know, and economists regularly remind us, that each segment of the business has a hand in producing value. Nevertheless, we rely on the simplifying convention that all the revenue is recognized on the date the goods or services are exchanged between seller and buyer—or, in accounting parlance, "when the earnings process is complete and collection is reasonably assured."

There are, however, exceptions to this generalization. When in-process manufacturing time is very long, and the value added at various process stages can be accurately and objectively measured, portions of the revenue may be recognized in advance of final delivery or exchange. Examples include large building or other construction projects, or the fabrication of complex equipment such as space vehicles or nuclear power generation plants. This so-called *percentage-of-completion* accounting for sales was devised more than a century ago for shipbuilders; regrettably, today it is used inappropriately by some companies to accelerate revenue recognition and therefore profits. Also, if there is considerable doubt that the buyer will ever pay—for example, because of financial failure or because the customer retains the right to return the goods following evaluation—revenue recognition should be delayed beyond the date of delivery, perhaps until the cash is finally received.

Note that, while the distinction between cash- and accrual-based accounting is only a matter of timing, timing is everything. Most valuation disagreements among accountants simply revolve around timing. Even under the accrual concept, reasonable persons frequently disagree as to just when goods or services are actually shipped or delivered or consumed. The accounting systems illustrated in this book all utilize the accrual concept.

Consistency

Since these principles are necessarily vague and general, accountants are left with wide discretion in recording particular transactions and events. Specific accounting rules only narrow, they do not eliminate, this discretionary area. The **consistency** principle requires that, having decided how to account for a particular transaction within a particular company, the company must consistently employ that method for all similar transactions: consistency over time as well as consistency among similar transactions. Consistency facilitates comparisons of financial data across accounting periods. If consistency is relaxed, comparability is lost and analysts end up comparing the apples of one accounting period to the oranges of another.

Of course, conditions change over time; to prohibit entirely any change in accounting policies and procedures would thwart efforts to present more useful and meaningful data. Consistency requires, instead, that when accounting practice is changed, in order to enhance the usefulness of the financial data, accountants must thoroughly disclose the nature of the change and estimate its effect on the financial reports.

Materiality

The principle of **materiality** instructs accountants to focus attention on the significant, or material, events in the business. Measuring, valuing, and recording financial data costs money; if the benefit of the resulting information is less than the cost of collecting, valuing, and processing it, then today's procedures should be abandoned and reasonable valuation and estimates and simpler procedures should be used.

For example, tracking the use of low-value parts in an assembly process— parts such as nuts, bolts, and washers–is typically inefficient and uneconomic. Estimating these inventory values for a particular job is typically quite satisfactory because the values are immaterial in the context of the total enterprise; even a substantial estimating error has an immaterial financial impact.

Another example: if you are billed monthly for certain services, such as utilities and telephone, it is typically unnecessary to allocate charges among the current accounting period, the past accounting period, and the next accounting period. If the bill is for approximately the same amount each month, and you are careful to record one bill in each monthly accounting period, that's good enough. Allocations are expensive and can be avoided if their effects are immaterial.

SPECIFIC RULES

As valuable as these principles are, they are necessarily subject to wide differences in interpretation and application. In some ways, that is their strength:

the principles provide accountants with useful guidance as they struggle to value the huge variety of transactions and conditions encountered in contemporary, complex, global business enterprises.

But the pressure for consistent financial reporting across firms (not just across time within a single firm) has led the accounting profession and agencies of the federal government to promulgate a great volume of specific rules. These typically relate to particularly troublesome and controversial accounting areas. While you are free, for internal management purposes, to develop data in any manner you deem useful, public reporting of financial information must follow these rules.

The **American Institute of Certified Public Accountants (AICPA),** an association of professional, or certified, accountants, is a strong voice in the United States for setting accounting policies and rules. For many years, the AICPA issued informal and nonbonding recommendations. In the late 1950s, the perceived need for more standardized procedures led the institute to organize the Accounting Principles Board (APB). From 1959 to 1973, this board issued 31 formal opinions that **certified public accountants (CPAs)** were required to follow. Controversy surrounding the APB's work grew steadily during its 14-year existence, as its opinions dealt with very sensitive and difficult issues.

The need for a more independent rule-setting body, one with industry and academic representatives as well as CPAs, led in 1973 to the formation of the **Financial Accounting Standards Board (FASB),** a board of seven full-time members appointed by a board of trustees. As with APB opinions, FASB pronouncements must be followed by professional accountants in the United States. The FASB continues to tackle many thorny and unresolved accounting issues. The FASB also coordinates with international rule-making bodies, particularly the International Accounting Standards Board (IASB), in its attempt to gain global consistency in financial reporting. Thus far, the FASB's rules have been met with a good deal of both opposition and support, a condition likely to continue. Since different audiences for financial statements have different needs and objectives, it is unrealistic to expect that any rule aimed at conclusively settling a highly controversial issue will be warmly embraced by all. Only if the FASB proves effective and efficient—and its record to date is not encouraging—will rule making remain in the private sector; otherwise, it is likely to be taken over by the government.

Indeed, a governmental regulatory agency, the **Securities and Exchange Commission (SEC),** in its role as watchdog for the investing public, is increasingly active in pressuring the FASB and requiring that certain standardized accounting rules be followed in financial statements issued to public shareholders. While only public companies whose securities are traded in interstate commerce are affected, virtually all major U.S. corporations fall into this category; thus, SEC requirements become de facto rules for the entire accounting profession. Reactions to recent accounting scandals have stimulated the SEC to become more active in rule making. In late 2003, a new body, the Public Company Accounting Oversight Board (PCAOB), legislated by the U.S. Congress, began issuing accounting standards. The respective roles of the FASB and the PCAOB, and the relationship between the two, will not be clear for several years.

Typically, the formal rules set forth by the FASB and the SEC are so specialized that we won't worry about them here. Among the areas these rules now cover are calculating earnings per share, accounting for the value of fluctuating international currencies, accounting for research and development expenses, accounting for leases, accounting for taxes, and accounting for stock options and other management incentive payments.

The full set of principles and specific rules serve to define what is widely referred to as **Generally Accepted Accounting Principles** (GAAP). Both auditors and the SEC require that companies in the United States conform to so-called GAAP accounting. You can be sure that new rules will continue to emerge in an attempt to reduce the diversity of accounting treatments, to improve comparability among financial statements, and to reduce accounting abuses.

Other agencies of the U.S. government also insist on certain standardized accounting procedures for those companies that they regulate or with whom they contract. For example, companies, universities, and research institutes that perform contract research for the government and defense and aerospace contractors must submit to certain specified accounting procedures.

ROLE OF THE CPA AUDITOR

The independent and certified accountant plays a key role in ensuring that accounting guidelines and specific rules are followed. Virtually all companies—large and small, privately or publicly owned—engage CPAs to audit their accounting records and systems and to certify their financial statements.

Certified public accountants must demonstrate, through formal education, on-the-job experience, and rigorous testing, a level of accounting competency that warrants certification, or licensing, by the state. As members of a profession, CPAs are required to observe a code of professional ethics. Similar procedures are followed in other countries; chartered accountants in England or Canada are generally equivalent in background and function to CPAs in the United States.

CPA firms offer their services for a fee. They are required to remain independent of the clients whose financial statements they certify, even though hired and paid by these same clients. Their independent status requires that, unlike lawyers, they not be advocates of the client. Rather, the CPA code of ethics underscores the CPA's obligation to the general public, while performing a service to the client. In certifying the correctness and fairness of publicly reported financial statements, the CPA asserts that he or she has performed sufficient review of the accounting procedures and individual records to be able to verify that the financial statements fairly represent the position and condition of the firm and that the firm has followed GAAP accounting. Should these assertions prove incorrect and, subsequently, a member of the general public (typically an investor) who relies on the accountant's certification suffers financial loss, the individual suffering the loss may bring legal action against the CPA for malpractice or incompetence. Therefore, the independent CPA is exposed to substantial potential liability by virtue of this public responsibility.

The process of certifying the accuracy and fairness of financial statements is called **auditing,** and the audit function is central to the independent CPA's public practice. CPAs do not prepare the companies' financial statements (indeed, a CPA who prepared the statements would no longer be independent); rather, CPAs review and test the accuracy of the statements prepared by the companies themselves. Further, auditing does not involve a review of each specific accounting entry, an effort that would be prohibitively time consuming and expensive. Rather, CPAs employ well-developed and standardized procedures to accomplish the following:

1. Review a random sample of routine transactions and valuations of assets and liabilities (such as inventory, accounts receivable, and accounts payable)
2. Test the accounting systems, including computer programs, to assure their reliability and completeness
3. Review other routine procedures and control processes of the accounting department
4. Check out in detail any unusual transactions or accounting entries
5. Make such other tests as may seem necessary or prudent.

To protect shareholders, creditors, and honest and competent company employees, the independent CPA also tests for conditions that might facilitate fraud.

As part of the audit, the independent CPA is expected to blow the whistle on any intentional or unintentional over- or understatement of profit or misstatement of financial position. He or she has responsibilities to the shareholders not to permit overstatements of profit, to the taxing authorities not to permit understatements of profit, and to the company's creditors not to permit an unrealistic presentation of the company's financial position. Surely by now, you see that many opportunities exist for disagreements between the CPA performing the audit and the company's managers who prepared the financial statements being audited. The client-CPA relationship is unusual, often delicate, and sometimes stormy.

Those who rely on audited financial statements often have the mistaken expectation that the auditors will root out any and all fraudulent practices. In fact, auditors specifically disclaim their ability to detect and expose fraud.

When fraud involves collusion among top executives, as has been the case in several high-profile instances in recent years, it is extraordinarily difficult to detect. The amount of work that would be required of the auditors to assure the absence of any fraud would be prohibitively expensive.

A CPA may practice professionally as an individual or as a member of a partnership. Partnerships vary in size, but the profession is dominated by four large firms, which practice throughout the world. In addition to performing audits, the larger accounting firms also provide advisory services, tax return preparation, and tax planning. Until 2002, the firms also did a great deal of general consulting; in that year, following a number of high-profile scandals, consulting was deemed to compromise the firms' independence, so the large firms divested their consulting practices.

HOW GOOD ARE THESE PRINCIPLES AND RULES?

Readers of financial statements, even sophisticated and experienced financial analysts, have too strong a tendency to accept formal financial statements as representing truth with only a small margin of error. Yet, accountants must utilize estimates and approximations if they are to provide useful financial data in a timely manner. The principles and rules discussed here help accountants exercise reasonable judgment, but they surely do not resolve all accounting dilemmas, particularly as to valuations.

As the number and complexity of accounting rules have grown, some managements have followed the questionable practice of dressing up their financial statements by pushing specific rules to the limit and ignoring the spirit of the rules and of the underlying principles. Chapter 11 addresses some of the resulting abuses.

The SEC has for years put great emphasis on full disclosure, with the result that the notes to the financial statements continue to expand, becoming ever more complex and difficult to decipher. As critically important as these notes are to the full understanding of reported financials, they have, unfortunately, become increasingly legalistic—even obfuscating in their complexity—and therefore less useful.

Since input data contain estimates and approximations, the output of the accounting system—the financial statements—are at best estimates: of the value of assets and liabilities at a moment in time (the balance sheet) and of profit performance (change in owners' equity) during the particular accounting period. To repeat, financial statements are only estimates. To be completely certain of the amount of profit and, thus, the value of owners' equity, a company would have to cease operations, liquidate its assets, and pay off all liabilities. Only then could one know for certain how much value remained for its owners.

Nevertheless, by following these rules and principles, particularly the principle of consistency, accountants provide extremely useful data to their audiences: to managers for making operating decisions, to stockholders for making investment decisions, and to lenders for making credit decisions.

As an interpreter of financial statements, you must strike a balance: use the financial data to make decisions, but retain a healthy skepticism. Financial statements simply cannot convey absolute truth about the business's financial position or profitability.

INCOME TAXES AND ACCOUNTING RULES

Getting right to the point, let me state that virtually all companies (and many individuals) maintain two sets of accounting records: one for reporting to external audiences, and one for calculating income tax liabilities. This "two sets of books" phenomenon should not be surprising, and it surely is not illegal or unethical.

Income tax laws and regulations, including specific instructions regarding how certain transactions or events are to be accounted for in determining tax liabilities, are developed for the purpose of raising revenue for governments—federal, state, and local. Also, at times legislatures will put in place, by means of the tax laws, certain incentives for business; for example, periodically the U.S. Congress passes tax laws to encourage businesses to invest in new capital equipment. All companies can and should seek to minimize their cash tax obligations; such action is fully consonant with looking out for their shareholders' interests.

But tax laws do not—and are not expected to—define good accounting policy. Because required tax accounting is frequently at odds with required GAAP accounting, two sets of books are necessary. Accountants must not let tax regulations color their thinking about how a particular transaction should be recorded for the most appropriate valuation of assets and liabilities and, thus, of revenues and expenses.

Minimizing tax expense drives tax accounting. Fair valuation drives financial accounting.

ACCOUNTING MECHANICS

We now turn briefly to some of the mechanics of operating an accounting system: the procedures for assembling, organizing, and presenting the vast stream of accounting data. Bookkeeping is emphasized in most accounting courses and books. By contrast, this book assumes that you are, or will be, a user of accounting information—a manager or an investor, for example—not a practitioner of the accounting trade. You seek to be a proficient user, not a skilled bookkeeper. Nevertheless, you need to know something of the mechanics and jargon of bookkeeping, enough to communicate and to visualize the key activities of an accounting department. To be a proficient user of financial statements, you need to know how an accounting system functions.

Everyone in an organization encounters and influences data that end up in the accounting department. Purchase orders, employee time records, invoices to customers, bills from vendors, checks received, and checks issued are all part of this data flow. Thus, you need a sense of the way these and other documents are utilized by the accounting department.

Recall the verbs in our definition of accounting: *observe, measure, record, classify,* and *summarize.* To this point, we have concentrated on the first two of those verbs: observe and measure. Once measured and valued, assets, liabilities, revenues, and expenses must be recorded in the accounting system, classified in some useful manner, and summarized for interpretation.

CHART OF ACCOUNTS

Classification is key to bookkeeping. Consider the very many transactions and the huge volume of data generated in any large organization. Perhaps hundreds or thousands of sales take place each day; hundreds or thousands of bills are received, checks paid out, and checks received. Legions of employees earn wages and salaries, and many types of fixed assets are owned. Money is borrowed and repaid; tax obligations are recorded and paid.

Accounting's task is to turn data into information. The accounting system must do more than simply collect and record the data; it must classify the data in ways useful to management, creditors, shareholders, tax collectors, and other financial statement readers. The chart of accounts, introduced briefly in Chapter 3, is a road map of the accounting system, listing all available accounts for classifying the entity's assets owned, liabilities owed, revenues earned, and expenses incurred. The chart of accounts defines the accounting system's classification scheme. Drawing up a chart of accounts is the first step in designing an accounting system.

The nature, size, and complexity of the business organization dictate the particular general ledger accounts needed (recall that "T-account" is shorthand for a general ledger account). For example, some companies classify sales extensively in order to provide management information for product line and pricing decisions; others enterprises may sell only a single product or service, and thus need only one sales account in the chart of accounts. An organization with many separate departments categorizes expenses not only by type of expense (such as salaries, telephone, travel, supplies, rent), but also by the department (machine shop, assembly shop, sales department, accounting department), since each departmental manager needs financial information relevant to his or her segment of the enterprise.

Bear in mind that information costs money, not only to prepare but also to assimilate and interpret. Resist the temptation to construct a very lengthy chart of accounts to provide elaborate data classification. Computerization makes this urge almost irresistible, since the computer can classify and re-classify data with great speed. Too often, however, computer-generated accounting reports are so voluminous as to be both intimidating and not terribly useful.

You want the data summarized—*summarize* is the fifth and penultimate verb in the definition of accounting. The more detailed the classification, the less the data are summarized. Yours is the difficult task of making the best cost-versus-benefit trade-offs: sufficient detail (classification) to be useful, but enough summarization that financial reports can be prepared and interpreted efficiently.

Table 6-1 lists the categories for the Zitar Corporation's chart of accounts; Table 6-2 gives in detail the account names and numbers for certain account categories. Zitar, a small company, does not need an elaborate or extensive chart of accounts, only about 100 general ledger accounts. A larger company with many product lines, departments, and offices might have thousands of accounts. Such a company might, for example, want not a single set of general and administrative (G&A) expense accounts, but further classification to each of the several departments comprising the G&A function.

Note in Table 6-2 that the Zitar Corporation uses 15 current asset accounts, including three categories of inventory, and a separation of receivables into "trade" and "other." The company has four classifications of sales: products A and B, special contracts, and the inevitable "other" category. The company's selling expenses are divided among 11 accounts, but more extensive categorization here might be useful. For example, field sales force salaries might be separated from office support staff salaries, and travel and entertainment expenses might be further divided into air travel, auto rental, hotel, meal, and entertainment expenses. The sales manager might desire this additional detail, or tax reporting requirements might demand it (for example, requiring that entertainment be separated from other travel expenses).

TABLE 6-1. The Zitar Corporation: Chart of Accounts—Table of Contents

Account Number (Range)	Account Category
101–149	Current assets
150–199	Noncurrent assets
200–249	Current liabilities
250–259	Long-term liabilities
261–299	Owners' equity
300–349	Revenue (sales)
350–399	Cost of goods sold
400–699	Operating expenses
400–499	Selling expenses
500–599	Research and development (R&D) expenses
600–699	General and administrative (G&A) expenses
700–799	Other income and expense

TABLE 6-2. The Zitar Corporation: Excerpts from Chart of Accounts

Account Number	Account Name
101–149	Current Assets
101	Petty Cash
105	Cash (Checking)—First Bank
107	Cash (Savings)—Provident Savings & Loan
111	Accounts Receivable—Trade
112	Allowance for Doubtful Accounts—Trade
113	Accounts Receivable—Other
115	Notes Receivable—Trade
117	Notes Receivable—Other
118	Travel Advances—Employees
121	Inventory—Raw Material
125	Inventory—In-process
131	Inventory—Finished Goods
141	Prepaid Expenses
145	Other Current Assets
147	Freight Clearing
300–349	Revenue (Sales)
301	Sales—Product A
305	Sales—Product B
321	Sales—Special Contracts
329	Other Sales
345	Sales Returns and Allowances
347	Sales Price Discounts
400–499	Selling Expenses
401	Sales Commission Expense
411	Sales Salaries Expense
413	Fringe Benefit Expense—Sales
415	Travel and Entertainment Expense
421	Advertising Expense
423	Promotional Literature Expense
425	Miscellaneous Supplies and Other Expense
429	Miscellaneous Outside Services Expense
431	Telecommunications Expense
433	Occupancy Expense
435	Depreciation Expense

THREE KEY ELEMENTS OF ACCOUNTING SYSTEMS

Accounting systems are composed of three primary elements: **source documents, journals,** and **ledgers.** Data flows from source document to journal to ledger. While the physical appearance of these elements varies with the sophistication of the system, the elements are present in a simple accounting system maintained by a single bookkeeper, as well as in elaborate, computerized systems.

Source Documents

As implied by the name, source documents are the original evidence of a transaction to be recorded. What are the primary, recurring transactions in a merchandising or manufacturing company? Essentially, there are only four: (1) sales to customers; (2) cash received from customers (concurrently with the sale in a cash-sale transaction); (3) incurrence of expenses; and (4) cash disbursement, or the payment of cash (again, sometimes concurrent with the incurrence of an expense). Surely, other transactions occur—money is borrowed and repaid, fixed assets are acquired and disposed of, and so forth— but the bulk of transactions that the accounting system must handle routinely are these four.

Sales. The primary evidence of a sale on credit is the invoice prepared for the customer. Cash sales may be evidenced by a copy of a customer receipt or by a cash register tape (or the equivalent in electronic or magnetic media). The customer invoice must contain all the information required to classify the sale, as well as simply to record it: the customer's name, the date, the amount of the sale, the type of product sold or service rendered, the freight charges, the sales tax, the terms of payments, and so forth. The invoice may be coded by geographic region or by responsible salesperson, if the accounting system is to provide sales data classified by sales region or salesperson. (A computer-based accounting system is, or course, capable of multiple classifications.) Assigning numerical codes to the various products, customers, regions, and so forth facilitates data processing.

Cash Receipts. Most cash is received in the form of checks rather than currency, so the check itself becomes a source document. Most organizations, however, are anxious to deposit checks without delay, and, thus, a satisfactory source document is either a copy of the check or the "voucher" attached to checks drawn by commercial or industrial organizations. Just as the cash register tape or copy of the customer receipt evidences a cash sales transaction, it also evidences the cash received in that transaction.

Incurring Expenses. An **invoice** (or **bill**) received from a vendor is evidence of an expense incurred, just as a customer invoice is evidence of a sales transaction. The accrual concept requires timely recording of each expense, typically well in advance of when the bill is paid; thus, the vendor's invoice, not the company's subsequent payment, triggers the accounting entry to record the expense.

The vendor's invoice alone does not provide assurance that the merchandise or services received correspond in fact with those ordered or authorized by the company. Furthermore, the invoice does not verify either that the merchandise has been received or that the prices charged are the agreed-on prices. Therefore, the accounting department collects and matches several documents to confirm that the vendor's invoice represents a bona fide expense to be recorded. A copy of the company's **purchase order** (forwarded from the purchasing department) confirms both the agreed-on price and the fact that the products or services were indeed ordered. Evidence is still required that the company received the merchandise. This verification document typically is the **packing slip** (or **packing list)** that the vendor enclosed with the shipment, indicating quantities of each item shipped; the receiving clerk notes on the packing slip any discrepancies between the quantities shown and the quantities actually received, and then forwards it to accounting. By matching data on the packing slip and the original purchase order to the vendor's invoice data, the accounting clerk verifies that the expense is appropriate without any further contact with either the purchasing department or the receiving department.

Wages and salaries expenses are evidenced by specialized documents, typically **time sheets,** showing not only the total hours an employee worked, but, when appropriate, the particular activity on which he or she worked. Time sheets are usually signed by the individual's supervisor to provide independent accuracy verification.

Cash Disbursements. As most payments are made by check rather than in currency, a copy of the issuing firm's check is the logical source document for cash disbursements (payments). Commercial and industrial organizations typically use a multipart check form, sending the original to the payee and retaining one or more (nonnegotiable) copies as a source document.

The preceding discussion of the four most prevalent transactions describes hard copy (paper) source documents for each. As organizations move to paperless environments, source documents increasingly come to the accounting department in electronic form.

Journals

Journals are the accounting system's so-called books of original entry: data from source documents are recorded first in one of the journals. (The term **register** is often used interchangeably with journal when referring to specialized journals.)

An accounting system could be operated with a single journal. All transactions would be listed in chronological order, referencing the appropriate source document. In an organization of even modest size, such a journal would be voluminous and simply a jumble. The initial step in classifying accounting data is accomplished through the use of specialized journals. While all accounting systems do maintain a **general journal** for recording nonroutine transactions, most systems use one or more specialized journals, often one for each category of source document: **sales journal, cash receipts journal, expense journal,** and **cash disbursements journal** (or **check register).** Even finer classification of transactions can be achieved with yet more specialized journals. For example, many companies utilize an expense journal and a cash payments journal for nonsalary expenses and payments, as well as a specialized **payroll journal** for all wage and salary payments. These payments are unlike other transactions: they are circumscribed by numerous, strict legal requirements, and wage and salary expenses are typically recognized at the same time payroll checks are drawn.

As one example of a specialized journal, Table 6-3 shows entries in Zitar's sales journal for the first several days of March. Note that the normal entry involves a debit entry to Accounts Receivable (account 111) and a credit entry to one of the sales accounts (accounts 301, 305, or 321). Essentially, all debit entries to Accounts Receivable occur in this journal. Sales are classified by type, and at month-end, the totals in these three columns indicate total sales for the month by product category.

This specialized journal also records certain ancillary transactions that are, by their nature, linked to the sales transactions. Zitar pays commissions to sales agents on certain sales. Proper matching of expenses and revenues demands that the commission expense be recognized in the same accounting period as the related sale. While the commission expense could be recorded in the expense journal, a more convenient approach is to record commission expenses concurrently with the related sales transactions, in the sales journal. When a sales transaction on which an agent earns a commission is recorded in the sales journal (for example, the sale on March 2 to General Mammoth), the appropriate debit to Commission Expense and credit to Commission Payable is recorded simultaneously. Careful design of specialized journals facilitates efficient bookkeeping.

General Ledger

The third element is the general ledger to which categorized and summarized data from the journals are transferred. The general ledger is the fundamental books of the accounting system, containing all accounts listed in the chart of accounts. All accounting data ultimately find their way to the general ledger, and financial statements are prepared from the summarized data contained therein.

At the end of the accounting period (for example, a month), data from the journals are transferred to the general ledger:

1. First, the journals are totaled and double-checked to be certain that they are in balance, that the sum of all debits equals the sum of all credits.
2. The totals of those columns in the specialized journals devoted to a single account are transferred to the corresponding general ledger account. (Note that the detailed individual entries are not transferred, only the totals.)
3. All entries in the general journal are transferred to the appropriate general ledger account.

TABLE 6-3. The Zitar Corporation: Sales Journal

| | | Debit | | | Credit | | | | |
| | | #401 Commission Expense | #111 Accounts Receivable | | #301 Product A | #305 Product B | #321 Special Contracts | #147 Freight Clearing | #225 Commissions Payable |
Date	Customer Name				Sales			Other	
Mar 1	Cox Supply Co.		4,050		4,000			50	
Mar 2	General Mammoth Corp.	150	3,200			3,200			150
Mar 4	Mansfield Distributors		3,160		2,480	650		30	
Mar 8	Foster Bros. Wholesalers		3,900		2,500	1,400			
Mar 8	Willamette Supply Co.	100	2,600		2,600				100

At this point, the general ledger should be in balance (debits equal to credits). All data from the source documents have flowed through one of the journals and now appear in classified and summarized fashion in the general ledger. Table 6-4 shows that the format of the general ledger is analogous to the T-account. General ledgers provide space to record both the date of the entry and the journal (specialized or general) from which it came.

A trail exists from the general ledger back through the journals to the source documents, so that, if required, the detail behind each general ledger entry can be examined and the resulting balance verified. This trail is needed both by the accounting staff and by the company's independent auditors. Suppose the sales manager or an auditor at Zitar wishes to know just what caused the promotional literature expense to be so high during the month of March. In general ledger account 423 (Table 6-4) on February 28 (that is, after the first two months of the year), the balance was $600. The entries for March net to a debit of $550, and thus indeed the March expenses do seem high in relation to the expense for the previous two months. The entries in March are as follows:

1. On March 12, a $100 debit entry was made from page 3 of the cash disbursements journal (CDJ 3).
2. Similarly, on March 21, a $150 entry was made from page 4 of that journal.
3. At the end of the month, the expense journal (EJ) was summarized, and the column total ($400) related to this expense was transferred to the general ledger.
4. Also on March 31, a $100 credit entry is detailed on page 2 of the general journal (GJ 2); this entry reduces the expense and is probably a correction.*

TABLE 6-4. The Zitar Corporation: General Ledger Account

			#423: Promotional Literature Expense		
Date	Explanation	Debit	Date	Explanation	Credit
February 28	Balance	600	March 31	GJ 2	100
March 12	CDJ 3	100			
March 21	CDJ 4	150			
March 31	EJ	400			

*Entries are not erased from either journals or ledgers; rather, corrections are made by reversing the original (incorrect) entry and then proceeding with the correct entry. A debit entry is reversed by an equal credit entry to the same account, and a credit entry is reversed by a debit.

The accountant can now go back to the pages indicated in these three journals—the expense, cash disbursements, and general journals—and determine the individual transactions that comprised the total $550 expense for the month. The entries in the journals, in turn, reference the particular source documents, and, if the auditor or sales manager wants still more detail, these individual source documents can be retrieved from the files.

Subsidiary Ledger

The general ledger provides insufficient detail for certain asset and liability accounts; often, you need to know more about, for example, what is owned and what is owed and by whom. The accounts receivable balance in the general ledger indicates the total amount owed by all customers, but does not show the amount owed by each customer. You obviously need this customer-by-customer information; you want to pursue late-paying customers, and you want to cease selling to customers who are excessively delinquent. Similarly, you need to know, not only the total amount owing to vendors (total accounts payable), but also the detailed amount owing to each.

The detailed backup data in support of a general ledger balance are maintained in a **subsidiary ledger,** a set of records that elaborates on a particular general ledger account and reconciles with its balance. The accounts receivable subsidiary ledger is the most common. It is organized by individual customer and records each sale as a debit, each payment as a credit, and any other adjustment. Any asset or liability account may be the subject of a subsidiary ledger. Some accounting systems provide subsidiary detail on inventory, on fixed assets, and even (if the company maintains accounts in numerous banks) on cash. Computerized computer systems maintain subsidiary ledgers with very little additional work.*

ACCOUNTING CYCLE

The bookkeeping just described goes on throughout the accounting period. Source documents arrive daily in the accounting department, and data contained therein are processed to the journals. All subsidiary ledgers are maintained on a current basis. However, no entries are made in the general ledger during the course of the accounting period.

At the end of the accounting period, a good deal of work remains to get the accounting books sufficiently in shape to construct financial statements for the period. Information from the journals now is transferred to the general ledger, as discussed earlier; in accounting parlance, data from the journals are

---

*I keep a personal, very simple accounts payable subsidiary ledger: all unpaid bills remain in one spot on my desk until I pay them!

"posted to the general ledger". A preliminary **trial balance**—a listing of all accounts in the general ledger and the debit or credit balance in each—is constructed. This preliminary trial balance verifies that the ledger is in balance, that the debits equal the credits. A balanced trial balance does not guarantee that all entries were made to the correct accounts, but if the trial balance fails to balance, a hunt for the error or errors commences.

The first trial balance is considered preliminary because additional entries are necessary to complete the accounting for the period. These additional entries—referred to as *adjusting entries* or *end-of-period adjustments*—are discussed in the next chapter. Suffice it to say here that adjusting entries are not triggered by transactions with parties outside the organization and, therefore, are not the subject of typical source documents; rather, the adjusting entries, initiated by the accountant, are necessitated by certain events (including simply the passage of time) that alter values of assets or liabilities. For example, the accountant may learn that a certain customer will be unable to pay, and thus the accounts receivable balance needs to be adjusted; or she may ascertain that outstanding borrowing will soon require an interest payment, and thus interest expense and current liabilities need to be adjusted. These adjusting entries are recorded in the general journal and then posted to the general ledger. After all adjusting entries have been made, the accountant constructs a final trial balance (a final check that debits equal credits) and then the financial statements for the accounting period.

SUMMARY

As accountants wrestle with the problem of valuing assets and liabilities, and thereby owners' equity (changes in owners' equity equal revenues minus expenses), they follow certain general principles and specific rules.

The key principles that assist in resolving day-to-day accounting dilemmas are the following:

Express in *monetary terms*

Carefully define the accounting *entity*

Assume the entity remains a *going concern*

State asset and liability values *conservatively*

Record revenues and expenses on their *realization* dates

Account *consistently* across periods

Focus accounting attention on *material* events and conditions

These principles improve comparability between financial statements emanating from *different* companies and ensure even greater comparability over time of a single company's financial statements. Valuation procedures assume that a firm will continue in its present business. Accountants must exercise care not to overstate either profits or the overall financial health of the firm. Several of the principles address the pivotal question of when revenues and expenses—and therefore profits—should be recognized.

In a further step to meet these laudable objectives, the FASB was organized in the United States within the private (nongovernmental) sector to establish specific rules to resolve (or at least help to resolve) crucial accounting dilemmas. The SEC, the federal agency watchdog for the investing public, also formulates guidelines and rules for financial statements published for shareholders and potential investors.

Independent CPAs are employed by major companies to audit the companies' accounting systems and to attest to the fairness of their published financial statements. In fulfilling their responsibilities to both client companies and the public, CPAs make certain that companies' accounting practices are in accord with GAAP accounting: compliant both with the general principles and with the specific rules issued by the professional and governmental regulatory authorities.

While this book emphasizes accounting concepts and the interpretation of accounting reports, rather than bookkeeping, some knowledge of accounting mechanics is essential for all who must interact with and gain information from an accounting system.

The basic road map of the accounting system is the chart of accounts, a listing of all available accounts; the extent and nature of classifications are codified in this account listing. The accountant must balance costs and benefits in deciding the amount of detail to maintain in the accounting books.

The basic flow of accounting data is from source document to journal to general ledger. While a wide variety of source documents are used, each typically evidences one of four basic transactions: a sale, incurrence of an expense, a receipt of cash, or a payment of cash. Specialized journals facilitate efficient processing, permitting similar transactions to be grouped. A general journal records both transactions that do not fit the specialized journals and various adjusting (end-of-period) entries.

At the end of the accounting period, data from the journals are posted to the general ledger, and a preliminary trial balance is constructed. End-of-period entries are then recorded in the general journal and posted to the general ledger before a final trial balance is constructed and the financial statements—income statement, balance sheet, and cash flow statement—are drawn up.

NEW TERMS

American Institute of Certified Public Accountants (AICPA). An association of professional certified accountants in the United States.

Auditing. The process, carried out by certified public accountants, of testing and verifying financial statements.

Bill. Another name for *invoice*.

Cash disbursements journal. A specialized journal to record cash payments. This journal is frequently referred to as a *check register*, as checks are typically recorded sequentially by check number within the journal.

Cash receipts journal. A specialized journal to record cash receipts.

Certified public accountant (CPA). An accountant who has, through formal education, on-the-job experience, and rigorous testing, demonstrated competency in accounting and been awarded state certification.

Check register. Another name for *cash disbursements journal.*

Conservatism. The accounting guideline that requires the accountant, when in doubt as to the value of assets and liabilities, to lean in the direction of understating revenues and assets and overstating expenses and liabilities.

Consistency. The accounting guideline requiring that similar transactions and conditions be accounted for in the same manner over time.

Expense journal. A specialized journal to record expenses.

Financial Accounting Standards Board (FASB). An independent rule-setting body, within the private (nongovernmental) sector of the United States, responsible for promulgating accounting regulations to which accountants and CPAs must adhere.

Generally Accepted Accounting Principles. The set of accounting principles and specific rules to which all U.S. companies must adhere in preparing financial statements for external audiences.

General journal. The journal to record all transactions and recognitions that do not fit one of the specialized journals.

Going-concern assumption. The accounting guideline that requires accountants to value assets and liabilities assuming the enterprise will continue its present set of activities, unless or until a decision to the contrary is made by the enterprise.

Invoice. The document issued by the seller specifying amounts owed by the buyer in connection with items provided or services rendered. Invoices are often referred to as *bills*.

Journals. Books of original entry, in which data from source documents are recorded and from which summary totals are posted to the general ledger. Accounting systems typically utilize both specialized journals and a general journal.

Ledger. The general ledger, the fundamental accounting records of the organization. In an accounting system, the flow of data is from source documents to journals to the general ledger.

Materiality. The accounting guideline that advises the accountant to focus attention on valuing the material, or important, events or changes in condition that affect assets and liabilities, and to rely on estimates and approximations as expedients in the valuation of immaterial events and changes in condition. Materiality is judged in relationship to the total values of the particular company.

Packing slip (packing list). The document prepared by the seller and included with shipments to the buyer. The packing list details the quantity and description of all items included in the shipment.

Payroll journal. A specialized journal to record wages and salaries expenses, together with transactions related to payroll (for example, employees' tax withholding).

Purchase order. The document, from the buyer to the seller, specifying the particular items(s) or service(s) ordered, quantities, prices, terms of purchase, required delivery date, and so forth.

Realization. The accounting principle inherent in accrual accounting, requiring that revenue be recognized as earned only at the particular time the goods or services are delivered or furnished.

Register. Another name for *journal*. For example, the cash disbursements journal is sometimes referred to as the *check register*.

Sales journal. A specialized journal to record sales transactions.

Securities and Exchange Commission (SEC). A federal regulatory agency responsible for regulating securities markets and certain relationships between publicly owned companies and their shareholders.

Source documents. The evidence of transactions with other entities. These documents (for example, invoices and checks) provide the data required to record the transactions.

Subsidiary ledger. A set of records providing detail on the composition of a particular general ledger account. For example, an accounts receivable subsidiary ledger details amounts owed by individual customers; the sum of the balances in the subsidiary ledger agrees with the single balance in the Accounts Receivable account in the general ledger.

Time sheet. The document maintained by individual employees detailing the hours worked and, in many instances, the jobs, tasks, or projects worked on. Time sheets are the source documents for salaries and wages expenses.

Trial balance. A listing of all accounts in the general ledger and their balances.

EXERCISES

1. Indicate whether each of the following statements is true or false:
   a. The principle of conservatism requires that accountants understate the company's asset valuations and reported profits.

    b. The materiality principle suggests that transactions involving minor dollar amounts need not be processed.

    c. The general ledger is that set of accounting records from which figures for the balance sheet and income statement are drawn.

    d. Copies of checks sent to vendors to settle accounts payable represent important source documents.

    e. The realization principle requires that orders from customers not be recorded in the sales journal at the time the order is received.

    f. The consistency principle requires that all companies within a particular industry follow consistent accounting procedures.

2. What useful accounting data might be captured in a specialized payroll journal, in addition to the net wages paid to each employee?

3. In addition to the invoice from the supplier, what other evidence does the accounting department need to collect—and where is this evidence found—in order to safely authorize payment of the invoice?

4. Why is the consistency principle important? To what audiences is it important?

5. Suppose you are confronted with the following accounting dilemmas. In each case, what decision would you make and what accounting principles are relevant to the resolution:

    a. An employee has been discharged and this month is being paid severance pay equal to two months' salary. Should this severance pay be considered an expense of this month, or should it be split between the next two months?

    b. Certain items have been in inventory for more than a year; there is only a 30 percent probability that they will ever be sold or used. Should their value be removed from the total inventory value?

    c. A manufacturer of sophisticated analysis instruments ships a new model to an important customer; the customer agrees to try the new model for two months and then either return the instrument or pay full price for it. Should this shipment be counted as a sale this month? If not, should you account for a decrease in inventory value and, if so, how?

    d. The company president purchases 1,000 shares of stock from a former employee of the company. How should the company account for this transaction?

    e. The company provides a $1,000 travel advance to the sales manager, who is about to depart on a business trip to Japan. What entries, if any, would you make?

    f. A major customer with a $400,000 outstanding account receivable declares bankruptcy. What entries, if any, would you make?

g. Your company purchases $500 of merchandise from a vendor who offers a 10 percent discount if your company pays the invoice within 15 days. In the past, your company has always taken such lucrative discounts for prompt payment. At what value should you record this inventory and the corresponding account payable?

h. Your company pays $120 for telephone classified advertising for the coming year. Should you treat that as an advertising expense of the current period? If yes, why? If not, how might you account for it?

i. Annual interest charges on your five-year loan are $1,200, payable at the end of each calendar quarter. Should you recognize any interest expense in February? If so, how?

j. Your company owns a computer for which it paid $8,000 two years ago. The computer is still carried at that value in your fixed asset valuations. You now believe that the computer will be worthless in two more years. Should you adjust the value of the computer at this time? If so, how?

6. Following are the column headings from a specialized journal. What is a logical name for this journal? Explain what each of the columns means:

Accounts receivable
  Debit
  Credit
Commission expense
Cost of goods sold
Date
Customer name
Invoice number
Revenue
  Product A
  Product B
  Services
  Sales returns
Commission payable
Inventory

CHAPTER 7

FURTHER REFINEMENTS
IN VALUATION

By now you are aware that accountants face many choices:

- What changes in conditions should be valued and recorded?
- Should values be determined by the time-adjusted, market, or cost value method?

This chapter outlines several more choices. While the possibilities are by no means exhausted in this short discussion, the common procedural choices included here illustrate the wide range of issues and dilemmas that the accountant must resolve. Note that *all* relate to valuing assets and liabilities and thus to the *timing* of the recognition of expenses and revenues (and thus of profit).

The need to refine valuations of assets and liabilities, and thus revenues and expenses, is driven primarily by the accrual concept and the principle of conservatism. These refinements, or adjustments, are typically recorded at the end of the accounting period. Frequently, expenses and revenues need to be transferred from one accounting period to another in order to achieve a better match and to guard against overstating earnings. These shifts trigger, of course, offsetting shifts in assets and liabilities to preserve the accounting equality (debits equaling credits). The key concern is timing: when shall revenues and expenses be realized?

Remember that a firm's interim profitability (monthly, quarterly, or yearly) can only be estimated. Total profitability can be determined with absolute certainty only after an unhoped-for event: the affairs of the firm are wound up, all assets have been converted to cash, and all liabilities have been paid.

At that point, all remaining cash is the property of the shareholders, and only then can they know the total return received from their investments. Of course, for most companies, this day of reckoning—when the enterprise owns nothing but cash and has no obligations other than to stockholders—never arrives. The company may be sold as a going concern or merged, but these events do not cause the ultimate cash reckoning.

Thus, to meet the informational needs of stockholders, creditors, and (most importantly) managers, profits must be estimated on an interim basis. Accordingly, accountants struggle at the end of each accounting period to adjust the general ledger for the proper timing of revenues and expenses and for appropriate changes in asset and liability values. This chapter focuses on the most common adjustments, typically referred to as **end-of-period adjustments,** or simply **adjusting entries.** Note that transactions do not trigger these adjusting entries; accountants initiate adjusting entries.

PREPAIDS AND ACCRUALS

Recall that, in accrual accounting, cash inflows and outflows often do not coincide with the flow of revenues and expenses, respectively. When this timing difference spans accounting periods—when the cash transaction occurs in one period and the revenue or expense recognition in another—an asset or liability must be recognized at the intervening period-end. These assets and liabilities then have to be adjusted or closed out in the appropriate future period.

The four categories of balance sheet accounts affected by these timing differences are shown in Table 7-1.

Liabilities and Revenues

First, consider revenue. Most business in industrialized countries is conducted on a credit, rather than a cash, basis; we expect customers to pay for goods or services sometime following their delivery. When a sale is made, an asset, Accounts Receivable, is increased (debited) to balance the increase in revenue (credit to Sales). When the customer pays the invoice, the reduction in the

TABLE 7-1. Balance Sheet Accounts Affected by Timing Differences

	Revenue	Expense
Cash flow occurs before revenue/ expense is to be recognized	Unearned revenues: liability	Prepaid expenses: asset
Cash flow occurs after revenue/ expense is recognized	Accrued receivables (or accounts receivable): asset	Accrued liabilities (or accounts payable): liability

asset account, Accounts Receivable, is matched by the increase in another asset account, Cash. Two transactions (delivery of merchandise or services, and receipt of cash) trigger two sets of entries. No adjustments are required.

Suppose O'Malley Heating and Air Conditioning Company contracts to provide preventive and emergency service for one year in return for a $24,000 annual fee. Typically, customers pay the annual service fee in advance, thus giving rise to **deferred income (deferred revenue),** a liability in O'Malley's general ledger. The customer's annual payment triggers this entry:

Cash		Deferred Income: Service Fee (liability)	
Balance 24,000			24,000

Then, O'Malley recognizes revenue from the service agreement month by month throughout the contract period, not all at once. O'Malley's accountants are careful to make the adjusting entry each month to recognize both the service revenue earned that month and the corresponding decline in the liability:

Deferred Income: Service Fees		Service Revenue	
2,000	Balance		2,000

Another example: O'Malley subleases part of its space for $800 per month; the tenant makes quarterly payments in advance. Let's assume O'Malley receives a check for $2,400 (to cover three months—April through June—at $800 per month) on March 20. The debit is again to Cash, but O'Malley has not earned $2,400 of rental income in March; the rental will be earned pro rata over the next three months. In the meantime, O'Malley has an obligation to the tenant to provide the space. Therefore, the appropriate entry at the time the cash is received is:

Cash		Unearned Revenue: Rental Income	
Balance 2,400			2,400

Now, step forward to the end of April. What end-of-period adjustments are necessary to state correctly the company's revenues and expenses for the month of April? In April, O'Malley earned $800 rental income from the sublease, even though no transaction occurred during the month. The following adjusting entry must be made both to recognize the revenue and to reflect the reduction in the obligation to the tenant ($2,400, less the $800 benefit already received by the tenant):

Unearned Revenue: Rental Income (liability)		Rental Income	
800	Balance		800

Accrued Assets and Revenues

Alternatively, suppose the terms of O'Malley's sublease are that the tenant pays at the end of each quarter for the use of the space during that quarter. Once again, O'Malley should recognize that, in April, it has earned $800 of rent, even though payment will not be received for two more months. The appropriate April month-end adjusting entry is:

Accrued Rent Receivable (asset)		Rental Income	
800			800

This entry recognizes that the tenant is obligated to O'Malley: O'Malley now has an asset equal to the rent for the time that the tenant has already occupied the space. As similar adjustments are made at the end of May and June, the Accrued Rent Receivable account will grow to $2,400. When, at the end of June, the tenant pays $2,400, Accrued Rent Receivable will be credited to return it to a zero balance, offset by a debit to Cash.

If O'Malley owns notes due from individuals or other firms, interest income on these notes will be similarly accounted. If interest payments are received in advance, O'Malley will recognize a liability that is then reduced as the interest is earned over time. If interest is received in arrears (at the end of the period), O'Malley will recognize an asset (accrued interest receivable) that builds over time until discharged by the borrower's payment.

Accrued Liabilities and Expenses

Concerns with timing of expenses parallel concerns with the timing of revenues. If O'Malley's borrowing agreement with the bank provides for quarterly interest payments at the end of the quarter, O'Malley needs an adjusting entry in each month of the quarter, in order to match the interest expense to the period and to recognize the company's liability to pay this interest one or two months hence; that is, O'Malley's accountant recognizes the **accrued liability.**

Thus, if O'Malley's outstanding bank borrowing in October is $40,000 and the interest rate is 9 percent per annum (0.75 of 1 percent per month), an interest expense of $300 should be accrued for October, by the following end-of-period entry:

Accrued Interest Payable	Interest Expense
300	300

Prepaid Expenses

If O'Malley pays in advance for goods and services to be received in a subsequent accounting period, a **Prepaid Expense** account is created. (Recall from Chapter 3 that prepaid expenses are also referred to as *deferred expenses*.) This asset is then adjusted as the expense is matched to the appropriate accounting period. For example, O'Malley purchases a one-year property insurance policy, providing comprehensive coverage from July 1 through June 30 of the following year. If the $6,000 premium is paid in advance in June, the reduction in cash must be accounted for, although no expense is then incurred. Thus, the asset cash is exchanged for another asset, a prepaid:

Cash	Prepaid Insurance Premium (asset)
Balance 6,000	6,000

Beginning in July 1, O'Malley's accountant recognizes, as an expense each month, a pro rata portion of the insurance premium, and makes the following adjusting entry:

Prepaid Insurance Premium	Insurance Expense
Balance 500	500

After 12 monthly adjusting entries, the prepaid asset will be reduced to zero.

REMINDER: PRINCIPLE OF MATERIALITY

Only material adjusting entries should be made. If the adjustment would make no material difference to expenses (or revenues) or to assets (or liabilities), then the adjustment should be ignored. Situations abound in which an adjusting entry is theoretically appropriate but practically unnecessary. If O'Malley pays $72 for an advertisement in the telephone directory yellow pages for the following year, it is theoretically correct to prorate this expenditure through the year, showing $6 of advertising expense each month. But for O'Malley, such adjustments are so immaterial as to be unwarranted; the $72 payment is simply recorded as an expense in the particular accounting period when it is made.

Monthly billings for utilities, telephone, and other services offer similar examples. The utility company's billing cycle may not coincide with a calendar month (for example, the cycle may be from the 14th day of one month through the 13th day of the following month); theoretically, the bill should be split between the two months. However, if O'Malley's utility bill does not vary greatly from month to month, this refinement is unnecessary. As long as one utility bill is included in each month's expenses, utility expenses are adequately matched to the accounting period.

Salaries, the largest expense category for most companies, are typically paid weekly, biweekly, semimonthly, or monthly. Nevertheless, when unpaid salaries are material, adjusting entries—debits to the Salary Expense account that are offset by a credit to Salaries Payable—may be required at month-end.

LIABILITIES CREATED BY TODAY'S OPERATIONS

Accountants need to be alert to obligations that a company (1) generates routinely in today's operations, but (2) will not discharge until months or years into the future. Noteworthy examples are employee pensions and product warranties. In both cases, the firm undertakes a contingent obligation to perform: expenditures for warranty repair are required only if the product proves defective, and pension payments are required only when the conditions specified in the pension agreement (such as age of employee and length of service) have been meet.

You might be inclined to ignore these contingent future liabilities, arguing that companies, after all, obligate themselves in a myriad of ways every day. For example, purchase orders imply a promise to pay; employment agreements with engineers or managers obligate the company to pay future salaries. These two obligations are not valued in an accounting sense, however, until, in the first case, the products or services are received and, in the second case, the engineers or managers actually perform services and thereby earn salaries.

How do pensions and warranties differ from employment contracts? Employment agreements obligate a company to pay *future* salaries for (and at the time of) *future* services. In contrast, future warranty repairs arise because of the past delivery of products or equipment. Thus, the warranty provided to the customer is a part of the total cost of the products or equipment delivered today. This warranty cost, then, should be matched to the period when the sale is made (since that is the period when all of the revenue is recognized) not to the period when the customer brings the defect to light. (Alternatively, a portion of the revenue could be deferred and recognized later, either as warranty repairs are effected or when the warranty period expires.)

The exact amount of the warranty obligation is, of course, not known in advance. Typically, companies must repair under warranty only a small percentage of the products they sell. Accountants exercise judgment to estimate, in a probabilistic sense, the future warranty expenditures that may arise from today's shipments. If, for example, warranty expenditures have historically amounted to about 1 percent of sales, then in a month when sales total $850,000, the appropriate end-of-period adjusting entry is:

Allowances for Warranty (or Warranty Reserve) (liability)	Warranty Expense
8,500	8,500

When the warranty work is performed sometime in the future, the associated expenditures are debited to a liability account—Allowance (or Reserve) for Warranty and not an expense account; thus, the future period's profit is not affected by the necessity to repair products shipped in earlier periods.

Accounting for pension obligations is similar. Although pension obligations are discharged in the future, they arise because of the past employment of the pensioner. The exact dollar amount of the future pension obligation, however, may be even more difficult to predict than warranty obligations. Pension payments are typically made years, rather than months, after they are earned by (become the right of) the employee, and they are a function of the employee's length of service, age at retirement or death, and final salary. Nevertheless, if an employee by working today earns the right to a future pension, then the employer should recognize, as an expense of the current period, the obligation to pay that future pension.*

Other employee benefits require similar recognition of liabilities for future payments. For example, most companies carefully define vacation and sick-leave benefits. A company that provides its employees with 20 days of paid vacation leave per year recognizes that an employee accrues 20/12 days of vacation for each month of work. The wages or salary paid to a vacationing employee in August is not an expense of August but, rather, an expense of the months during which the vacation leave was accumulated. Proper matching requires that these obligations to pay vacation wages and salaries be recognized month by month by debiting Vacation Expense and crediting Va-

---

*Many employers pay insurance companies to provide employees with pensions. By doing so, the employer is paying today to relieve itself of these future obligations by shifting the obligation to the insurance company. Pension expenses are thereby matched to the appropriate accounting periods.

cation Wages and Salaries Payable. Then, in an accounting period when an employee with a weekly salary of $1,000 takes a two-week vacation, no expense account is affected and the vacation salary payment is recorded as follows:

Cash		Vacation Wages and Salaries Payable	
Balance	2,000	2,000	Balance

Proper accounting for another employee benefit illustrates the difficulty of matching. Many companies make bonus or profit sharing payments to key employees at year-end. Are these payments appropriately recognized as expenses of the final month of the year, or should they be recognized pro rata throughout the year? There is no right answer; the accountant must exercise judgment. On the one hand, if the accountant is relatively certain that bonuses of at least a certain amount will be paid at year-end, failure to spread the expense impact of these bonuses throughout the year results in an overstatement of profit in each of the first 11 months of the year, offset by a drastic understatement of profit in the 12th month. On the other hand, early in the year the accountant may not have a clear view of what bonus amounts, if any, will be paid at year-end, since bonuses are a function of the company's profits. The accountant faces a dilemma, the resolution of which will almost surely not satisfy everyone. Typically, if the profit-sharing payments or bonuses are highly predictable because of the company's size, stability, and past practice, such year-end payments are accrued throughout the year; by year-end, the general ledger contains an amount in the Bonuses Payable liability account approximating the amount of the payments. In other situations, the accountant may have no practical alternative but to wait until year-end to recognize the bonus expense.

ACCOUNTING FOR ACCOUNTS RECEIVABLE

Discounts for Prompt Payment

Standard terms of sale or purchase frequently permit the buyer to deduct a small percentage of the invoice amount if the invoice is paid within a specified number of days; the discount is a reward for prompt payment. For example, the terms "2%, 10 days/net, 30 days" mean that the invoice may be reduced by 2 percent if payment is made within 10 days of the invoice date and, alternatively, the full amount of the invoice is due in 30 days. The **discount for prompt payment** is also called a **cash discount.**

The timing of revenue and expense recognition is affected by how these allowed or earned discounts are handled. When shall the discount be reflected—at the time of purchase or sale, or when payment is finally made or received? The accountant is guided by when he or she *expects* payment to occur.

A buyer who foregoes a cash discount is, in essence, borrowing from the supplier for a period of days in return for giving up the discount. Thus, a buyer who does not promptly pay an invoice subject to "3%, 10 days/net, 30 days" terms is paying 3 percent to borrow the invoice amount for an additional 20 days—from the 10th to the 30th day.* Because passing up a discount this high is a costly financial policy, a large majority of buyers will raise sufficient funds through other borrowing or sale of equity stock to take advantage of such attractive cash discount terms.

A company that offers its customers a 3 percent discount for paying within 10 days can expect most customers to avail themselves of this attractive discount; therefore, the company may assume that in normal circumstances, it will receive only 97 percent of the invoice amount. Under these conditions, the so-called **net method** of accounting for discounts allowed should be used, as illustrated in Table 7-2.

TABLE 7-2. Net Method of Accounting for Cash Discounts

Sale: $1,000	Terms: 3%, 10 days/net, 30 days

(1) Recording the sale:

A/R	Sales
970	970

(2a) Recording the receipt of cash, if the discount is taken:

A/R	Cash
Balance 970	Balance 970

(2b) Recording the receipt of cash, if the discount is not taken:

A/R	Cash	Other Income
Balance 970	Balance 1,000	30

*This is expensive borrowing: as there are approximately eighteen 20-day periods in a year, paying 3 percent for 20 days of borrowing is equivalent to a 54 percent (18 × 3 percent) annual interest rate.

However, not all cash discounts are attractive. Suppose the discount terms offered to customers are "1/2%, 10 days/net 30 days"—equivalent to annual interest cost of about 9 percent. Under these circumstances, few customers will pay within 10 days and take the 0.5 percent discount, and thus the **gross method** of accounting for sales should be utilized: any discounts taken by a customer are recorded only when the customer's payment is, in fact, received. The gross method of accounting for cash discount terms is illustrated in Table 7-3.

Accounting for cash discounts for prompt payment on *purchases* of goods or services is exactly parallel. If the cash discount available is so attractive that the company will typically pay promptly, then the company should use the net method of recording these purchases or expenses. Otherwise, the gross method is more appropriate. Note that, while the net method of recording sales (*revenues*) is more conservative, the gross method of recording sales purchases (*expenses*) is more conservative, since the benefit of the discount is deferred until a future accounting period.

Cash discount accounting illustrates again that interim financial results are affected by accountants' choices. However, the consistency accounting principle significantly reduces the problem. Note that the profit for a particular month will be virtually identical for both the gross and net methods as long

TABLE 7-3. Gross Method of Accounting for Cash Discounts

Sale: $1,000	Terms: 1/2%, 10 days/net, 30 days

(1) Recording the sale:

A/R		Sales	
1,000			1,000

(2a) Recording the receipt of cash, if discount is taken:

A/R		Cash		Discounts Allowed[a]	
Balance	1,000	Balance 995		5	

(2b) Recording the receipt of cash, if discount is not taken:

A/R		Cash	
Balance	1,000	Balance 1,000	

[a] This is an expense account or an account that offsets sales.

as no change in methods is undertaken during the period and the aggregate value of transactions doesn't vary much between periods.

Allowance for Doubtful Accounts

Extending credit to customers almost inevitably exposes a company to bad-debt losses. The extent of those losses depends on the nature of the customer group, the care with which the company screens for customer creditworthiness, and the diligence and tenacity with which it pursues slow-paying customers.

The accrual and conservatism principles suggest that the expense (or loss) on accounts that ultimately prove to be uncollectible should be matched to (that is, included in the accounting period of) the original sale. Of course, a seller who knows in advance that a certain customer will not pay will not extend credit to that customer; the customer will be required to pay cash or will simply be turned away. Thus, we cannot identify in advance exactly *which* accounts will prove uncollectible, and yet, on a probabilistic basis, we know that *some* accounts receivable will never be collected. With this knowledge, we can adjust the Accounts Receivable balance. Failure to do so overstates an asset and accelerates profit recognition inappropriately.

Again, the accountant exercises judgment in order to develop a decision rule for this adjustment. The company's history of bad-debt losses is a useful guide, but this historical pattern should be tempered with such considerations as the state of the economy, changes in credit-granting policies of the firm, and changes in the mix of customers.

Suppose the accountant determines that a conservative estimate of bad-debt losses is 0.5 of 1 percent of credit sales. Then, if credit sales for February are $460,000, the estimated value of those receivables that will ultimately prove uncollectible is $0.005 \times 460,000 = \$2,300$, and the end-of-period adjusting entry is as follows:

Allowance for Doubtful Accounts	Bad Debts Expense
2,300	2,300

The account **Allowance for Doubtful Accounts** is used to preserve the gross value of all accounts receivable in the accounting records. Note that the allowance for doubtful accounts is a *pooled,* (overall) *adjustment,* not an adjustment of amounts owed by particular customers. Note, too, that the effect of this adjustment is to include a bad-debt expense in February that is a function of that month's credit sales, even though the company won't know which of the credit sales proved uncollectible until months later.

Allowance for Doubtful Accounts is called a **contra account,** which offsets the value of its associated asset or liability account while preserving the values in that associated account. This particular contra account appears in the chart of accounts among the assets, right after the Accounts Receivable account (the account to which it is "contra"), even though it carries a credit balance. When a balance sheet is constructed, a single amount for Accounts Receivable, Net is the gross accounts receivable less the allowance.

If bad-debt losses are estimated conservatively high, the credit balance in the contra account Allowance for Doubtful Accounts could grow to unreasonable levels. Therefore, decision rules typically provide that the total balance in Allowance for Doubtful Accounts not exceed a certain percentage of the Accounts Receivable balance.

Now, how and when is the balance in the contra account reduced? What triggers debit entries? Recognition that a certain receivable is uncollectible. Suppose you learn in February that a customer who had purchased $1,700 of merchandise late in the previous year (that is, several months previously) has encountered financial difficulties and probably will never pay the $1,700 account currently outstanding. The customer's promise to pay is, therefore, valueless and should be eliminated from the assets. Here is the entry to recognize this write-off:

Accounts Receivable			Allowance for Doubtful Accounts	
Balance	1,700		1,700	Balance

Note two facts about this entry: (1) the *net* value of the Accounts Receivable in your general ledger (the gross amount less the allowance for doubtful accounts) is unaffected; and (2) no expense account is debited in this entry, since both the debit and credit are to asset accounts (one of them being a contra asset account). Thus, the profit in February is not reduced by the current discovery that a particular receivable created in an earlier period is a bad debt.

Occasionally, a customer whose account was previously written off does, in time, pay. Obviously, the payment will be accepted, but what account should be credited to balance the debit to Cash? Typically, the credit is made to Allowance for Doubtful Accounts, but you could reasonably argue that the credit should go, instead, to the Bad Debts Expense account.

ACCOUNTING FOR INVENTORY

Inventory accounting also presents some interesting dilemmas for accountants. Just as accounts receivable values are adjusted to allow for probable noncollection in the future, so inventory values may be adjusted for spoilage, shrinkage, obsolescence, and similar phenomena that affect its value.

Consider obsolescence, one of the inevitable costs of carrying inventory. The proper matching of expenses to accounting periods suggests that, when the risk of obsolescence is high, some inventory obsolescence expense should be included in each accounting period, even though it's impossible to verify exactly which items are obsolete until some time in the future. As in the case of Allowance for Doubtful Accounts, the solution is to create a contra account, perhaps entitled Allowance for Obsolete Inventory. This account is built up by employing a decision rule that relates this period's obsolete inventory expense to the level of purchases during this period and to inventory balances. (This decision rule should provide for larger increases in Allowance for Obsolete Inventory when the firm anticipates that some event—for example, introducing a new model of a product—will increase the risk of obsolescence.) When, in a future accounting period, particular inventory items are determined to be obsolete, the Inventory account and its contra account, Allowance for Obsolete Inventory, are both adjusted, with no effect on the firm's profitability during that future period.

Deriving Cost of Goods Sold by End-of-Period Adjustment

Some companies derive their cost of goods sold expense only once per accounting period by adjusting inventory values at the end of the period. (This end-of-period adjustment approach eliminates the opportunity to maintain detailed, day-to-day inventory values in the accounting records.)

You are well aware how important it is to match cost of goods sold to sales in each accounting period so that the gross margin (or gross profit) indicates accurately the margin earned on the revenue of the period. Suppose a $4,000 credit sale of equipment having a finished goods inventory value of $2,500 is accounted for as follows:

Accounts Receivable		Finished Goods Inventory		Sales		Cost of Goods Sold	
Balance 4,000		Balance	2,500		4,000	2,500	

This method of determining the cost of goods sold, referred to as the **perpetual inventory method,** requires that a specific cost be associated with each sale. This requirement is no problem in the illustration above: the equipment has a large dollar value, and relatively few sales are made. Consider, alternatively, a variety store or supermarket grocery store that sells thousands (or tens of thousands) of items each day. Tracking the cost of each item would be an onerous chore. Moreover, the tracking costs might be more than the worth of the information derived.*

Alternatively, the cost of goods sold can be derived only once per accounting period. One can reasonably assume that the cost of items sold in a particular period is equal to the amount purchased into inventory, adjusted for any increase or decrease in that inventory. Stated another way, the total value of merchandise available for sale in the period equals the amount in inventory at the beginning of the period plus the amount purchased during the period; this total is either sold or remains in inventory at period-end. Thus,

$$\text{Cost of goods sold} = \text{beginning inventory} + \text{net purchases} - \text{ending inventory}.$$

Of course, this method will result in cost of goods sold including also the value of goods lost, stolen, misplaced, or discarded. While such inventory shrinkages are typically small compared with the amount actually sold, the end-of-period adjustment method does not allow such shrinkages to be isolated and valued.

The advantage of this method of determining the cost of goods sold is that it requires less accounting; the cost of goods sold is determined only once each accounting period, rather than at the time of each sale. There are three disadvantages: (1) The ability to determine the margin on individual sales is lost; only the aggregate margin for sales during the period can be determined. (2) As just noted, inventory shrinkages cannot be isolated. (3) The method requires that the inventory be physically counted and valued at the end of each accounting period, a tedious and time-consuming job.† Note that the ending inventory for one accounting period becomes the beginning inventory for the following period.

This method of determining cost of goods sold is outlined for the Tollini Company in Table 7-4, with the adjusting entries shown in bold type. Throughout the month of October, the Tollini Company recorded no entries in either the Inventory or Cost of Goods Sold accounts; it debited the Purchases account (rather than the Inventory account) for all merchandise acquired. Think of Purchases as a temporary expense account—temporary in the sense that, after the end-of-period adjustment, it will have a zero balance.

---

*The advent of low-cost computers, universal product codes, and laser-based code-reading devices makes it feasible to maintain perpetual inventory values in accounting records; however, to date, most such systems have been used to control physical inventory, not to maintain accounting records.

†Even when the perpetual inventory method is used, physical inventories must be taken from time to time to determine the amount of inventory shrinkage due to breakage, pilferage, and similar causes. However, items may be counted on a rotating basis (referred to as *cycle counting*), and they need not all be counted at the end of each accounting period. Estimates of inventory shrinkage are typically adequate for interim financial statements.

TABLE 7-4. The Tollini Company: General Ledger at October 31 with Adjusting Entry Recorded

Cash	Receivable	Inventory	All Other Assets
10,000	30,000	19,800 | **19,800** **21,000**	33,400

All Liabilities	Owners' Equity
50,000	40,000

Sales	Cost of Goods Sold
86,000	**53,800**

Purchases	Purchase Returns	All Other Expenses
58,000 | **58,000**	**3,000** | 3,000	28,000

Note, too, that during the month, Tollini returned to its suppliers items valued at $3,000, the amount in the Purchase Returns account. Net purchases for the month, therefore, were $55,000. The amount in Inventory on the trial balance (before adjusting entries are made), then, is the inventory value at the end of the last accounting period—that is, the value at the close of business on September 30 and the start of business on October 1.

The accounting records do not reveal the inventory value on October 31. Rather, Tollini personnel must physically count and value the inventory on that date. Let's assume this value to be $21,000; then, the value of merchandise sold must have been $19,800 + 58,000 − 3,000 − 21,000 = $53,800. (Reminder: the sum of the debits in this adjusting entry equals the sum of the credits!)

Since Tollini's inventory grew during October, had the accountant not made the end-of-period entry but simply treated net purchases as equal to the cost of goods sold, the October 31 inventory values on the balance sheet would have been understated, with a corresponding overstatement of the cost of goods sold and, thus, an understatement of profit for the month.

Conventions to Handle Price Fluctuations

When identical inventory items are purchased at different times and at different prices (because of price-level changes, quantities purchased, or other reasons), what value should be used to credit Inventory and debit Cost of Goods Sold when the item is sold? The cost value method presents us with more than one cost.

The **first-in, first-out (FIFO)** convention assumes that the oldest costs, the first to arrive in inventory, are the first to flow to the cost of goods sold; the corollary assumption is that the last prices (the most recent prices) remain in inventory. The **last-in, first-out (LIFO)** convention assumes just the opposite: that the most recent (last-in) prices appear first in the cost of goods sold, while the oldest (first-in) prices remain in inventory.

During inflationary periods, the LIFO convention reflects the increasing prices sooner in cost of goods sold than does the FIFO method. Thus, the costs of goods sold are better matched to sales by the LIFO convention, since the cost-of-goods-sold value more closely approximates the replacement, or current, cost of the inventory being sold. This improved matching is advantageous in making various production, investment, and marketing decisions, particularly pricing. It follows, then, that the LIFO convention states profits more conservatively in times of inflation than does the FIFO convention. The negative is that "old " prices remain in the inventory valuation; therefore, in periods of high inflation, stated inventory values can become substantially lower than replacement costs, and thus be somewhat unrealistic.

Accordingly, high inflation rates encourage the use of the LIFO convention for two reasons. First, management typically relies more on profit-and-loss data than balance sheet data in making operating decisions. Second, LIFO leads to lower profits, lower income tax expense, and thus higher cash flow during periods of inflation. Bear in mind that the conventions have the opposite effect on profits and taxes during periods of deflation (declining prices).

These conventions do not abandon the use of historical cost. The difference between FIFO and LIFO is simply the assumption about the flow of these costs. Note, too, that the convention refers only to the monetary flow, not to the physical flow of inventory items. Good inventory management practice requires that the older inventory be used up before the newer inventory. The physical flow of inventory, then, is generally first in, first out, but we still have the choice of using either FIFO or LIFO to account for the monetary flows.

The last section discussed the two basic methods of accounting for cost of good sold and inventory: (1) end-of-period adjustment and (2) perpetual inventory records. FIFO and LIFO conventions can be used with both methods.

Generally accepted accounting principles permit the use of either the FIFO or LIFO convention. When interpreting financial results, particularly during periods of rapidly changing prices, be aware of which convention is used, as specified in the financial statement footnotes.

The current U.S. income tax laws also permit either the FIFO or LIFO convention but with restrictions. First, a company may not switch freely between the two conventions in an attempt to take advantage of periods of increasing and decreasing prices. Second, a company that uses the LIFO convention for income tax purposes must use the same convention for reporting profits to shareholders. This requirement that tax and financial accounting be the same is unusual. In virtually all other instances, one is free to select tax accounting methods that will minimize tax obligations consistent with the tax laws and to select alternative accounting methods for public reporting. Some managers forgo cash savings (in actuality, a postponement of cash tax payments, rather than outright savings) associated with the LIFO convention because they are unwilling to report the lower profits that LIFO yields in inflationary times.

A third convention, *average prices,* results in valuations on the income statement (cost of goods sold) and balance sheet (inventory) falling somewhere between FIFO and LIFO. This convention requires that a new average price be calculated after each new purchase of an inventory item—a trivial calculation chore for a computer. This moving average price is then used in valuing all cost of goods sold transactions until the next purchase occurs, at which time a new average price is calculated.

ACCOUNTING FOR PRICE-LEVEL CHANGES

The terms inflation and deflation refer to increases and decreases in general price levels in an economy. In periods of inflation or deflation, the prices of specific assets or liabilities also typically increase or decrease, although sometimes at widely varying rates. The historical cost method of valuation takes no account of these general or specific price changes. When the rate of inflation or deflation in an economy is low—say, 3 percent per year or less (the rate in the United States for many decades prior to about 1960 and after about 1990)—valuation distortions are minor. However, rates of inflation in countries around the world have varied widely, sometimes reaching very high levels. As the inflation problem, once thought to be unique to less-developed and unstable economies, has become a periodic or perpetual problem for many countries, the inadequacies of historical cost valuations have become more apparent.

Distortion arising from price-level changes is potentially most significant for those assets and liabilities that are held for a considerable time—fixed assets, long-term liabilities, and, to a lesser extent, inventory. The distortion affects not only the balance sheet but also the income statement:

1. If inflation has been high since an inventory item was purchased some months ago, the balance sheet value of the asset (inventory) is now low in relation to current replacement prices. When the item is sold, its price to the customer reflects the intervening inflation, but its cost does not: today's sale price is matched to the historical purchase cost. The company's reported gross margin is higher than the difference between today's sale price and the replacement cost of the inventory. The use of historical costs, then, undervalues the inventory owned and overstates profits during periods of rapid inflation. (The problem is mitigated, but not eliminated, by the use of LIFO in preference to FIFO accounting.)

2. A similar situation pertains to fixed assets. Historical costs undervalue the asset on the balance sheet; the write-down (depreciation, to be discussed in Chapter 8) based on these historical costs is less than it would be based on replacement costs. Arguably, then, in periods of inflation, profits are overstated.

3. Long-term liabilities are affected in the opposite manner. In periods of rapid inflation, the purchasing power of specified repayments on long-term liabilities declines each year. In effect, the obligation becomes progressively less painful to the borrower as inflation occurs. The lender is harmed by the effects of inflation, but the borrower benefits. The financial statements of neither borrower nor lender reflect these changes.

Accounting for price-level changes is a very complex subject, the detailed discussion of which is beyond the scope of this book. However, a very brief discussion of two approaches to price-level accounting is appropriate here.

Specific Price Adjustment accounting adjusts for specific price changes of nonmonetary assets, primarily inventory and fixed assets. Cash and accounts receivable—both monetary assets—are unaffected. Making specific price adjustments appears easier than it is.

The time-adjusted value method, discussed in Chapter 2, provides just such a specific adjustment. If the stream of net benefits of ownership is expressed in today's dollars and the benefit stream is adjusted for time, the result will be an inflation-adjusted value. Recall, however, that this valuation method is unreliable and subject to bias because the expected future benefits of ownership are so difficult to estimate. For certain other assets, particularly fixed assets, price indices may be available to inflate historical costs to a reliable current value. The market value method also offers some possibilities.

But how would one value assets, particularly aging equipment, that will not be replaced in kind? If technological improvements have occurred since the original acquisition, the owner takes advantage of the improvements at the time of replacement. Thus, the appropriate specific price adjustment would not necessarily reflect the current cost of duplicating the exact physical asset now owned. Determining equivalent productive capacity is frequently difficult, time consuming, unreliable, and subject to bias.

If satisfactory specific price adjustments for inventory and fixed assets can be determined, then these adjustments can be reflected in balance sheet valuations. However, then we face the dilemma of how these revaluations—gains or losses from owning the inventory and fixed assets—should be reflected in the income statement. Specific price adjustment accounting in periods of inflation leads to an increase in owners' equity. Is this increase a profit? Yes, but a profit that arises from inflation and not from operations and, therefore should be separated and so identified. If the problems associated with making the specific price adjustments could be overcome, this approach would improve the matching of revenues and expenses.

General price-level adjustment accounting attempts to reflect in the financial statements the effects of inflation on equivalent purchasing power, without concern for changes in prices of individual assets and liabilities.

A prime objective of financial accounting is to provide comparable data for successive accounting periods in order that financial trends can be observed and comparisons can be struck. Our emphasis on consistency arises from this objective. In periods of inflation, the purchasing power per monetary unit declines. An asset purchased for $1,000 five years ago is not comparable to an asset purchased for $1,000 today, as more purchasing power was spent to acquire the five-year-old asset than to acquire the new one. Or, viewed another way, if the five-year-old asset were to be replaced today at the same cost in purchasing power, its cost would be substantially more than $1,000. Some argue that financial statements ought to be expressed in terms of equivalent purchasing power.

The general price-level adjustment approach recomputes financial statements in equivalent monetary units (in the United States, in constant dollars). Adjustments are not restricted solely to inventory and fixed assets, but are made to all assets and liabilities and to revenues and expenses. The adjustments can take the form of restating prior years' financial statements in today's dollar values, or restating the current financial statements in the value of the dollar in some base year.

The index used to make the price adjustments, rather than one relevant to the specific asset or liability being adjusted, is one of several readily available **general price indices;** these include the Gross National (or Domestic) Product price deflator and the Consumer Price Index. Their application to the revaluation of an asset or liability is straightforward, objective, verifiable, and essentially free from bias. As a result, auditors tend to be more comfortable with this approach to accounting for inflation and deflation than with the various methods for specific price adjustments.

In practice, readers of financial statements are often more confused than enlightened by inflation-adjusted statements. After a flurry of activity in the decade of the 1970s, when U.S. inflation was high, the accounting profession (the FASB, the SEC, and the AICPA) has shown less interest in the subject of inflation accounting. Over the past two decades, much debate concerning the most effective method of incorporating inflation in accounting valuations has resulted in no consensus on the best choice between these two approaches. Inflation-adjusted accounting reports have, in some instances, been provided as supplementary information, but the primary financial reports continue to be those prepared on the basis of historical costs.

ACCOUNTING FOR CURRENCY FLUCTUATIONS

Another problem related to, and arising from, price-level change is that of accounting for currency exchange-rate fluctuations. The rate of exchange between two currencies typically changes when the rates of inflation in the two countries are different. If country A is experiencing 5 percent inflation per year while inflation in country B is 10 percent per year, the currency of country B is likely to lose value (become devalued) in terms of country A's currency.

All global firms, and many smaller companies, have operations—or at least own assets and liabilities—in countries other than their home country. If a company operates a foreign subsidiary, the foreign government insists that the subsidiary's financial statements be maintained in the foreign currency. Yet, such companies maintain consolidated financial records in the currency of a single country, typically their home country. Thus, assets, liabilities, incomes, and expenses (and thereby owners' equity) denominated in another currency must be translated into the home country's currency for purposes of reporting consolidated financial results of worldwide operations.

Over the last several decades, international currencies have fluctuated widely. Companies based in countries with strong currencies, such as Switzerland, have seen their foreign assets (denominated in the foreign currency) lose substantial value in terms of the Swiss franc, even while the assets have maintained their value in terms of that foreign currency.

Suppose your company is located in a strong currency country (call this country SCC) and sells machinery to a customer in a weak currency country (WCC), with one-year credit terms. At the time of the sale, the exchange rate between SCC and WCC is 1:3—that is, one unit of SCC currency purchases three units of WCC currency. The customer receivable is denominated as 30,000 units of WCC currency, which translates to a value of 10,000 units of SCC currency. Stepping forward a year to the time when the customer pays, suppose that the prevailing exchange rate is now 1:4—that is, the currency of WCC has devalued vis-à-vis that of SCC. One unit of SCC currency now purchases four units of WCC currency. The customer pays the outstanding receivable: 30,000 units of WCC currency. When this payment is converted into SCC currency, your company receives only (30,000/4), or 7,500 units of SCC currency. Thus, during the course of the year, this account receivable declined in value from 10,000 to 7,500 in terms of SCC currency, although to the customer the account payable continued to be 30,000 units of WCC currency.

Alternatively, suppose your company, located in SCC, borrowed 150,000 units of WCC currency (when the exchange rate was 1:3), promising to repay at the end of one year. The loan proceeds were converted into SCC currency immediately (50,000 units), and your company used the funds productively. A year later, when your company repays the loan and the exchange rate is 1:4, it needs to convert only 37,500 units of SCC currency into WCC currency to repay the loan: 37,500 units × 4 = 150,000 of WCC currency. Your company has fared very well on this transaction, while losing on the receivable transaction described in the previous paragraph.

These two brief examples illustrate one way to hedge foreign-currency positions (that is, minimize the financial impact of currency fluctuations): owe as much in the foreign currency as you own. Many companies follow this hedging strategy. Highly organized foreign-exchange futures markets permit still other approaches to hedging. Nevertheless, many companies experience exchange-rate gains and losses that must be accounted for.

EXCEPTIONS TO THE ACCRUAL CONCEPT AND THE REALIZATION PRINCIPLE

While generally accepted accounting principles require accrual-basis rather than cash-basis accounting, certain businesses, particularly small businesses and professional service firms, do utilize the cash-basis method. Recall that this method recognizes sales or revenue only when cash is received, regardless of when merchandise was delivered or service rendered, and recognizes expenses only when cash is paid out. Firms for which cash-basis accounting is appropriate have few fixed assets and little inventory, and they typically do not have public shareholders to whom they must report. The method is simple and, although it is conservative in recognizing profit, it is permissible for income tax reporting under certain circumstances. While emphasis in this book is on the accrual concept, remember that certain types of enterprises use the cash-basis method quite appropriately.

Still other businesses find the realization principle not applicable to their circumstances. Recall the discussion in Chapter 6 of the **percentage-of-completion method** of accounting for revenues. Rather than recognizing revenue only at the end of the project—and then recognizing an enormous amount of revenue—the contractor on a large, prolonged project (such as the construction of a dam) recognizes a portion of the revenue in each of the accounting periods throughout the project. The great challenge here is to match expenses to revenues as they are being recognized, and to do so accurately and without bias.

Some businesses operate under **cost-plus contracts;** that is, the amount charged to the customer is a function of the costs incurred, typically the cost plus a fixed fee or the cost plus a percentage of the cost. Consulting firms, auditing firms, law firms, research centers, and defense contractors frequently contract with their customers on a cost-plus basis. In these circumstances, revenue for interim accounting periods while the contract is in process is relatively easy to determine. The recognition of revenue, therefore, is typically not delayed until the end of the contract but is recognized period by period, as the costs are incurred.

SUMMARY

As further evidence that accounting is an imprecise science requiring judgment, this chapter reviews both end-of-period adjusting entries and several accounting dilemmas requiring accountants to choose among acceptable accounting procedures that relate to the timing of expense and revenue and, thus, of profit.

End-of-period, or adjusting, entries are required to record revenues and expenses in a timely manner and to adjust accordingly the asset and liability values. These entries are not triggered by transactions but are initiated by accountants as they apply the accrual concept. Adjusting entries are essential in order to be faithful to the principle of conservatism: anticipating expenses and not accelerating revenues.

The precise matching of revenues and expenses is unattainable; judgment must be exercised. However, two other accounting principles mitigate distortions that may arise from differences in judgment. First, the principle of materiality tells us to be concerned only with timing adjustments that have a material effect on profit or on the balance sheet. Second, the consistency principle, dictating that recurring timing problems be resolved consistently across accounting periods, assures that financial statements, even if not precisely accurate, are comparable from period to period.

To state revenues, expenses, assets and liabilities satisfactorily, various general ledger accounts must be adjusted to reflect changed conditions or simply the passage of time. These end-of-period adjustments fall into two categories:

1. Adjustments to unearned revenue (liability) accounts and prepaid asset accounts.
2. Recognition of liabilities created by today's operations. Examples: allowances for warranty expenses, pensions and other employee benefits.

Cash discounts for prompt payment of invoices—whether applicable to sales to customers or purchases from vendors—can be accounted for by the net or gross method. The choice is driven largely by whether the cash discounts are typically taken; if they are, the net method is preferable. The net method of accounting for sales is more conservative than the gross method, while just the opposite is true of cash discounts for purchases.

A provision for bad-debt losses on credit sales is accrued so as to match the bad-debt expense (on a probabilistic basis) to the time period when the credit sale occurs. The account Allowance for Doubtful Accounts is a contra account for the Accounts Receivable account. When, in a subsequent period, an individual customer account is found to be uncollectible, the reduction in Accounts Receivable is offset with a debit entry to the contra account rather than to an expense account.

Similarly, inventory contra accounts are often used to accrue provisions for obsolescence, spoilage, and other hazards associated with holding inventory. When perpetual inventory records are not maintained within the accounting system, the cost of goods sold is derived and recorded as an end-of-period entry.

Chapter 8 focuses on end-of-period adjustments to a particular asset category, long-term assets.

Accounting for inventory and cost of goods sold, in the typical situation where prices for physically identical items vary over time, requires the choice between FIFO (first-in, first-out) and LIFO (last-in, first-out) conventions. In inflationary periods, FIFO leads to higher reported profits and higher values for inventory, while LIFO, the more conservative convention, provides more useful data for operating decisions and also minimizes income tax expenses.

As countries have periodically experienced high inflation rates in recent decades, the dilemma of accounting for these price-level changes has taken on added importance. Matching expenses with revenues is difficult when inventory and fixed assets are purchased in one period and used in a subsequent period when price levels are sharply lower or, more typically, higher. Comparability among financial statements from different years is weakened when the monetary unit is unadjusted for changed price levels. While consensus has not developed on the best method for effecting these adjustments, the two primary methods are specific price adjustments and general price-level adjustments. The first seeks to reflect price changes in specific assets such as inventory and fixed assets. The second seeks to reflect, in the financial statements, the changed purchasing power of the monetary unit by adjusting historical costs through the use of broad indices such as the Consumer Price Index.

Price levels changing at varying rates in different countries lead to changes in currency exchange rates, in turn presenting problems in valuing assets and liabilities denominated in other than a company's home-country currency.

Even such fundamental accounting principles as the accrual concept and the realization principle are subject to exceptions. Certain companies should use cash-basis rather than accrual-basis accounting; similarly, the percentage-of-completion and the cost-plus-contract methods of accounting for revenue (and associated costs of revenue) are, in certain circumstances, logical and appropriate exceptions to the realization principle.

NEW TERMS

Accrued Liability. An obligation that arises when an expense is recognized in one period and the cash outflow occurs in a subsequent period. Examples are accrued tax, interest, and rent liabilities.

Adjusting entries. An alternate name for *end-of-period adjustments*.

Allowance for Doubtful Accounts. A contra account to adjust the value of the asset account Accounts Receivable for the probability that receivables from some customers will prove uncollectible.

Cash discounts. Small discounts (typically 2 percent or less) allowed by some suppliers to customers who pay invoices within a prescribed number of days of the invoice date. An alternative name is *discounts for prompt payment.*

Contra account. An account that serves to reduce the value of its associated asset or liability account (with a liability or asset balance, respectively) while preserving the values in that associated account.

Cost-plus contract. A contract providing that the price charged is a function of the cost of performing the contract—cost + a fixed fee or cost + a percentage of cost; revenues and costs on such contracts are typically recognized in each accounting period rather than recognizing the full amount only at completion of the contract.

Deferred income (deferred revenue). A liability that arises when cash is received in an accounting period preceding the period when the corresponding income (revenue) is to be recognized. Alternative names are *unearned revenue* or *unearned income.*

Discounts for prompt payment. An alternative name for *cash discounts.*

End-of-period adjustments. Entries occurring at the end of the accounting period and initiated by the accountant (rather than being triggered by transactions); these entries are necessary to match revenues and expenses and, accordingly, to adjust asset and liability values. An alternative name is *adjusting entries.*

First-in, first-out (FIFO). A convention that values inventory and the cost of goods sold by assuming that the oldest purchase prices included in inventory are the first prices reflected in costs of goods sold.

General price indices. Indices such as the Consumer Price Index or the Gross National (Domestic) Product price deflator that are designed to reflect the impact of inflation and deflation on general price levels; these indices are used in general price-level adjustment accounting.

General price-level adjustments. A method of accounting for price-level changes (inflation and deflation) by reflecting in the valuation of assets and liabilities the changed purchasing power of the monetary unit (such as dollars).

Gross method. In contrast to the net method, this method of accounting for cash discounts recognizes sales or purchases at their face (full) value, accounting for any cash discounts allowed or earned in the period when cash is finally received or paid out.

Last-in, first-out (LIFO). A convention that values inventory and the cost of goods sold by assuming that the most recent purchase prices included in inventory are the first prices reflected in costs of goods sold.

Net method. In contrast to the gross method, this method of accounting for cash discounts recognizes sales or purchases at their net value (net of available cash discount) with any discount not taken by customers or forgone on purchases accounted for (revenue and expenses respectively) in the period when cash is finally received or paid out.

Percentage-of-completion method. A method of recognizing sales (or revenue) on an extended contract (for example, research or construction) in proportion to the percentage of the contract completed, rather than recognizing the full amount only at the completion of the contract.

Perpetual inventory method. A method of accounting for each inventory increase (receipt) or decrease (shipment) as it occurs. Alternatively, inventory values may be adjusted only at period-end as cost of goods sold for the full period is derived and recorded.

Prepaid Expense. An asset that arises when cash outflow occurs in an accounting period preceding the period when the corresponding expense is to be recognized.

Specific price adjustments. A method of accounting for price-level changes (inflation and deflation) by reflecting the changed prices of individual assets.

Unearned revenue (unearned income). An alternative name for *deferred income* (or *deferred revenue*).

EXERCISES

1. Why are end-of-period adjusting entries necessary?

2. Describe the end-of-period adjustment method of determining cost of goods sold for an accounting period.

3. What accounting principle suggests that it is unnecessary to divide the monthly utility bill (for gas and electricity) between the two calendar months to which it applies?

4. A company signs a two-year employment agreement with its new president. Explain why this action does not trigger the recording of a liability equal to the two years of salary promised in the contract.

5. How would you account for the receipt of payment on a receivable that you had previously written off as uncollectible?

6. Most automobile manufacturers offer extended warranties on the cars they sell. Describe why and how entries are made to account for fulfillment of these warranty services at the time of the sale and later as the warranty repair work is performed.

7. An insurance carrier specializes in 12-month fire insurance policies. All of its clients pay premiums in advance, in cash, for these policies. How should the insurance company record these cash payments when received? What adjusting entries should be made at the end of each month?

8. Do you think there are any manufacturers in the United States that should set up reserves for (that is, accrue for) future legal liabilities arising from

the products they sell today? If so, which companies, and why would you recommend this accrual?

9. This month Clark & Sons purchased, by mistake, enough pencils to last five years at a total cost of $73. What adjusting entries, if any, should Clark's accountants make at the end of this accounting period?

10. Marones, Inc., is a manufacturer of specialized instruments. When a customer places an order, Marones requires a 20 percent nonrefundable down payment, with the balance of the sale price payable on delivery.

  a. Al Smith placed in October an order for a $25,000 instrument with a $5,000 down payment. How should Marones record this payment?

  b. Two months later, Smith receives the new instrument and pays the $20,000 balance of his order. How should Marones record these transactions?

  c. How is this accounting different from the one for an auto dealer that sells a car with a $4,000 down payment and $16,000 to be paid in monthly installments over 48 months?

11. Explain why the failure to pay by a customer who bought $400 of merchandise on credit four months ago does not affect profits in the current month, if the company uses the contra account Allowance for Doubtful Accounts.

12. Indicate whether each scenario below is true or false. During periods of high price inflation, higher reported profits (or, the equivalent, earlier reporting of profits) occurs if the following accounting method is used:

  a. LIFO, rather than FIFO

  b. The net method of accounting for cash discounts on purchases, rather than the gross method

  c. The specific price adjustment method, rather than the use of simple historical costs

  d. Use of an Allowance for Doubtful Accounts contra account, rather than expensing bad debts as they occur

  e. The net method of accounting for cash discounts offered to customers, rather than the gross method

13. Accountants are frequently called on to choose between alternative accounting methods. Which of the following choices do you think is most true to the conservatism principle? Explain.

  a. FIFO or LIFO

  b. Net or gross method of accounting for cash discounts allowed to customers

  c. Using an Allowance for Doubtful Accounts account or not

14. Does the preliminary trial balance of the general ledger as it exists before end-of-period entries are made typically indicate a higher operating profit than the final trial balance? Discuss your answer.

15. Can you think of an end-of-period entry that would affect the Cash account in the general ledger?

16. A company provides a stock option to its treasurer that permits her to purchase, at today's market price, 1,000 shares of the company's common stock at any time over the next five years. Does this action call for an adjusting entry? If so, what would it be?

17. Two years ago, company A purchased 5,000 shares of common stock in company B (20 percent of the outstanding shares of B) at $20 per share. The market price per share of B's stock is now $30 per share. Should A make an adjusting entry to reflect this new higher value? Discuss your answer.

18. The Electronics Outlet, a television and electronics appliance store, starts the month with $137,000 in inventory. The following transactions take place during the month:

    · Receives $21,230 worth of stereo equipment purchased for credit
    · Returns a $300 television to a supplier because it did not work
    · Purchases $330 of promotional material for the store
    · Acquires and receives three used television sets for $90 per unit, a total of $270
    · Discovers that one of the three televisions purchased in the previous entry is defective and throws it in the trash
    · Merchandise sales during the month total $33,000
    · Determines that the month-end inventory at the store has a value of $132,000

    What is the cost of goods sold for the month for the Electronics Outlet?

19. Refer again to exercise 18 and assume that, after the appropriate adjusting entries were made, Electronics Outlet discovered another $700 of inventory that hadn't been included in the $132,000 end-of-period valuation. Show the appropriate correcting entry.

20. In T-account form, record each of the following end-of-period adjustments. Indicate for each account whether it is an asset, a liability, a revenue, or an expense account.
    a. Allow for estimated bad debts of $3,500 on credit sales that occurred during the current accounting period.
    b. Record the fact that three months of coverage have expired on the annual insurance policy for which the company paid $8,000 six months ago.
    c. Record the cost of goods sold for the period, during which purchases were $36,500, purchase returns were $2,100, and inventory grew by $1,000 from the beginning to the end of the period.

d. For a second company that maintains perpetual inventory valuations and uses the contra account Allowance for Inventory Shrinkage, record the loss of $1,200 of inventory, apparently to theft.

e. A six-year-old computer that is currently valued as a $2,000 asset is determined now to be valueless.

f. A company that maintains an Allowance for Doubtful Accounts account determines that a six-month-old receivable of $3,000 is unlikely ever to be collected.

g. In light of ongoing negotiations with a disgruntled customer, a service company determines that it most likely will have to refund $2,000 to that customer.

h. A company owes $900 in interest on a loan at the end of each calendar quarter. If the next payment is due in two months, make the adjusting entry for this month.

i. The company's board of directors has declared $10,000 of dividends, payable early in the next accounting period.

j. Next month, the company will receive a check for $3,000 from the U.S. Treasury representing six months of interest on a government bond owned by the company.

21. Make appropriate entries for the following transactions at the Franke Company or Thompson Products. Indicate for each account used whether it is an asset, a liability, a revenue, or an expense account.

a. Franke purchases $7,500 of merchandise for resale; Franke uses the end-of-period adjustment method for determining the cost of goods sold for the period.

b. Thompson purchases $5,000 of parts to be incorporated into the products it manufactures; Thompson maintains perpetual inventory valuations.

c. Franke sells $10,000 of merchandise and offers all of its customers payment terms of "2%, 10 days/net, 30 days"; Franke accounts for cash discounts offered on the net basis.

d. Thompson sells $10,000 of its products and offers all of its customers payment terms of "2%, 10 days/net, 30 days"; Thompson accounts for cash discounts offered to customers on the gross basis.

e. Franke receives a $400 payment from a bankrupt customer in full payment of the customer's $800 receivable (originally recorded 10 months ago). Franke uses an Allowance for Doubtful Accounts contra account.

f. Thompson receives a $500 payment from a bankrupt customer in full payment of the customer's $1,000 receivable (originally recorded in the last fiscal year). Thompson does not use an Allowance for Doubtful Accounts contra account.

g. Franke pays an employee $2,000 for a vacation leave that the employee is taking this accounting period. Franke accrues the liability for vacation salaries as employees earn vacation days throughout the year.

h. Thompson pays $2,000 of dividends on its common stock; these dividends were declared and accrued during the last accounting period.

i. Franke pays $1,400 for an insurance policy that provides protection beginning next month.

j. Thompson pays $5,000 in an out-of-court settlement of a lawsuit for which it had previously accrued (reserved) $10,000 of probable liability.

22. The Dewey Corporation's preliminary trial balance (before month-end adjusting entries) on June 30 is shown below.

PRELIMINARY TRIAL BALANCE (JUNE 30)

	Debit	Credit
Cash	$ 600	
Short-term Investments	7,000	
Accounts Receivable	9,300	
Allowance for Doubtful Accounts		$ 450
Inventories	11,200	
Allowance for Obsolescence		710
Prepaid Expenses	2,300	
Fixed Assets, Net	22,500	
Accounts Payable		4,900
Salaries and Wages Payable		1,500
Vacation and Sick Leave Wages Payable		2,100
Interest Payable		300
Bank Loans Payable		6,000
Other Accrued Liabilities		1,000
Long-term Debt		5,000
Capital Stock		3,000
Retained Earnings		32,270
Sales		37,000
Cost of Goods Sold		
Purchases	27,000	
Salaries and Wages Expense	9,800	
Rent and Other Occupancy Expenses	1,850	
Bad-debt Expense		
Utilities Expense	880	
Insurance Expense		
Miscellaneous Expenses	600	
Interest Expense	1,500	
Interest Income		
Cash Discounts Earned		300
Total	$94,530	$94,530

a. Because some employees left the company before taking vacation leave, the Vacation and Sickness Leave Wages Payable account is overaccrued by $200.

b. Dewey needs to add $450 to its Allowance for Doubtful Accounts in light of the $37,000 credit sales during the month.

c. In reviewing the detail of its outstanding customer receivables, Dewey determines that $250 is due from a customer who has gone out of business and is therefore unlikely to pay.

d. Inventories increased by $1,300 from the beginning to the end of June.

e. Dewey earned $300 on its short-term investments during the month, with payment to be received in August.

f. Dewey paid $1,500 in bank interest during June; however, only one-third of this amount is an appropriate expense of June, the balance being interest prepayments for July and August.

g. Dewey has not yet received the invoice for, nor paid, the $1,000 rent for the month of June.

h. Included in the Prepaid Expenses account is $1,400 of prepaid insurance for the seven months of June through the end of the calendar year.

i. Dewey's accountant determined that the company's fixed assets have declined in value by $600 during the month of June.

j. After the month-end inventory was counted and valued, the accountant discovered $100 of inventory that is obsolete and must be scrapped.

k. In the course of negotiations with a supplier who provided services to Dewey in April, the supplier agreed to reduce his bill (not yet paid by Dewey) by $300. Dewey expects to pay the reduced balance next month.

l. Because of good operating results in the period January through June, Dewey's president has decided to pay employee bonuses totaling $1,200, with payment to be made in early July.

CHAPTER 8

ACCOUNTING FOR LONG-TERM ASSETS

Long-term assets, as contrasted with current assets, have a useful life of more than one year. A company's purpose in owning long-term assets—as contrasted with, say, inventory—is to use them for a number of years and not to earn a profit by purchasing and reselling them. Examples of long-term assets are land and buildings used by the firm; production and office equipment; software used in the firm's operations; autos, trucks, and forklifts; and intellectual property, particularly patents and trademarks.

The accrual concept requires that long-term assets be **capitalized** when acquired—that is, the expenditure is treated as an increase in an asset, not as an expense. While valuing a long-term asset when it is acquired is relatively straightforward, that initial valuation cannot remain unchanged over time, since most long-term assets decline in value over their lives. To avoid distorting reported income, either at the acquisition date or when the asset is finally disposed of (probably for a fraction of its original cost), its value must somehow be reduced periodically and in steps throughout its life.

A retail store values purchased inventory at its acquisition cost throughout the weeks or months the inventory is owned, because the store expects to sell the inventory at a price higher than cost. The same store, however, cannot purchase a delivery truck, use it for several years, and expect to sell the truck for more than a fraction of its original cost. Thus, while the store typically does not record a decline over time in its merchandise inventory value, it does account systematically for the decline in the delivery truck's value.

The impact of long-term asset accounting is particularly great in firms that are fixed asset intensive. For example, electric power utilities invest far more in property, plant, and equipment (fixed assets) than in accounts receivable and inventory (current assets). The procedures they use to reflect, as expenses over the years, the decline in value of those fixed assets have a major impact on the utilities' year-to-year reported profits. By contrast, a professional service firm is a labor-intensive enterprise; since the firm owns few fixed assets and the primary expenses are personnel salaries and related expenses, long-term asset accounting has little influence on profits or asset balances.

VALUING FIXED ASSETS

For most companies, by far the largest long-term asset category is fixed assets: property, plant, equipment, computers, and vehicles that are used in the conduct of the business and have multiyear lives. Thus, we turn first to accounting for fixed assets, leaving until later in the chapter the challenge of accounting for long-term intangible assets and long-term investments.

Note that the accounting principles discussed in Chapter 6 must undergird fixed-asset accounting procedures. For example, the matching principle is key here: we need a way to match the cost of owning a fixed asest, including what was paid for it initially, to the years over which it is used—that is, when the benefits of ownership are realized. Other key principles to bear in mind as you consider fixed-asset accounting are as follows:

- Going-concern assumption: these fixed assets typically have the greatest value if they are both used and useful.
- Conservatism: lean in the direction of understating both asset values and profits.
- Consistency: use the same procedures from accounting period to accounting period.
- Materiality: don't bother to capitalize assets—such as hand tools, cell phones, and the like—below a threshold value (perhaps $1,000 for a small company and $25,000 for a large company), even though they will be useful for more than one year; instead, follow the conservative route and expense these low-value assets when acquired.

Recall the three valuation methods: the time-adjusted value, the market value, and the cost value. The first two of these methods automatically recognize a decline in the value of fixed assets throughout their years of use. The time-adjusted value method recomputes the remaining stream of ownership benefits at the end of each accounting period. As the asset ages, the remaining benefit period shortens and its usefulness may decline; thus, the time-adjusted value decreases. Similarly, the market value method redetermines the asset's then-current market value at the end of each accounting period. As the asset ages or wears, its market value typically declines. Under both of these methods, the decline in the value during each accounting period is recorded as an expense of that period.

Appealing as these two methods of valuation are in concept, however, they are, you will recall, extremely difficult to implement. We are left, then, with the cost valuation method, currently the only method acceptable under GAAP accounting in the United States. We need some procedure to record the decline in the capitalized value of fixed assets during their years of use. That is, we must allocate, over the years of ownership, the difference between the original cost and the anticipated amount to be realized when the asset is ultimately sold or scrapped. This procedure is called **depreciation accounting,** the rational, equitable, and systematic allocation over the estimated years of its use of the difference between an asset's acquisition cost and its estimated salvage value.

Bear in mind that depreciation is a process of allocation, not a method of estimating market values: the fixed asset's value on the firm's balance sheet (assuming depreciation has been properly recorded) bears no necessary relationship to the asset's market value—nor is it intended to. The presumption is that the owner will not sell the asset until the end of its useful life, and thus interim market values have little relevance. The asset's estimated market value at the end of its useful life—its disposition or salvage value—is, however, relevant to depreciation accounting.

Calculating Depreciation Expense

Depreciation accounting requires you to estimate (1) the **initial cost** of the asset (its acquisition cost); (2) its **useful life** (the years of use), (3) the asset's **salvage value** (value at the end of its estimated useful life); and then to choose (4) the most rational, equitable, and systematic method of allocating this difference between the acquisition cost and the salvage value—the depreciation method. The fourth factor, the depreciation method, requires the most discussion, but the first three factors also present some challenges.

Initial Cost. The initial cost includes, in addition to simply the asset's cash purchase price, any other outlays required to get the asset into a position and condition where it is productive. Thus, freight charges and installation costs are normally included in the initial cost of the asset, so they can be depreciated over the asset's life and not expensed in the accounting period when they are incurred.

Major overhauls or rebuilding of existing fixed-assets should be treated as fixed-asset expenditures if the result of the refurbishing is to substantially increase the asset's useful life. Thus, a major overhaul of a machine tool or the replacement of the factory roof is capitalized and depreciated.

Useful Life. What determines the period over which the asset will be depreciated, that is, its life? We are concerned only with its useful life, the productive years for its present owner. This useful life may be a good deal shorter than the asset's total life; a truck may have several owners during its total life, but the depreciable life for a particular owner is the time during which that owner uses the truck.

The life of an asset is typically dictated by wear, obsolescence, or change in requirements. A truck may truly wear out. The motor, transmission, and other mechanical parts may deteriorate to the point that the truck is no longer reliable or not worth repairing; if so, the owner ceases to use it, and its useful life to the present owner terminates. A second owner—perhaps one more tolerant of breakdowns—may now purchase the truck, thus commencing a useful life for this second owner.

Unlike a truck, a computer seldom wears out, but is likely to obsolesce. It may be operable for 20 or 30 years, but its useful life may be five years or less, after which technological improvements will give the owner strong motivation to replace it.

The life of a specialized production tool may be determined by neither wear nor technological change, but by changes in demand for the end product or service produced. For example, an injection molding die used to make a heavily promoted child's toy may have a short useful life simply because the demand for the toy will evaporate when the promotional program ends. The tooling is neither worn out nor technologically obsolete, but its useful life is over.

Income tax regulations provide guidelines on the depreciable lives for certain classes of assets. Beware of routinely accepting these guidelines. While they govern income tax calculations, frequently shorter or longer lives are more appropriate for profit reporting, that is for the equitable distribution of the difference between original cost and salvage value. Remember: income tax laws should not determine accounting policy. More on this subject later.

Salvage Value. If the useful life is difficult to estimate, salvage value—the asset's probable market value at the end of that useful life—is that much more difficult. Fortunately, for most (but not all) assets, estimated salvage value is a small percentage of initial cost (for example, 10 percent or less), and estimation errors do not have a material effect on depreciation expense and asset valuation. When in doubt about salvage value, remember that the lower the salvage value, the more conservative the depreciation accounting.

Salvage values are influenced by many unpredictable factors. The market value at the end of a commercial building's estimated 40-year useful life will depend not only on the condition of the building itself but the condition of its neighborhood. If the building happens to be in a part of the city that has developed into a high-rent, retail-commercial area, the building may have a market value well in excess of its initial cost. (The cost value method prohibits using an estimated salvage value greater than the asset's initial cost.) On the other hand, in 40 years the building's undesirable location may drive away prospective buyers. Rates of inflation (or deflation) also influence salvage values.

In many situations, zero salvage value is the only sensible assumption. A specialized asset, useful to its present owner but unlikely to be useful to others, should carry a zero salvage value. **Leasehold improvements**—that is, lighting, electrical equipment, and other improvements installed in leased facilities—normally are fully depreciated (that is, reduced to zero salvage value) over the term of the lease, as the leasee will obviously derive no continuing benefit from the improvements after moving out of the facility.

Depreciation Method. The depreciation method selected determines how the total depreciation expense is spread over the accounting periods comprising the asset's useful life. Depreciation methods fall into three fundamental categories: straight line, units of production, and accelerated.

The **straight-line method** is explained by its name: the value of the asset is reduced from its initial cost to its estimated salvage value in a straight line over the asset's estimated useful life. If the asset has a 10-year life, depreciation expense during each year of life is 1/10 of the depreciable value (or 1/120 during each month of life). Straight-line depreciation is the most prevalent method.

The **units-of-production method** assigns depreciation expense in proportion to actual usage. The asset's life is defined in usage terms—for example, estimated hours of machine operation, units produced, or vehicular miles traveled. The amount of depreciation expense in any accounting period is a function of usage during that period. For example, if a particular tool's life is defined as the production of 450,000 units, the depreciation expense on this asset during a year when 75,000 units are produced will be 16.7 percent (75,000/450,000 = 0.167) of the difference between the asset's initial cost and its estimated salvage value. The units-of-production method of depreciation fits certain assets well, but it is not widely used.

Accelerated methods assign greater depreciation expense to the asset's early years, and lower expense to later years. Given that straight-line depreciation is simple and apparently equitable, what are the arguments for accelerated depreciation? Recall the objective: to allocate, rationally and systematically, over the asset's life the difference between initial cost and estimated salvage value. To reflect the asset's declining market value is not an objective. Although accelerated depreciation frequently comes closer than straight-line depreciation to approximating market values, this cannot be an argument for its use when that is not the objective.

There are, however, two rational arguments for accelerated depreciation. First, the asset may well be more useful during its early life, before wear and technological obsolescence have taken their toll. Toward the end of its life, the asset may serve only a standby role, with the primary productive load assumed by newer, technologically superior assets. A second argument is that maintenance expenses are likely to increase over the asset's life. The total cost of owning and operating an asset (including both depreciation and maintenance) may be spread more equitably if more depreciation is recognized early in the asset's life and less later, when maintenance expenditures are likely to be greater.

Moreover, accelerated depreciation is conservative. If the company is growing and, thus, continually adding to its stock of fixed assets, accelerated depreciation results in conservatively stated profits throughout the growth phase of the company. In my opinion, accelerated depreciation should be used more widely.

One appealing method of accelerated depreciation is the **declining-balance method.** The declining balance method applies a specified fraction to a declining value or declining base. The base each year is the initial cost less the depreciation accumulated to date—that is, the remaining book value of the asset. The fraction is set by reference to the straight-line method; for example, the *double-declining-balance method* sets the fraction at twice (double) the straight-line rate.

Thus, an asset with a 10-year estimated life is depreciated 1/10 per year under the straight-line method and 2/10 (1/5) under the double-declining-balance convention. However, the depreciation expense declines year by year because the one-fifth fraction is applied to the declining book value. For example, depreciation expenses for an asset with an initial cost of $8,000 and a useful life of 10 years, assuming double-declining-balance depreciation, would be as follows:

Year 1: $\frac{1}{5}$ ($8,000) = $1,600
Year 2: $\frac{1}{5}$ ($8,000 − $1,600) = $\frac{1}{5}$ ($6,400) = $1,280
Year 3: $\frac{1}{5}$ ($8,000 − $1,600 − $1,280) = $\frac{1}{5}$ (5,120) = $1,024

Two refinements to this convention should be noted. First, since the unswerving application of the declining-balance convention precludes an asset ever being depreciated to zero, we switch from the declining-balance convention to the straight-line method during that year when the straight-line method would result in greater depreciation expense, and the remaining book value is depreciated accordingly over the remaining life. Second, estimated salvage values are ignored when determining depreciation expenses under the declining-balance convention. When the switch is made to the straight-line method, however, the asset is then depreciated to its anticipated salvage value.

Table 8-1 compares the straight-line depreciation method with the double-declining method for an asset having an initial cost of $26,000, a 10-year estimated useful life, and an estimated salvage value at the end of its useful life of $2,000. Total depreciation over the asset's life is, of course, the same under both methods: $24,000, the difference between initial cost and estimated salvage value. The two methods differ solely in the timing of expenses, but timing is all important.

Contra Account: Allowance for Depreciation

So much for calculating depreciation expense for each year of the asset's life. What are the accounting entries to record the expense and the corresponding

TABLE 8-1. An Illustration of Alternative Depreciation Methods

	Initial Cost of Asset		$26,000 (includes freight and installation)	
	Useful Life:		10 years	
	Estimated Salvage Value:		$2,000	

Method	Year	Depreciation Calculation	Annual Depreciation Expense	End-of-Year Book Value
Straight-line	1	$\frac{1}{10}(26,000 - 2,000)$	$ 2,400	$23,600
depreciation	2	$\frac{1}{10}(26,000 - 2,000)$	2,400	21,200
	3	$\frac{1}{10}(26,000 - 2,000)$	2,400	18,800
	4	$\frac{1}{10}(26,000 - 2,000)$	2,400	16,400
	5	$\frac{1}{10}(26,000 - 2,000)$	2,400	14,000
	6	$\frac{1}{10}(26,000 - 2,000)$	2,400	11,600
	7	$\frac{1}{10}(26,000 - 2,000)$	2,400	9,200
	8	$\frac{1}{10}(26,000 - 2,000)$	2,400	6,800
	9	$\frac{1}{10}(26,000 - 2,000)$	2,400	4,400
	10	$\frac{1}{10}(26,000 - 2,000)$	2,400	2,000
		Total	$24,000	
Double-declining-	1	$\frac{2}{10}(26,000)$	$ 5,200	$20,800
balance depreciation	2	$\frac{2}{10}(20,800)$	4,160	16,640
	3	$\frac{2}{10}(16,640)$	3,328	13,312
	4	$\frac{2}{10}(13,312)$	2,662	10,650
	5	$\frac{2}{10}(10,650)$	2,130	8,520
	6	$\frac{2}{10}(8,520)$	1,704	6,816
	7	$\frac{2}{10}(6,816)$	1,363	5,452
	8[a]	$\frac{1}{3}(5,452 - 2,000)$	1,151	4,301
	9	$\frac{1}{3}(5,452 - 2,000)$	1,151	3,150
	10	$\frac{1}{3}(5,452 - 2,000)$	1,151	2,000
		Total	$24,000	

[a]Change to the straight-line method and depreciate to estimated salvage value over remaining life. If the double-declining-balance method were used in the eighth year, the depreciation expense would have been ($\frac{2}{10} \times 5,453$) = $1,091, somewhat less than the straight-line amount shown.

decline in the asset's book value? The procedure could parallel the accounting for Prepaid Assets, discussed in Chapter 7. That is, we could debit the Depreciation Expense account, and credit the Fixed Assets account. If we did so, however, the general ledger would no longer carry information about the initial cost of the fixed asset. The Fixed Asset account would instead show the book value: initial cost less all depreciation recorded to date.

In order to preserve the unadjusted original value of assets (or, indeed, of liabilities) in the general ledger, we once again use a contra account. Remember that contra accounts permit values of assets (or liabilities) to be adjusted in a separate, but related, account and without disturbing the existing asset (liability) balance. Contra asset accounts carry credit balances, and contra liability accounts carry debit balances. Depreciation is thus accumulated in the contra account **Allowance for Depreciation.** Here is the ledger entry for depreciation for the third year of the asset illustrated in Table 8-1, assuming double-declining balance depreciation:

Depreciation Expense	Allowance for Depreciation
$3,328	Balance $3,328

After this entry is made, the general ledger *balances* with respect to this asset are as follows:

Fixed Assets	Allowance for Depreciation
26,000	12,688

The Allowance for Depreciation account balance is the sum of the first three years' depreciation expenses, and the difference between these accounts balances, $13,312, is the book value of the asset, as shown in Table 8-1.

When the company's balance sheet is constructed, it may contain both amounts or simply the net of the two accounts:

Fixed assets, net $13,312

or

Fixed assets $26,000
less: Allowance for depreciation   12,688 $13,312

The second construction provides the financial statement reader with more information: what the company paid for its assets and how much depreciation has been accumulated to date.

The contra account Allowance for Depreciation is sometimes incorrectly referred to as **Reserve for Depreciation.** Obviously, the balance in this contra account is not a reserve in the sense of cash available for the purchase of a replacement asset. Note that this entry is not triggered by a transaction. Rather, it is one of the adjusting, or end-of-period, entries that accountants must initiate to state properly the accounting period's profit and ending asset values.

Accounting for the Disposition of Fixed Assets

Seldom does actual ownership of an asset conform exactly to estimates of life and salvage value. What accounting entries need to be made when the original life and salvage-value assumptions do not work out precisely?

Sometimes, an asset is owned for longer than its estimated life. When the asset has been fully depreciated—that is, when the allowance for depreciation is built up to the point that the asset's book value equals its estimated salvage value, if any, no additional depreciation expense is recognized. The balances in both the asset and the contra accounts simply remain unaltered as long as the company owns and uses the asset.

Refer again to the example in Table 8-1. Both methods depreciated the asset over 10 years to a $2,000 estimated salvage value. If, after 12 years, the asset is finally salvaged for exactly $2,000, the accounting entry is:

Cash		Fixed Assets		Allowance for Depreciation	
Balance 2,000		Balance	26,000	24,000	Balance

The initial cost and accumulated depreciation balances contained in the Fixed Assets and contra accounts with respect to this particular asset are driven to zero; the income statement (and thus owners' equity) is unaffected by this transaction.

Frequently, an asset is sold prior to the end of its estimated useful life. Suppose a cattle rancher owns a piece of ranching equipment with useful life and initial and salvage values conforming to the assumptions in Table 8-1. Now, suppose the rancher sells this equipment after six years. Sales of productive assets are fundamentally different from sales of the ranch's products. The cattle rancher is in the business of selling cattle, not buying and selling ranching equipment. Thus, when the ranching equipment is sold, the sale is viewed as an "occasional sale," not in the normal course of business, and is not recorded as revenue as the sale of cattle would be. Almost inevitably, the sale occurs at a price above or below the asset's current book value; thus, typically the owner must recognize some gain (profit) or loss upon disposing of the asset.

Gain (or Loss) on the Disposition of Fixed Assets

Assume that the equipment in Table 8-1 is sold after six years, for $8,000. The owner incurs a loss on this sale, if the straight-line depreciation method was used, but a gain if the double-declining balance depreciation convention was used. Any gain or loss is recorded in an account entitled Gain (or Loss) on Disposition of Fixed Assets. Assuming straight-line depreciation was used, the appropriate entry is

Cash	Fixed Assets		Allowance for Depreciation		Gain (or Loss) on Disposition of Fixed Assets	
Balance 8,000	Balance	26,000	14,400	Balance	3,600	

If the double-declining-balance convention was used, the entry would be as follows:

Cash	Fixed Assets		Allowance for Depreciation		Gain (or Loss) on Disposition of Fixed Assets	
Balance 8,000	Balance	26,000	19,184	Balance		1,184

Note that neither the Sales account nor the Cost of Goods Sold account is used. Gain (or loss) on the disposition of fixed assets is nonoperating revenue or expense and, accordingly, is typically shown below the operating profit line on the company's income statement.

Trade-In

The accounting entry to reflect the sale of an asset for cash is straightforward. Complications arise when the disposition is part of a barter transaction, for example, when one asset is *traded in* on another (when part of the payment is the transfer of the old asset to the seller of the new asset).

Suppose a restaurant trades in a piece of equipment having an initial cost of $13,000 and a current book value of $5,000, as partial payment for a new piece of restaurant equipment having a list price of $21,000. After much negotiation, the equipment dealer agrees to accept the restaurant's old equipment plus $14,000 in cash. Apparently, the dealer is ascribing a $7,000 value to this trade-in, the difference between the list price of the new equipment and the cash balance to be paid by the restaurant. If so, the restaurant has realized a $2,000 gain on the disposition of the old asset, as its book value is only $5,000.

Now let's assume that, in the absence of any trade-in, the restaurant could have negotiated a discount off the $21,000 list price. If so, was the equipment dealer really ascribing a value of $7,000 to the trade-in? Or was the $7,000 difference between the list price and the cash to be paid the sum of a price discount and the trade-in value? The restaurant's accountant must separate these two effects in order to record this transaction. For example, assume the restaurant believes the dealer would have accepted $19,500 in cash and no trade-in; if so, the dealer actually valued the trade-in at only $5,500, and the restaurant realized a correspondingly smaller gain on the trade-in of the old equipment. These assumptions lead to the following entries:

Cash		Fixed Assets		Allowance for Depreciation		Gain (or Loss) on Disposition of Fixed Assets	
Balance	14,000	Balance 19,500*	13,000†	8,000	Balance		500

Note that the new equipment is recorded at its equivalent cash purchase price, somewhat less than its list price. The gain on the disposition is now substantially lower. The basic accounting principle of conservatism is better served by factoring into the accounting entries the apparent cash discount, since doing so values both assets and profits more conservatively.

Timing the Recognition of a Loss

When an asset is no longer useful to its owner, the asset's value should be adjusted accordingly. Suppose, for example, that the Cody Company purchased computer equipment in 1998 for $85,000, and has been depreciating this equipment over a 10-year estimated useful life. Five years later, in 2003, Cody upgrades its computer network so that the 1998 computer is no longer used; Cody's management, after determining that only a very nominal amount could be realized by selling this now-obsolete computer, decides to retain it in storage. Without debating the wisdom of this decision, consider what accounting entry, if any, should now be made. Depreciating this computer in the normal fashion is inappropriate, as the computer is no longer productive; with respect to this single asset, the going-concern assumption no longer applies. If the computer has no market value (or scrap value), its remaining book value should be written off—that is, expensed. The conservative approach is to recognize the expense at the time the network is revamped, for it was that event that rendered the 1998 computer valueless to Cody.

INCOME TAX LAWS

Income tax laws and regulations are too complex to discuss in detail. But you should be aware of the types of provisions you may encounter. First, the law requires that assets meeting certain tests be capitalized and depreciated, not expensed—that is, accounted for as fixed assets. The law requires that related freight, installation, and similar expenditures also be capitalized. Inevitably, disputes arise: taxing authorities argue for capitalization, the company argues

---

*New asset
†Old asset

that the expenditure should be deductible immediately. Second, the law establishes guidelines for the useful lives of various classes of assets. While the company seeks as short a depreciable life as possible in order both to maximize and to accelerate in time the depreciation tax shield, tax authorities expect careful justification for use of a life shorter than the guidelines provide. Specific tax provisions fill many volumes and apply to a broad range of new and used assets in various industries and under alternative circumstances. Management should obtain sound tax advice from experts regarding these issues.

Income Tax Considerations

Tax laws in the United States contain certain incentives to spur investment in fixed assets. Thus, tax considerations are important in decisions regarding the acquisition, depreciation, and disposition of fixed assets. While you should understand these considerations, you also need to remember that the tax laws should not dictate a company's accounting policy, as tax law provisions are frequently inconsistent with sound accounting practice. Tax laws are enacted to generate tax revenue and implement overall government economic policy. They are not meant to dictate how companies should account for fixed assets. As a result, the accounting for fixed assets in reports of company performance directed to shareholders and creditors (financial accounting) may be quite different from fixed-asset accounting in the company's income tax returns (tax accounting). Most companies end up maintaining two sets of books—appropriately, legally, and openly.

Recall the objective of sound depreciation accounting: to allocate in a rational, systematic, and equitable manner the difference between initial cost and estimated salvage value over the estimated life of the asset. That's not the objective when calculating the depreciation expense deduction for income taxes, where the objective is to minimize tax payments and postpone them as long as possible. Every taxpayer—individual or corporate—is expected to take reasonable steps to minimize income tax payments. To do otherwise is both foolish and inconsistent with management's obligations to its corporate shareholders. Remember, too, that deferring tax payments is always desirable, as the money can be used during the period of postponement; money has a time value.

Therefore, for tax purposes, a company typically seeks to accelerate depreciation to the extent possible, thereby decreasing its taxable income (by increasing expenses) and decreasing its tax liability. Remember from our discussion of the cash flow statement (in Chapter 5) that depreciation expense is a **noncash expense.** Therefore, an increase in depreciation expense has no cash cost; instead, it produces a cash savings by reducing income taxes. Thus, we speak of depreciation as a **tax shield**—higher depreciation expenses shield more of the corporation's earnings from taxation. Understandably, therefore, the income tax laws and regulations are filled with provisions limiting the extent to which depreciation can be accelerated.

Deferred Income Taxes

When corporations use one depreciation method in financial accounting and another in tax accounting, timing differences in expense recognition result. These differences give rise to so-called **deferred income taxes.** Suppose that the Wong Corporation uses straight-line depreciation for financial reporting purposes and the double-declining-balance method for tax purposes. In 2002, its depreciation expense using the straight-line method is $250,000; its double-declining-balance expense is $360,000; its operating profit before depreciation and income tax expenses is $780,000; and its effective income tax rate is 40 percent. Under these assumptions, Wong's reported net income (financial accounting) and related income tax expense are as follows:

Operating profit before depreciation and taxes	$780,000
less: Depreciation expense	250,000
Taxable income (internal reporting purposes)	530,000
less: Income tax expense (at 40 percent)	212,000

But is Wong currently liable for $212,000 of income tax payments? No: using the double-declining-balance method, its income tax currently payable (tax accounting) is calculated as follows:

Operating profit before depreciation and taxes	$780,000
less: Depreciation (double-declining balance)	360,000
Taxable income (tax reporting purposes)	420,000
Income tax currently payable (at 40 percent)	168,000

This difference is obviously in Wong's favor; the company has postponed the cash payment of $44,000 ($212,000 less $168,000). How should this discrepancy—this postponement of taxes—be handled in Wong's accounting records? For financial reporting purposes, the proper income tax expense is $212,000; of this amount, $168,000 is currently payable (a current liability) and the balance, $44,000, is treated as a deferred income tax (a long-term liability):

Income Tax Liability*	Deferred Income Taxes†	Income Tax Expense
168,000	44,000	212,000

*Current liability
†Long-term liability

If Wong acquires no new or replacement assets (an unlikely prospect unless the company falls on sustained hard times), eventually the deferred income taxes will become payable when, late in the assets' lives, straight-line depreciation expenses exceed those calculated by the double-declining-balance method. Then cash payments for income taxes will be greater than the income tax expense reported on Wong's income statement, and the deferred tax liability will be reduced. Remember that the differences here are solely in timing; the full difference between initial cost and estimated salvage value of the fixed assets will ultimately be depreciated by both methods.

Interestingly, if a company grows and regularly adds fixed assets, the amount of its deferred income taxes will typically continue to grow. As a result, many growing companies view deferred income taxes as semipermanently deferred; although they are accounted for as long-term liabilities, the company does not anticipate having to pay them in the foreseeable future.

Incidentally, timing differences between financial and tax recognition of expenses (and revenues) can also run the other way, resulting in prepaid rather than deferred income taxes. For example, warranty expenses are typically not allowable deductions for tax purposes until the actual expenditures occur, although for financial reporting purposes the company accrues an allowance for warranty (with a corresponding warranty expense) at the time of delivery.

Limits on the Use of Tax Shields

Tax laws and regulations are replete with provisions, including some that are quite obscure, to limit the deductible expenses and tax credits available to individuals and corporate taxpayers. Careful tax planning is required to be certain that anticipated tax savings are, in fact, realized.

Keep in mind one key limitation. Tax shields are of no value to an individual or corporation who, in the absence of such shields, would not be required to pay tax. Thus, accelerated depreciation is of no benefit to an unprofitable corporation. A tax deduction is only of value if and when the deduction reduces actual cash tax payments. The U.S. tax law does, however, permit most taxpayers to carry back and carry forward any taxable losses for a limited number of years. For example, if a company incurs a $1 million pretax loss in the current year after several years of large pretax profits, the company can carry back and apply for a refund of some income taxes paid previously. If these earlier pretax profits did not total $1 million, the company can then carry forward the loss and apply it against future years' profits, reducing income tax payments in those future years.

ACCOUNTING FOR OTHER LONG-TERM ASSETS

As developed economies around the world have increasingly relied on knowledge as key economic drivers, intangible assets have assumed greater importance. Our traditional accounting procedures were developed a century or more ago for firms focused on these aspects:

- the production of goods, not services
- semiskilled or unskilled labor, not knowledge workers
- tangible, "hard" assets, not intellectual property and software
- dividend payments, not the accumulation of monetary investment assets

Accordingly, our accounting procedures are imperfect—or inadequate—for accounting for long-term intangibles: patents, know-how, software, trademarks, goodwill, and marketable securities. We need to review both how we now account for these long-term assets and how these methods are deficient.

Patents, Trademarks, and Other Intellectual Property

Suppose Glen Partners purchases a patent from an inventor for $45,000, presumably the market value of this patent. Since, by U.S. law, a patent provides protection of the invention for 20 years, the new owner, Glen Partners, should recognize a decline in the patent's value over its estimated useful life, a period that might be shorter than, but certainly is not longer than, its legal life. Let's assume that Glen Partners, after assessing the rate of technological change in its industry, determines that the patent's probable useful life is five years, or 60 months. Glen Partners then **amortizes** the value of the patent, month by month, again utilizing a contra account. The month-end adjusting entry to recognize patent amortization is as follows:

Allowance for Amortization		Amortization Expense	
	Balance 750	$750	

The acquisition and subsequent amortization of a trademark asset or of confidential know-how would be treated similarly.

Now let's suppose Glen Partners itself invents another novel device that it will use (that is, not sell to another company), a device for which the company applied and ultimately received a patent. How do we account for this intangible asset? Imperfectly, at best! Because we are stuck with the cost valuation method, the market value of this internally developed intellectual property is irrelevant. And, because of current conservative practices in accounting for engineering and development expenses, these expenses are written off (expensed) as incurred, not capitalized. Typically, legal and application fees associated with obtaining patent protection are capitalized (and amortized), but such fees are generally trivial compared with the out-of-pocket salary and other expenses incurred in creating the intellectual property.

Thus, unfortunately, the accounting treatments for acquired intellectual property and internally created intellectual property are vastly different. Accounting for trademarks faces a similar dilemma: arguably the most valuable trademark in the world is Coca-Cola; with a market value of billions of dollars, the trademark is nevertheless valued at $1 by the company that created that enormous value.

Software

The dilemma just discussed for patents and trademarks also applies to the differences in accounting for acquired versus internally developed software.

But consider the peculiar accounting environment for a company that produces and markets a line of software products. If the cost of creating the software is expensed as incurred, the company has virtually no cost of goods sold (except the minor cost of reproducing the software package) to match against future years of sales revenue. Its gross margin is very large, but so also are its development expenses—with most of these development expenses related to future products, not to those being sold today. Thus, the principle of matching revenues and expenses is not well served in such companies. The FASB attempted to address this problem with its standard FAS #86, issued in the early1990s, but for various complicated reasons, this accounting rule has done little to improve the problem of accurately matching of revenues and expenses in software companies.

Investments

How should we account for a company's long-term investments in *marketable securities,* (securities that the company does not expect to sell during the next 12 months)? Interestingly, the accounting treatment differs with the intent of the company.

Suppose the Meier Company buys a 25 percent equity interest (in other words, 25 percent of the outstanding common stock) in Optek Industries, a company in an industry related to Meier's. Meier makes this investment for strategic reasons: to learn more about the technology and industry dynamics with which Optek deals. Meier accounts for this investment using the *equity method:* 25 percent of Optek's profit (or loss) is included in Meier's income statement and added to (or subtracted from) the value of its asset, Investment in Optek.

Alternatively, suppose Meier is simply a passive investor in the common stock or bonds of General Globalwide, a large company traded on major security exchanges. Meier must *mark to market*—that is, reflect on its books the current exchange price (market value) of the security. As the value of General Globalwide increases or decreases on Meier's balance sheet, Meier's nonoperating income is similarly increased or decreased. Incidentally, Meier's short-term investments, including cash-equivalent investments, must also be marked to market.

Goodwill

The intangible asset Goodwill, arises often when one company acquires another. Suppose Welch Optical acquires all of (that is, 100 percent of the common stock of) the Fletcher Corporation for $286 million. Welch adds to its own balance sheet, at the estimated fair market value, all of the assets and liabilities of Fletcher. Let's assume that the market values of these assets and liabilities, at the time of acquisition, are, respectively, $431 million and $263 million, a difference of $168 million. Why would Welch pay $286 million for "net assets" (assets minus liabilities) of only $168 million? Presumably, because Welch views very positively the future business prospects of Fletcher, it is willing to pay "extra" to acquire Fletcher and its attractive future. This "extra," called **goodwill,** is treated as a long-term intangible asset.*

Is this intangible asset amortized? The answer over recent years has been yes, no, and maybe, as the FASB accounting rules have changed. Over those years, goodwill has, at different times, been viewed as (1) a permanent asset, not to be amortized; (2) an amount to be expensed (reducing owners' equity) at the time of the acquisition; and (3) an asset to be amortized over its useful life, defined as 40 years or less. Currently (2003), goodwill is capitalized at the time of acquisition. It is then revalued each time a balance sheet is constructed, and any reduction (referred to as any "impairment") in value is reflected with a corresponding expense. I leave to you to contemplate how difficult this "impairment" in goodwill's value is to assess!

SUMMARY

While the time-adjusted and market value methods account routinely for the decline in value of a fixed asset over its life, the historical cost method requires that some arbitrary but systematic procedure allocate, over the asset's life, the difference between the initial cost of the asset and its estimated salvage value. Depreciation accounting does not attempt to reflect, in the accounting records, the market values of assets but seeks simply to match, rationally and equitably, the costs of owning the fixed assets over the years the assets are used.

Depreciation accounting requires the estimation (in some instances with the potential for substantial error) of (1) the initial cost, (2) the estimated useful life, and (3) the estimated salvage value at the end of the useful life, and then (4) choosing a depreciation (or allocation) method. Three alternative methods of depreciation are straight line, accelerated, and units of production. The most common convention for accelerated depreciation is the declining-balance method. Depreciation is accumulated in a contra account so as to

---

*If the acquiring company pays less than the market value of the net assets of the acquired company, it does not account for the difference as "illwill" but, rather, simply adjusts downward the value of the assets acquired.

preserve information in the accounting records regarding the asset's initial cost.

At the time of disposition, any difference between the amount realized in the sale (or scrapping) of the asset and the asset's book value (equal to the initial cost less the accumulated depreciation) is recorded as other income or expense, typically in an account entitled Gain (or Loss) on Disposition of Fixed Assets.

Depreciation expense is a noncash expense but nevertheless deductible for income tax purposes (that is, a tax shield). Therefore, taxpayers seek to maximize and accelerate depreciation for tax purposes, while taxing authorities place limits on the use of this tax shield. As a result, tax laws in the United States are replete with provisions regarding the capitalization, depreciation, and disposition of fixed assets. The use of straight-line depreciation for financial reporting purposes and accelerated depreciation for tax reporting gives rise to deferred income taxes.

Other long-term assets—for example, intellectual property—are amortized over their useful lives. Goodwill, which may arise when company A acquires company B, must be devalued as and when its value is "impaired," and the extent of the impairment must be recognized as an expense.

Long-term investments in securities are either marked to market (reflecting current market prices) or accounted for on the equity basis, depending on the company's purpose in owning the investment.

NEW TERMS

Accelerated depreciation. Methods of depreciation that result in greater depreciation expenses in the early years of the asset's useful life and lower depreciation expenses in later years.

Allowance for Depreciation. The contra account to the Fixed Assets account reflecting the accumulated depreciation of the assets to date.

Amortization; to amortize. The accounting process to reduce systematically the book value of a long-term asset (other than fixed assets) to zero or an appropriate terminal value at the end of its predicted life.

Capitalization; to capitalize. The accounting process to record an expenditure as an asset (to be subsequently depreciated or amortized), rather than as an expense.

Declining-balance depreciation. A method of accelerated depreciation that results in a depreciation expense equal to a specified fraction of the book value (initial cost less accumulated depreciation to date) of the asset, an amount which declines each year.

Deferred income taxes. An income tax liability, not payable currently, typically resulting from differences in accounting procedures between financial reporting and tax reporting.

Depreciation accounting. The rational, equitable, and systematic allocation, over the years an asset is expected to be used (its *useful life*) of the difference between the asset's initial cost and its estimated salvage value.

Goodwill. The difference between the (1) acquisition price paid for a company, and (2) the market value of the assets obtained in the acquisition less the market value of the liabilities assumed.

Initial cost. An amount that includes the purchase price of the fixed asset and all ancillary costs required to get the asset into a productive condition.

Leasehold improvements. Those fixed assets that arise when a lessee invests in lighting, partitions, or other building improvements within a facility (typically a building) that is leased and not owned. These improvements are owned by the lessee and not the lessor.

Noncash charge; noncash expense. An expense, such as depreciation or amortization, that does not involve an expenditure of cash.

Reserve for Depreciation. An alternative, but somewhat inappropriate, name for Allowance for Depreciation, a contra account for fixed assets.

Salvage value. The value of an asset at the end of its useful life; the amount received by the present owner from the sale of an asset to a subsequent owner or from scrapping the asset.

Straight-line depreciation. A method of depreciation that results in equal depreciation expense in each accounting period of the asset's estimated useful life.

Tax shield. A *noncash expense (or charge)* that is deductible for income tax purposes. Such an expense does not consume cash but rather, because it is deductible, reduces cash outlays for income taxes.

Units-of-production depreciation. Methods of depreciation that result in the depreciation expense being a function of usage of the asset during the particular accounting period.

Useful life. The period of time during which an asset is useful to, and used by, its present owner; the useful life is often only a portion of the asset's total life.

EXERCISES

1. What is the objective of depreciation accounting?

2. How would you respond to someone who asked if the amounts shown in Allowance for Depreciation account could be used to purchase replacement fixed assets?

3. Can you name an asset for which the units-of-production depreciation method might be particularly appropriate?

4. One often hears the statement "By accelerating depreciation expenses, the company achieves greater cash flow." Do you agree? Why or why not?

5. Why is land not depreciated? Can you think of other assets that should not be depreciated?

6. Would you expect the market value to be above or below the book value for each of the following long-term assets? Explain your answer.
   a. A three-year-old personal computer
   b. A one-year-old instrument that was supplied by a manufacturer who is now quoting a two-year wait for delivery of the same instrument to new customers
   c. A seven-year-old asset owned in a country that has experienced high inflation for the past decade
   d. A 10-year-old commercial office building that is fully leased to financially strong tenants
   e. Leasehold improvements installed two years ago and being depreciated over eight years (the remaining term of the lease at the time of installation)

7. How difficult would it be to estimate the useful life of (and therefore the appropriate amortization period for) a patent acquired by your company from the inventor? What factors would you consider in estimating the useful life?

8. Under what circumstances does a tax shield provide no current benefit to the owner of the asset?

9. Calculate the depreciation expense in the third year of the life of an asset having a $70,000 original cost, 10 percent estimated salvage value, and an eight-year estimated life, assuming each of the following depreciation methods:
   a. Straight line
   b. Double-declining balance
   c. Declining balance at a 150 percent rate (rather than a double rate)

10. What are the primary accounting (not tax) arguments in favor of using accelerated depreciation rather than straight-line depreciation?

11. The principle of conservatism is best met by which of the following?
    · Using straight-line or accelerated depreciation?
    · Assuming short or long asset lives?
    · Expensing installation costs of a new fixed asset or capitalizing them?
    · Assuming low or high salvage values?

12. Might a computer costing $3,000 be appropriately expensed by one corporate owner and appropriately capitalized by another? Explain.

13. What is goodwill? Do you think it should be amortized? If so, over a short life or a long life?

14. Might it be appropriate for company A to depreciate an instrument over six years while company B appropriately chooses a depreciable life of eight years? Explain.

15. Intellectual property (generally in the form of patents) is critically important in certain industries. Assume that a company is debating whether to acquire a certain patent from another company or "design around" the patent in its own laboratories. Describe the accounting methodology and implications for each of these two alternatives.

16. Record, in T-account format, the following transaction: The Song Company receives delivery of a large punch press purchased for $19,300, payable in 30 days. Song pays the trucking company $1,200 in cash for delivering the press, and $2,200 in cash to the riggers and electricians for installing the press.

17. For a fixed asset having an original capitalized cost of $10,000, and a depreciable life of five years, how much greater is the depreciation expense on this asset in the second year of its life if its owner uses double-declining-balance depreciation rather than straight-line depreciation? Show your work.

18. A fixed asset that the Shores Corporation has owned for five years is sold for $14,000 cash.
    a. If the current book value of the asset is $12,000, show in T-account format how this sale should be accounted for.
    b. Assume this asset was being depreciated on a straight-line basis over 10 years and has a $1,000 salvage value. What did Shores originally pay for this asset (including installation costs)?

19. If an asset with an original cost of $5,000 and a useful life of five years is being depreciated on the double-declining method, in what year of the life of the asset should the depreciation method be changed to straight line? Show your work.

20. The Blueridge Corporation uses straight-line depreciation for financial reporting and the double-declining balance method for tax accounting. If the company acquires a fixed asset having an original cost of $40,000, an eight-year estimated useful life, and a 10 percent estimated salvage value, what will be the balance in the deferred income tax account attributable to this asset after two years, assuming Blueridge's tax rate is 35 percent.

21. The Kobori Corporation sends out for a major overhaul a large forklift truck that the company has owned for three years. The forklift had an

original cost of $12,000; it has been depreciated on a straight-line basis for three years, assuming a six-year life and a 20 percent salvage value. The overhaul costs $4,200 (paid in cash) and is expected to extend the life of the forklift for an additional three years. Record the following in T-account format:

a. The cash payment for the major overhaul

b. The appropriate depreciation expense on the forklift for the year after its major overhaul

22. Assume a company purchases a depreciable asset for $1 million, and that the asset has an eight-year estimated life and a zero salvage value. How much more cash will the owner of the asset have at the end of the second year if the owner uses double-declining-balance depreciation rather than straight line for tax accounting, assuming the company is profitable and pays income taxes at a 40 percent rate?

23. Palm Enterprises trades in a five-year-old tourist bus on a new model. The old bus had an original cost of $37,000, and its current book value is $16,000. Palm pays $23,000 in addition to the trade-in for a new bus having a list price of $40,000.

a. Account for this transaction in T-account format.

b. Assume now that Palm's accountant believes Palm could have negotiated a 5 percent discount on the list price if Palm had not traded in the old bus. Would you account for the transaction differently than in part a? Explain.

CHAPTER 9

WHAT CAN FINANCIAL STATEMENTS TELL US?

The financial statements that have been our focus to this point—the balance sheet, income statement, and cash flow statement—must offer more than data; they must provide useful information to the firm's various audiences, particularly managers, shareholders, and creditors. We turn now to techniques used to glean from the statements meaningful observations about an enterprise's financial health and operating results.

Readers of financial statements are concerned primarily with the company's progress in satisfying the triple objectives of (1) earning profits, (2) generating cash, and (3) maintaining a sound financial position. One reader may put more emphasis on profitability and cash flow, with less concern for financial position, while another may emphasize financial stability and security, with less concern for current profitability. If financial soundness is ignored, the company's risk of failure escalates; if the firm's profitability is ignored, its financial soundness may eventually become impaired; if cash flow is ignored, a liquidity crisis may eventually ensue.

We learn about a company by comparing financial data. It is the relationships between amounts on the income statement, the balance sheet, and the cash flow statement that convey information, much more so than isolated totals plucked from any one statement. If you are told that a particular company has $1 million in assets, you know that the sum of its liabilities and owners' equity is also $1 million, but you learn nothing about its financial condition or profitability. If you are also told that this company's owners' equity (or net worth) totals $500,000, you can deduce the relationship among equity, liabilities, and assets: borrowing has financed one-half of the company's assets. You have learned something about the company's financial position. If you are also told that the company earned $75,000 in profit in the year just ended, you learn something about its profitability: a 7.5 percent return on the $1 million asset investment, and a 15 percent return on its owners' equity.

Suppose you know that another company has $800,000 in accounts receivable. That number alone means little. If you also know that sales in the most recent year were $10 million, you can calculate that accounts receivable are about 8 percent of annual sales, and thus you can estimate the average time the company's customers take to pay their bills. Again, the comparison—not the absolute number—is key.

RATIO ANALYSIS

Therefore, the fundamental method of analyzing financial statements is by studying ratios. This chapter discusses the most common ratios and categorizes them by type of information revealed.

Actually, the discussions thus far have prepared you for ratio analysis. In discussing the balance sheet, Chapter 3 defined *current assets* and *current liabilities* in a parallel manner. This parallelism was purposeful, since the resulting ratio of current assets to current liabilities is meaningful. In Chapter 4, *gross margin* was defined; the ratio between gross margin and sales—the *gross margin percentage*—is another relevant ratio. Chapter 7 stressed that both inventory and the cost of goods sold are valued at cost, not sales, value. Thus, the ratio between inventory and COGS indicates the rate at which a company is using, or turning over, its inventory.

Financial analysts learn to think in ratios. Current assets are instinctively compared with current liabilities. Cost of goods sold is automatically compared with total sales to measure gross profitability, and to total inventory to judge inventory turnover.

Calculating ratios is straightforward. Their interpretation, however, requires judgment, and judgment is sharpened by experience. An analyst who has studied a variety of financial statements—statements of companies in different industries, and in both prosperous and recessionary periods—is able to glean more reliable conclusions from a particular set of statements than is an inexperienced analyst. This chapter provides a first step in gaining that experience.

CATEGORIES OF RATIOS

There are about as many financial ratios as there are analysts calculating them. Here, we explore only the most commonly used ratios. There are five primary categories of information that they provide:

1. *Liquidity:* How able is the company to meet its near-term obligations?
2. *Working capital utilization:* How efficiently is the company using the various components of its current assets and current liabilities?
3. *Capital structure:* What are the company's sources of capital?
4. *Profitability:* How profitable is the company in light of both its sales and its invested capital?
5. *Cash adequacy:* How successful is the company at generating cash?

To illustrate how to calculate and interpret the various ratios, this chapter analyzes the financial statements of Merck & Company, Inc., for the 2000 and 2001 fiscal years ending December 31. The income statements (statements of earnings), balance sheet, and cash flow statements for these two years and the income statement for 1999 are shown in Tables 9-1, 9-2, and 9-3. The financial notes accompanying these statements have been omitted, but not because they are unimportant. They are, in fact, invaluable for providing background information and explaining many of the entries on these financial statements. They are omitted simply because they are so extensive: 12 pages of densely packed narrative and tables. Arthur Andersen LLP audited these financial statements, of which the notes are an integral part. In its 2001 annual report, Merck describes itself as

TABLE 9-1. Merck & Co., Inc.: Consolidated Statements of Earnings for 1999, 2000, and 2001

	Years Ended December 31 ($ in millions except per share amounts)		
	2001	2000	1999
Sales	$47,715.7	$40,363.2	$32,714.0
Costs, expenses, and other:			
Materials and production	28,976.5	22,443.5	17,534.2
Marketing and administration	6,224.4	6,167.7	5,199.9
Research and development	2,456.4	2,343.8	2,068.3
Equity income from affiliates[a]	(685.9)	(764.9)	(762.0)
Other (income) expense, net	341.7	349.0	54.1
	37,313.1	30,539.1	24,094.5
Income before taxes	10,402.6	9,824.1	8,619.5
Taxes on income	3,120.8	3,002.4	2,729.0
Net Income	$ 7,281.8	$ 6,821.7	$ 5,890.5
Basic earnings per common share	$ 3.18	$ 2.96	$ 2.51
Earnings per common share with dilution	$ 3.14	$ 2.90	$ 2.45

[a]Note: the amounts in this row appear in brackets because they are, in effect, negative expenses: that is, income listed in a column of expenses.

TABLE 9-2. Merck & Co., Inc.: Consolidated Balance Sheet for 2000 and 2001

	December 31 ($ millions)	
	2001	2000
Assets		
Current assets:		
Cash and cash equivalents	$ 2,144.0	$ 2,536.8
Short-term investments	1,142.6	1,717.8
Accounts receivable	5,215.4	5,262.4
Inventories	3,579.3	3,021.5
Prepaid expenses and taxes	880.3	1,059.4
Total current assets	12,961.6	13,597.9
Investments	6,983.5	4,947.8
Property, plant and equipment (at cost):		
Land	315.2	311.6
Buildings	6,653.9	5,514.2
Machinery, equipment, and office furnishings	9,807.0	8,576.5
Construction in progress	2,180.4	2,304.9
	18,956.5	16,707.2
Less: Allowance for depreciation	5,853.1	5,225.1
	13,103.4	11,482.1
Goodwill and other intangibles (net of accumulated amortization of $2,224.4 in 2001 and $1,850.7 in 2000)	7,476.5	7,374.2
Other assets	3,481.7	2,752.9
Total Assets	$44,006.7	$40,154.9
Liabilities and Stockholders' Equity		
Current liabilities:		
Accounts payable and accrued liability	$ 5,108.4	4,605.8
Loans payable and current portion of long-term debt	4,066.7	3,319.3
Income taxes payable	1,573.3	1,244.3
Dividends payable	795.8	784.7
Total current liabilities	11,544.2	9,954.1
Long-term debt	4,798.6	3,600.7
Deferred income taxes and noncurrent liabilities	6,776.3	6,746.7
Minority interests	4,837.5	5,021.0
Stockholders' equity:		
Common stock, one cent par value[a]	29.8	29.7
Other paid-in capital	6,907.2	6,265.8
Retained earnings	31,489.6	27,363.9
Accumulated other comprehensive income	10.6	30.8
	38,437.2	33,690.2
Less treasury stock, at cost[a]	22,387.1	18,857.8
Total stockholders' equity	16,050.1	14,832.4
Total Liabilities and Stockholders' Equity	$44,006.7	$40,154.9

[a] Authorized 5,400,000,000 shares:

	2001	2000
Issued shares	2,976,129,820	2,968,355,365
Treasury stock, shares	703,400,499	660,756,186

TABLE 9-3. Merck & Co., Inc.: Consolidated Statement of Cash Flow for 2000 and 2001[a]

	Years Ended December 31 ($ millions)	
	2001	2000
Cash flows from operating activities:		
Income before taxes	$10,402.6	$ 9,824.1
Adjustments to reconcile income before taxes to cash provided from operations before taxes:		
Depreciation and amortization	1,463.8	1,277.3
Other	(359.5)	(222.8)
Net changes in working capital:		
Accounts receivable	(9.2)	(885.8)
Inventories	(557.5)	(210.1)
Accounts payable and accrued liabilities	458.3	(37.7)
Other	(15.3)	110.0
	11,383.2	9,855.0
Income taxes paid	(2,303.3)	(2,167.7)
Net cash provided by operating activities	9,079.9	7,687.3
Cash flows from investing activities:		
Capital expenditures	($2,724.7)	($2,727.8)
Purchase of securities, subsidiaries, and other investments, net	(1,397.4)	(969.6)
Other	(190.2)	56.1
Net cash used by investing activities	(4,312.3)	(3,641.3)
Cash flows from financing activities:		
Net change in short-term borrowings	$259.8	$905.6
Proceeds from issuance of debt	1,694.4	442.1
Payments on debt	(11.6)	(443.2)
Purchase of treasury stock	(3,890.8)	(3,545.4)
Dividends paid to stockholders	(3,145.0)	(2,798.0)
Proceeds from exercise of stock options	300.6	640.7
Other	(279.2)	1,350.8
Net cash used by financing activities	(5,071.2)	(3,447.4)
Effect of exchange rate changes	(89.2)	(83.7)
Net (decrease) increase in cash and cash equivalents	($392.8)	$514.9

[a] Some items have been combined to simplify the presentation.

a global, research-driven pharmaceutical company that discovers, develops, manufactures and markets a broad range of human and animal health products, directly and through its joint ventures, and provides pharmaceutical benefit services through Merck-Medco Managed Care.

. . . Merck sells its human health products primarily to drug wholesalers and retailers, hospitals, clinics, government agencies and managed health care providers.

In 2002, Merck was ranked by *Fortune* magazine as the 24th largest industrial company in the United States.

Merck is growing, profitable, and conservatively financed. It is also a complex company, and thus I will explain some of the entries on these statements as we proceed. We'll also modify some ratio definitions to fit better Merck's particular circumstances.

The ratios discussed here are generally applicable to manufacturing enterprises, more specifically to mature and diversified pharmaceutical companies. Companies engaged in different industries find some of these ratios irrelevant; other ratios, not discussed here, may be highly relevant to them. For example, a commercial bank is concerned with the ratio between loans outstanding and customer deposits—the loan-to-deposit ratio—since customer deposits provide the funds (liabilities) that the bank lends to its borrowers (the bank's primary asset).

Liquidity

A company unable to meet its obligations, typically by paying out cash, as they come due runs the risk of bankruptcy. Trade suppliers and employees must be paid on time. Interest and principal payments on borrowed money must be made when due. A company that has substantial liquid assets in relationship to its near-term obligations has strong *liquidity*. Note that a company may be very nonliquid (*illiquid*), and thus risk failure, even while making substantial profits. Liquidity tells part of the story of the company's financial position or condition, but it says nothing about company performance.

Current Ratio. The most widely quoted financial ratio is the **current ratio:**

$$\text{Current ratio} = \frac{\text{current assets}}{\text{current liabilities}}$$

Recall that current assets are those assets that are either presently the equivalent of cash or within the next 12 months will be turned into cash. Cash,

cash equivalents, accounts receivable, and inventory are the primary current assets, listed in order of decreasing liquidity. Current liabilities are obligations that must be met within the following 12 months, including primarily accounts payable, wages and salaries payable, short-term bank borrowing, the current portion of long-term debt, and miscellaneous accruals.

The higher the current ratio, the greater the margin of safety—that is, the more likely the company will have sufficient liquid assets to meet its obligations as they come due.

Note that current assets do not include all the cash that the company will receive during the next 12 months, nor do the current liabilities include all the obligations that must be met over the next 12 months. Sales occurring tomorrow, next week, and next month will result in cash receipts well before a year from now, and over the coming weeks and months, employees and vendors will have to be paid amounts not now included in current liabilities. Thus, the current ratio does not define comprehensively the ability of the company to meet all near-term obligations; it is only an indicator, albeit the most important one.

Merck's current ratio at the end of fiscal year 2001 was

$$\text{Current ratio} = \frac{12{,}962}{11{,}544} = 1.1*$$

This ratio is down from (13,598/9,954) = 1.4 in 2000, and is low compared with that of other manufacturing companies. But let's withhold judgment on Merck's liquidity until we have looked at some other ratios.

Acid-Test (or Quick) Ratio. Another liquidity measure is the **acid-test ratio,** often referred to as the **quick ratio:**

$$\text{Quick ratio} = \frac{\text{cash} + \text{cash equivalents} + \text{accounts receivable}}{\text{current liabilities}}$$

Note that the denominator here is the same as for the current ratio, namely, current liabilities. The numerator, however, includes only the most liquid of the company's current assets. No asset can be more liquid than cash! In addition, many companies own assets that are **cash equivalents:** short-term, interest-bearing investments of temporary or permanent cash reserves. These investments in, for example, U.S. Treasury securities are highly marketable: they can be converted into cash at a moment's notice. Thus, **marketable**

---

*Input data for this ratio are shown to five significant digits to assist you in tracing their source. However, most ratios are most usefully restricted to two or three significant figures.

securities are considered part of the company's quick assets.* Accounts receivable are also quick assets, since they will normally be collected (converted to cash) soon, typically within 90 days or less.

The primary difference between the numerators of the current and quick ratios is inventory; inventory is excluded from the quick ratio. A service company that holds little or no inventory might therefore have a quick ratio that approximates its current ratio. Also, a company that sells primarily for cash (for example, certain retail stores), has few (or no) accounts receivable in comparison with another that sells on credit. Typically, the sale-for-cash company has a lower quick ratio, but this fact does not necessarily mean it is illiquid. Recall the earlier admonition: financial ratios must be interpreted in light of the company's particular business and circumstances.

The quick (or acid-test) ratio at Merck for December 31, 2001 is

$$\text{Quick ratio} = \frac{2,144 + 1,143 + 5,215}{11,544} = 0.7.$$

Merck's quick ratio for 2001 is only about 36 percent below its current ratio. Moreover, the notes spell out that agreements with its banks permit Merck to borrow substantial funds should it need extra cash on short notice.

Working Capital Utilization

Current assets, particularly accounts receivable and inventory, are major investments for many companies. Often, although not at Merck, current assets represent a very large percentage of total assets. The more efficiently the company uses its current assets—that is, the faster it collects from customers and the less inventory it requires to accomplish its sales—the less capital the company will require. Recall that working capital is the difference between current assets and current liabilities. The following ratios indicate how efficiently the company is using the primary elements of working capital.

Accounts Receivable Collection Period. Most manufacturing companies extend customer credit. The longer customers take to pay their bills, the more the manufacturer must invest in accounts receivable. Comparing sales volume with outstanding accounts receivable indicates how promptly customers, on average, are paying:

---

*The company may own other marketable securities that are not equivalent to cash. For example, when company A purchases, on the open market, shares of company B's stock, company A is presumably making a long-term investment in company B; thus, this asset is not included in company A's current assets.

$$\textbf{Accounts receivable collection period} \text{ (in days)} = \frac{\text{accounts receivable}}{\text{average sales per day}}$$

where

$$\text{Average sales per day} = \frac{\text{annual sales}}{365} \quad \text{or}$$

$$\text{Average sales per day} = \frac{\text{quarterly sales}}{91}$$

The collection period is the number of equivalent days of sales remaining uncollected (and therefore in accounts receivable) at the end of the accounting period. Stated another way, the ratio equals the average number of days between the customer invoice date and the date that the payment is received. Averages, of course, contain limited information; actual times between invoice dates and collection dates vary widely by customer—some pay very promptly, while others take a distressingly long time to pay. The collection period, expressed in days, approximates the mean of a frequency distribution of days from invoice date to payment date.

You need to be cautious when interpreting the accounts receivable collection period. First, annual sales should include only credit sales; if the company has substantial cash sales, adjust the sales data accordingly. Second, if a business is seasonal, the collection period ratio is substantially distorted at certain times of the year. For example, suppose the ratio is calculated just after the conclusion of the company's busiest season. Average sales per day are higher during this busy season than for the year as a whole; moreover, the recent high rate of sales causes high accounts receivable balance at the end of the busy season. Seasonal distortion of the collection period is illustrated in Table 9-4. Note that each quarter's collection period, based on sales for just that quarter, is consistent at 45.6 days. However, the year-end accounts receivable is pushed up by the fourth quarter's high sales; when this year-end

TABLE 9-4. Effect of Seasonality on the Accounts Receivable Collection Period

Fiscal Period	Average Sales		Accounts Receivable	
	Sales for Period	Sales per Day	A/R at end of Period	Collection Period
Quarter 1	$400	$4.38	$200	45.6 days
Quarter 2	400	4.38	200	45.6 days
Quarter 3	400	4.38	200	45.6 days
Quarter 4	800	8.77	400	45.6 days
Full Year	$2,000	5.48	400	73.0 days

balance is compared with the average sales per day for the year (5.48), the resulting collection period is quite misleading.

A similar distortion results when a company is growing rapidly. The accounts receivable collection period is the first (but not last) ratio we encounter that draws data from both the income statement and the balance sheet. Recall that the income statement is for a period, while the balance sheet is a snapshot at the end of a period. The balance sheet, therefore, is a function of the activity level at the end of an accounting period, while the income statement records performance throughout the period.

Consider the hypothetical company illustrated in Table 9-5. It grew in sales volume by $3,000 per month—from $100,000 last December to $136,000 this December. The accounts receivable balance at each month end was just equal to sales in that month; the average collection period was obviously one month, or about 30 days. However, comparing average sales dollars per day for the entire year with the year-end accounts receivable balance yields an overstated collection period: the collection period appears to be 34.6 days instead of 30 days.

A simple adjustment compensates adequately for the effect of steady growth: use an average accounts receivable balance instead of the year-end

TABLE 9-5. Effect of Growth on the Accounts Receivable Collection Period

Month	Sales ($000)
December (2003)	$ 100
January (2004)	$ 103
February	106
March	109
April	112
May	115
June	118
July	121
August	124
September	127
October	130
November	133
December	136
Total for 2004	$1,434

Accounts receivable balances:

December 31, 2003 $100
December 31, 2004   136

$$\text{Collection period, based on year-end } A/R = \frac{136}{1{,}434 \div 365} = 34.6 \text{ days}$$

$$\text{Collection period, based upon average } A/R = \frac{(100 + 136) \div 2}{1{,}434 \div 365} = 30 \text{ days}$$

balance. If you have the balance sheet for the previous year-end as well as the current balance sheet, as we do for Merck, an average can be struck between these two numbers. Thus, the better formulation of the accounts receivable collection period ratio is

$$\text{Collection period} = \frac{(\text{opening A/R} + \text{ending A/R})/2}{\text{Annual sales}/365}.$$

(An even more accurate average is obtained by averaging the 13 month-end accounts receivable balances from the last month of one fiscal year through the last month of the next, data seldom available in published statements.)

In 2001, Merck grew by 18 percent in sales, and thus the two calculations of collection period, described above, might be quite different. Let's compare them. The ratio, based on *year-end* accounts receivable, is

$$\text{A/R collection period} = \frac{\text{accounts receivable}}{\text{annual sales}/365} = \frac{5,215}{47,715/365} = 40 \text{ days.}$$

The ratio based on *average* accounts receivable, is

$$\text{A/R collection period} = \frac{(\text{opening A/R} + \text{ending A/R 2}}{\text{Annual sales}/365}$$

$$= \frac{(5,215 + 5,262)/2}{47,715/365} = 40 \text{ days.}$$

In fact, the two ratios are the same because Merck accelerated A/R collections while growing sales by 18 percent, with the result that the accounts receivable balance actually fell from $5,262 million to $5,215 million. This rapid collection, and therefore decreasing collection period, is one way Merck maintains adequate liquidity even though the current ratio is low.

Inventory Turnover. Most manufacturing and merchandising companies invest substantially in inventory to serve customers and to assure the uninterrupted flow of manufacturing processes. Carrying inventory is expensive, when one includes the costs of storage, insurance, obsolescence, and tied-up capital. Thus, in deciding on inventory levels, every company trades off (explicitly or implicitly) the costs with the benefits of carrying inventory.

When inventory is sold, inventory values are reduced (credit) and a debit is made to Cost of Goods Sold. A comparison of inventory and the cost of goods sold, then, indicates the rate at which inventory is used—that is, the speed with which inventory is moving from receipt to final sale.

This **inventory turnover** ratio parallels the accounts receivable collection period in the following sense: the collection period ratio compares sales and accounts receivable, both valued at sales prices, while the inventory turnover ratio compares the cost of goods sold and inventory, both valued at cost values. The inventory turnover ratio is typically expressed in times per year, but an alternate form of the ratio is the **inventory flow period** ratio, expressed in number of days:

$$\text{Inventory turnover (times per year)} = \frac{\text{cost of goods sold}}{\text{inventory}}.$$

$$\text{Inventory flow period (days)} = \frac{\text{inventory}}{\text{cost of goods sold}/365} \quad \text{or}$$

$$\frac{365}{\text{inventory turnover}}$$

The inventory turnover ratio is subject to the same distortions from seasonality and growth as the accounts receivable collection period. For example, a retail store enjoying substantial holiday trade builds inventories in the fall in anticipation of high sales in November and December. Its apparent inventory turnover at the end of October is very different from an inventory turnover calculated on January 31, when inventories are depleted. Growth in rate of sales typically necessitates an increasing investment in inventory; for a growing company, inventory turnovers should thus be calculated with average inventory balances, rather than year-end balances.

Merck is engaged in many businesses, including both manufacturing and the resale of products manufactured by others. The inventory turnover for each discrete business would be informative, but we have data to calculate only the companywide average turnover:

$$\text{Inventory turnover} = \frac{\text{materials and production}}{(\text{opening inventory} + \text{ending inventory})/2}$$

$$= \frac{28,977}{(3,579 + 3,022)/2} = 8.8 \text{ times per year}$$

$$\text{Inventory flow period} = \frac{(\text{opening inventory} + \text{ending inventory})/2}{(\text{materials and production})/365}$$

$$= \frac{(3,579 + 3,022) \div 2}{28,977 \div 365} = 41.6 \text{ days}$$

Interpreting inventory turnover is difficult because the nature of the particular business so greatly affects the inventory flow. At one extreme, a dairy had better have a very rapid turnover of inventory if its milk, butter, and cheeses are to remain fresh until purchased by the consumer. At the other extreme, a manufacturer of heavy equipment requiring a year to complete the fabrication and assembly necessarily has very large inventories in relationship to annual sales. At Merck, inventory turnover is very rapid. Its manufacturing activities probably have substantially slower turnover, while the resale business—a fast-growing part of Merck in 2001—turns its inventory quickly.

Some published financial statements do not separate product and period costs. If cost-of-goods-sold data are unavailable, the inventory turnover ratio can be calculated using in the numerator sales rather than cost of goods sold. The numerator and denominator are now not expressed in comparable terms: the numerator is given in sales prices, including the company's gross margin, while the denominator (the inventory) is in cost values. While the resulting ratio does not, in fact, indicate inventory turnover in times per year, a comparison of this ratio over a number of years may be useful in assessing trends in inventory turnover over time. (But bear in mind that changes in profit margins also affect the ratio calculated in this way.)

Accounts Payable Payment Period. A major determinant of working capital is the amount of trade or vendor credit utilized. Just as a comparison of sales and accounts receivable reveals the collection period, a comparison of credit purchases and accounts payable yields the **accounts payable payment period**—the average time a company takes to pay suppliers. Comparing this payment period with the normal terms of purchase tells us how well the company is meeting its obligations to suppliers.

Unfortunately, total credit purchases seldom are either revealed in published financial statements or readily available within the company. However, we can use a *proxy* for credit purchases—that is, another value, more readily available, that tends to increase or decrease coincident with credit purchases. Cost of goods sold is a reasonable proxy, if the large majority of credit purchases is goods and merchandise for resale.

We do not have data on Merck's credit purchases in 2001 and 2000, and we know that the cost of sales (referred to in Table 9-1 as "materials and production") includes production labor and overhead in addition to materials purchased on credit. Thus, cost of goods sold here is quite an imperfect proxy for credit purchases. Moreover, accounts payable is combined with accrued liabilities on the Merck balance sheet. Nevertheless, Merck's payment period is

$$\text{A/P payment period} = \frac{\text{accounts payable}}{\text{cost of sales}/365}$$

$$\text{A/P payment period (2000)} = \frac{4{,}606}{22{,}444/365} = 75 \text{ days}$$

$$\text{A/P payment period (2001)} = \frac{5{,}108}{28{,}977/365} = 64 \text{ days}$$

Bearing in mind the caveat that we used proxies for both the numerator and the denominator, we see that Merck is apparently paying its vendors more rapidly in 2001 than in 2000.

For many companies, borrowing from trade creditors (that is, accounts payable) is a major source of capital. While there are practical limits to this source, the more this noninterest-bearing source is used, the less the company must obtain from sources to whom it must pay a return. Thus, many companies seek to stretch their A/P payment period. Some stretch is typically tolerable, but the risk is that vendors will lose patience and cease to supply the company, thus interrupting production processes and service to customers.

Working Capital Turnover. The **working capital turnover** ratio summarizes the efficiency with which the company is using its net investment in current assets less current liabilities:

$$\text{Working capital turnover} = \frac{\text{sales}}{\text{average working capital}}$$

This ratio, typically expressed in times per year, shows the dollars of annual sales achieved per dollar invested in working capital. If the ratio decreases, the company is investing more in working capital per dollar of sales; if the ratio increases, the company is making more efficient use of its working capital. Again, this ratio is subject to the same distortions from seasonality and growth that plagues all ratios that draw data from both the balance sheet and the income statement.

Given Merck's low current ratio, its 2001 working capital turnover must be quite high:

$$\begin{aligned}
\text{Working capital} \atop \text{turnover (2001)} &= \frac{\text{sales}}{(\text{opening working capital} + \text{ending working capital})/2} \\
&= \frac{47,716}{[(13,598 - 9,954) + (12,962 - 11,544)]/2} \\
&= 19 \text{ times per year}
\end{aligned}$$

Other Asset Turnover Measures. Still other asset turnover measures can be useful. A comparison of sales and fixed-asset investments indicates the number of dollars of annual sales realized from each dollar invested in productive assets. The higher the **fixed-asset turnover** ratio, the more effectively the company is utilizing its fixed assets. The most global measure of asset utilization is the **total-asset turnover** ratio: the relationship between annual sales and total assets. These two ratios, although unrelated to working capital, are mentioned here because we will use them later in discussing the linkage among ratios.

For Merck in 2001, these ratios are as follows:

$$\text{Fixed asset turnover} = \frac{\text{sales}}{\text{Average net fixed assets}}$$

$$= \frac{47,716}{(13,103 + 11,482)/2} = 3.9 \text{ times per year}$$

$$\text{Total-asset turnover} = \frac{\text{sales}}{\text{average total assets}}$$

$$= \frac{47,716}{(44,007 + 40,155)/2} = 1.1 \text{ times per year}$$

Don't misinterpret this ratio. It does *not* indicate that Merck must, this year, invest $1/1.1 = 91$ cents to achieve every dollar of sales. Rather, it says that, as Merck grows, it will need to make additional investments in total assets at the rate of $0.91 of asset investment for every dollar of *incremental annual sales volume*.

Some companies (such as electric power generating utilities) are asset intensive and thus have low total-asset turnover, while others (such as software producers) are the opposite. Accordingly, lots of investment capital is required to get into the electric power generating business, and very little is required to get into the software business.

Capital Structure Ratios

Capital structure ratios help us evaluate the liabilities and owners' equity side of the balance sheet—that is, how the company is financed. These ratios assess the financial riskiness of the business as well as the potential for improved returns through the judicious use of debt.

When a company borrows money, it undertakes a firm, ironclad obligation to pay interest and principal repayments on schedule. Failure to make these various payments when due (that is, default on the provisions of the loan agreement) subjects the company to the risk of bankruptcy. The higher the debt, the greater the risk. If the company's operating performance is erratic or encounters difficulties, the company may not be able to comply with its borrowing agreements. By contrast, when a corporation obtains additional funds by the sale of new capital stock, it undertakes no such firm obligations to its new shareholders; these shareholders will receive dividends only if and when the corporation's board of directors declares dividends. Failure to pay dividends does not constitute default and does not subject the company to the risk of failure, although it may subject the management to other pressures. Thus, from the point of view of the corporation, common stock financing is much less risky than borrowing.

However, the judicious use of borrowing benefits shareholders. If a company can borrow funds at an interest rate of x percent, and invest those funds to earn consistently at a rate of greater than x percent, then this incremental return boosts the return on shareholders' invested capital. We will return to this phenomenon, known as **debt leverage** in the next chapter.

For now, bear in mind that using borrowed funds is inherently neither good nor bad. The greater the borrowing—that is, the greater the use of debt leverage—the greater the company's financial risk, but also the greater the potential return to shareholders. Decisions on the amount of debt a corporation should use—that is, capital structure decisions—are matters of judgment to be exercised by the corporation's board of directors and management.

Total Debt to Owners' Equity. The debt of most corporations is composed of both current and long-term liabilities; owners' equity consists of both capital invested by the owners and earnings retained in the business. The ratio between the two sums indicates the relative contribution of creditors and shareholders to the company's financing. For Merck, the **total debt to owners' equity** ratios in 2001 and 2000 show a small reduction in debt leverage.

$$\text{Total debt to owners' equity} = \frac{\text{current liabilities} + \text{long-term liabilities}}{\text{total shareholders' equity}}$$

$$\text{For 2000} = \frac{9,954 + 6,747}{14,832} = 113\%$$

$$\text{For 2001} = \frac{11,544 + 4,799}{16,050} = 102\%$$

Note that Merck holds a large amount of treasury stock. More on that subject in the next chapter.

Total Debt to Total Assets Ratio. A closely related ratio compares the total debt to the company's total assets, thus indicating the percentage of total assets financed by liabilities. Since assets are equal to the sum of liabilities and owners' equity, the **total debt to total assets** ratio provides only the same information that was inherent in the total debt to owners' equity ratio. For Merck, this ratio in 2001 and 2000 confirms the reduction in debt leverage.

$$\text{Total debt to total assets} = \frac{\text{current liabilities} + \text{long-term liabilities}}{\text{total assets}}$$

$$\text{For 2000} = \frac{9,954 + 6,747}{40,155} = 42\%$$

$$\text{For 2001} = \frac{11,544 + 4,799}{44,007} = 37\%$$

If we ignore the deferred taxes and minority interests, Merck's owners' equity provided the financing for the remainder of its assets (63 percent in 2001).

Long-Term Debt to Total Capitalization. This ratio requires careful definition of both the numerator and denominator. Long-term debt is that portion of total borrowings having a maturity more than one year out (and therefore not included in current liabilities). **Total capitalization** is defined as the total permanent capital, that is long-term debt plus owners' equity. Current liabilities are not a permanent source of capital; since they arise spontaneously from operations. Amounts owing to trade creditors and employees and miscellaneous accruals would not be present if the company were not actively doing business. The ratio **long-term debt to total capitalization,** then, indicates the percentage that permanent, or long-term, borrowed funds are of total permanently invested capital. For Merck in 2001 and 2000, (again ignoring deferred taxes and minority interests) this ratio was

$$\text{Long-term debt to total capitalization} = \frac{\text{long-term debt}}{\text{long-term debt} + \text{owners' equity}}$$

$$\text{For 2000} = \frac{3,601}{3,601 + 14,832} = 20\%$$

$$\text{For 2001} = \frac{4,799}{4,799 + 16,050} = 23\%$$

Recall from Chapter 8 that deferred taxes typically appear on the balance sheet between long-term debt and owners' equity. Some analysts include deferred taxes in total capitalization, while others omit them entirely. Since most deferred taxes do not require payment in the foreseeable future, deferred taxes have been omitted here, as have their companion, noncurrent liabilities.

The amount labeled "minority interests" represents the investment, by shareholders unrelated to Merck, in Merck's less-than-wholly-owned subsidiaries consolidated into these financial statements.

Times Interest Earned. One measure of a corporate borrower's ability to service its debt—that is, to pay interest when due—is the **times interest earned** ratio, which compares the company's annual interest expense with its earnings before the payment of either interest or income taxes. Thus, the ratio's numerator is **earnings before interest and taxes,** often abbreviated **EBIT.** (Note that pretax operating profit is used, since current U.S. tax law permits interest to be deducted when computing taxable income.)

In Merck's case, I am assuming that interest expense approximates "Other (income) expense, net" which in 2001 was $341.7 million (see Table 9-1). This amount is the net of interest earned on almost $3.3 billion of interest-bearing assets (cash and cash equivalents plus short-term investments) and interest paid on almost $9.9 billion of short- and long-term debt. The net of these assets and liabilities is about $6.6 billion, and the net interest paid amounts to about 5 percent of $6.6 billion. This average interest rate is low, suggesting that our assumptions are probably somewhat off.

Therefore, note again that the trend in the times interest earned ratio may be more useful than the absolute number. The higher this ratio, the greater the safety margin, and the lower the risk that the company will be unable to service its debt. As this ratio declines, the greater the risk that some untoward event such as an economic recession will cause the company to be unable to meet its interest payment obligations. Merck's times interest earned ratio is very high:

$$\text{Times interest earned} = \frac{\text{earnings before interest and taxes}}{\text{annual interest expense}}$$

$$= \frac{10{,}403 + 342}{342} = 31 \text{ times}$$

Of course, interest must be paid in cash, not "earnings." We saw in Chapter 5 that a company's cash flow may, for any number of reasons, be quite different than its reported earnings. Accordingly, a variation of this interest coverage ratio uses cash flow from operations from the cash flow statement, instead of EBIT, as the numerator. For Merck in 2001, this *cash times interest earned* is also very high:

$$\text{Cash times interest earned} = \frac{\text{cash flow from operations}}{\text{annual interest expense}} = \frac{9{,}080}{342} = 27 \text{ times}$$

Public utilities such as electric power and natural gas distribution companies rely heavily on borrowed capital. For them, interest coverage ratios are critical for judging the financial risk inherent in their capital structures. For example, Edison International, the parent company of Southern California Edison, in 2001 had a times interest earned ratio of 3.4 and long-term debt that was 3.9 times shareholders' equity.

Profitability

The ratios dealing with liquidity, working capital utilization, and capital structure focus primarily on the company's financial position. By contrast, the profitability ratios measure the company's performance, the rate at which it is earning financial returns.

Two types of profitability ratios are useful. The first measures profit in relation to sales levels, and is obtained by comparing data solely within the income statement; the second measures profit in relation to investment, and involves comparisons of income statement and balance sheet data.

Percentage Relationships on Income Statement. The ratio of net income to total sales is useful in measuring the company's profitability. In addition, the percentage that each line item on the income statement is of total revenue (or sales) also provides useful insights. For example, the gross margin percentage shows the relationship between sales revenue and product cost. Sales expense as a percentage of sales revenue shows what percentage of the revenue dollar the company spends on selling and marketing activities.

Table 9-6 analyzes Merck's income statements for 2000 and 2001. Merck's business revenues shifted in favor of resale products during 2001, accounting for some significant shifts in percentages. For example, while expenditures on research and development increased very slightly, as a percentage of sales they went down quite sharply. Also, we would expect that gross margins on resale products to be substantially less than on proprietary (manufactured) products, and that seems to be the case.

As is the case for the pharmaceutical industry in general, Merck's net margin on sales (15 percent in 2001) is quite handsome.

These percentages vary greatly by industry. Consider the probable percentages for a large food retailer: the cost of goods sold is a very high percentage of sales, research and development expenses are essentially nonexistent, and net earnings as a percentage of sales is very small, typically 1 to 2 percent.

TABLE 9-6. Merck & Co., Inc.: Percentage Analysis of Income Statements for 2000 and 2001

	2001	2000
Sales	100%	100%
Cost, expenses, and other		
Materials and production	61	56
Marketing and administration	13	15
Research and development	5	6
Equity income from affiliates	(1)	(2)
Other (income) expense, net	1	1
Total cost expenses and other	78	76
Income before taxes	22	24
Taxes on income	7	7
Net income	15%	17%

Related ratios for 2001:

$$\text{Gross margin} = \frac{\text{Sales} - (\text{materials} + \text{production})}{\text{sales}} = \frac{47,716 - 28,977}{47,716} = 39\%$$

$$\text{Tax rate} = \frac{\text{Taxes on income}}{\text{Income before taxes}} = \frac{3,121}{10,403} = 30\%$$

Return on Sales. Note particularly the last of these percentage relationships, net income to total sales, or **return on sales (ROS)**.

Return on Equity. We turn now to profitability as it relates to investment. The most fundamental ratio is net income to total owners' equity, the **return on equity (ROE)**. To compensate for growth, the ratio compares earnings for the year with average equity:

$$\text{Return on equity} = \frac{\text{net income}}{(\text{opening equity} + \text{ending equity})/2}$$

Merck's return on equity in 2001 was very handsome indeed:

$$\text{Return on equity} = \frac{7{,}282}{(16{,}050 + 14{,}832)/2} = 47\%$$

The ROE ratio compares net income, after payment of all expenses including interest and taxes, with the total book value of the shareholders' investment, including both invested capital and earnings retained by the business. (Bear in mind that total company *market* value may be very different from *book* value. More on this in Chapter 10). Since all shareholders invest for a return, ROE is one ratio that can be compared across different industries.

Return on equity does not, however, indicate the degree of risk inherent in the return. Shareholders are typically willing to accept a lower ROE in exchange for a lower risk. In Chapter 10, we will see how the use of greater debt leverage leads to both higher risk and potentially a higher ROE.

To repeat, this ratio compares net income to book shareholders' equity. An investor typically has to pay more or less than equivalent book value for a share of stock, as market values of securities bear no necessary relationship to book values. Thus, the ROE generally does not indicate, for a particular investor, the rate of return on his or her particular securities investment.

Return on Assets. A company's return on equity is influenced by its capital structure. A company employing high debt leverage is typically subject to wider swings in ROE than another company obtaining its permanent capital primarily through shareholders' equity. To factor out the influence of capital structure when appraising the company's earnings on investment, total assets can be used as the measure of investment, that is, the total amount owned by the company, whether financed through debt or owners' equity.

The **return on assets (ROA)** ratio compares earnings to total assets. However, net income is inappropriate as the numerator, since net income is profit after payment of interest, and the amount of interest is a function of the capital structure, that is, a function of the amount of debt. Thus, we use earnings before interest and taxes (EBIT) in calculating return on assets:

$$\text{Return on assets} = \frac{\text{earnings before interest and taxes}}{(\text{opening assets} + \text{ending assets})/2}$$

At Merck in 2001, return on assets (ROA) was

$$\text{ROA} = \frac{10,403 + 342^*}{(44,007 + 40,155)/2} = 26\%$$

Bear in mind that the return on assets percentage and the return on equity percentage are not comparable, since ROA is a before-tax return and ROE is an after-tax return.

Earnings per share (EPS). Although Chapter 10 will discuss this ratio extensively, I introduce it here because it is the final important piece of information on the income statement:

$$\textbf{Earnings per share (EPS)} = \frac{\text{Net income}}{\substack{\text{Average number of shares} \\ \text{of common stock outstanding}}}$$

This ratio shows the net income attributable to each share of common stock held by outside shareholders (thus Treasury stock is excluded). Merck's balance sheet tells us the total shares outstanding (issued shares less Treasury shares), and that number divided into net income yields Merck's 2001 earnings per share of $3.18 (see Table 9-1).

It should be intuitive to you that the ratio of market price per share and earnings per share is of keen interest to investors. Chapter 10 will elaborate on this point.

Like most complex companies, Merck shows a second earnings per share, this one taking account of future dilution: $3.14 for Merck in 2001 (see Table 9-1). Companies typically have outstanding contracts that grant rights to acquire additional common stock at favorable prices under certain conditions. The most important of such contracts are incentive stock options granted to the company's managers. If conditions come to pass such that the rights are exercised and new common shares are issued, present shareholders will be diluted: each will own a slightly smaller percentage of the total. The diluted earnings per share reflects, in a rather complex calculation, the dilution that will result from the exercise of these rights to acquire common shares.

Cash Adequacy

I have emphasized repeatedly that cash is the lifeblood of every company. Accordingly, ratio analysis is useful in comparing a company's ability to

---

*Income before taxes plus other (income) expense, net

generate cash with its need for cash in order to make capital investments, acquire other companies, pay dividends, repay borrowings, and for a host of other purposes. Of the many possible cash flow ratios, let's consider just four.

Cash Flows from Operating Activities to Net Income. Recall that Merck's cash flow statement divides cash flow into three parts (as do all cash flow statements, but sometimes the parts carry slightly different names):

- Cash flows from operating activities
- Cash flows from investing activities
- Cash flows from financing activities

A quick check of the degree of difference between "net income" from operations and "cash flow" from operations is provided by the **cash flow from operating activities to net income** ratio. For Merck, in 2001, this ratio is

$$\text{Cash flow to net income} = \frac{\text{Cash flows from operating activities}}{\text{Net income}}$$

$$= \frac{9{,}080}{7{,}282} = 125\%.$$

While many companies generate less cash than net income, Merck's situation is the reverse: The cash flow statement shows that the company in 2001 had nearly $1.5 billion of noncash expenses (depreciation and amortization), that taxes paid in cash were about $800 million less than taxes accrued in its income statement, and that Merck was able to grow substantially while increasing working capital only modestly.

Dividends to Net Cash Flow Before Financing. An investor considering purchasing Merck shares might be interested in determining how "safe" the dividend is; that is, might Merck's board cut the dividend in the near future? Comparing net cash flow before financing (that is, the cash flow from operating activities less net cash used by investing activities) to the cash dividends paid should help answer this question:

$$\begin{array}{l}\textbf{Dividends to net cash flow} \\ \textbf{before financing}\end{array} = \frac{\begin{array}{c}\text{(cash flows from operating activities} \\ \text{less cash flow used for investing activities)}\end{array}}{\text{Dividends}}$$

$$= \frac{9{,}080 - 4{,}312}{3{,}145} = 152\%$$

This ratio appears a bit low. But, look at the details: Merck in 2001 used $3,891 million to purchase its owns shares in the market (treasury stock) and a net of $1,397 million to add to its portfolio of other securities and acquire other companies. One can reasonably assume that the board of directors at Merck would decide to curb both of those activities before it would cut dividends to present shareholders.

Cash Flows from Operating Activities to Cash Flow for Investments.
Is Merck able to generate sufficient cash from its ongoing activities to finance its desired investments without having to resort to additional external financing (stock or debt)? A look at the ratio **cash flows from operating activities to cash flow for investments** should anwer this question:

$$\frac{\text{Cash flows from operating activities}}{\text{Cash flow used by investing activities}} = \frac{9,080}{4,312} = 2.1$$

Clearly, Merck can, at least in 2001.

Many variations on this ratio might prove useful. An analyst inside the company or on Wall Street might want to modify the denominator to include also one or both of the following:

- dividend payments
- required repayments of loan principal

Cash Flows from Operating Activities to Owners' Equity. If cash is, indeed, the lifeblood of the company—and the medium of paying dividends—shareholders should be interested in a ratio that focuses on cash and parallels ROE. This ratio is **cash flows from operating activities to owners' equity.** For Merck in 2001,

$$\frac{\text{Cash flows from operating activities}}{\text{Owners' equity}} = \frac{9,080}{16,050} = 57\%.$$

In anyone's world, 57 percent is a very handsome return, even higher than Merck's ROE of 47 percent.

INTERPRETING RATIOS

Experienced financial statement analysts instinctively observe ratios while reviewing financial statements. They learn to think in ratios. They scan financial statements, searching for strength and weakness in the company's financial condition and operating performance. Of course, not every ratio provides useful information in every situation. For example, if you observe that a particular company's current ratio conforms to typical industry averages, you may conclude that liquidity is neither a problem nor strength. Further analysis may reveal a relatively long A/R collection period and a relatively low total debt to owners' equity ratio. These observations suggest areas for further inquiry: Why are customers paying slowly? Could the company benefit from further debt leverage?

Ratios should be calculated accurately, but not with undue precision. A collection period calculated to the nearest whole day, rather than to the second decimal point, is quite adequate to suggest customer payment patterns. Since income statement, balance sheet, and cash flow values are affected by a host of variables such as seasonality, price-level changes (such as inflation), and random events occurring just prior to the balance sheet date, don't attribute much significance to minor changes in ratios.

No Absolute Standards

By now, you undoubtedly are wondering: What are appropriate, or target, values for each of these various ratios? What would an analyst consider a "good" current ratio, or total debt to total assets ratio, or operating cash flow to net income ratio, or collection period ratio, or return on equity?

Unfortunately, your questions cannot be answered definitively. Adequate liquidity or appropriate debt leverage is a function of industry characteristics, the company's stage of development, philosophy of the management and owners, and many other factors. While conventional wisdoms do exist—such as, a manufacturing company should have a current ratio of 2.0 and a quick ratio of 1.0—these wisdoms are dangerous. Some companies (including Merck) enjoy very adequate liquidity with a current ratio closer to 1.0—for example, service companies with low inventories or companies that sell for cash and, thus, have no accounts receivable. Other companies—for example, those whose manufacturing processes require large inventories—need a current ratio well in excess of 2.0 to assure adequate liquidity.

So you will be frustrated in seeking absolute standards for the various ratios. Indeed, for most ratios, you cannot conclude that *the higher the better* or *the lower the better.* You might be inclined to think that the higher the current ratio, the more sound the company's financial condition. While a high current ratio indicates strong liquidity, an excessively high current ratio suggests an inefficient use of current assets or a failure to utilize credit available from vendors (accounts payable). Or, you may be inclined to conclude that the shorter the collection period, the better. However, if a company achieves rapid collection by refusing to sell to any but the most creditworthy customers and then harasses customers to pay quickly, the company may suffer a loss of sales volume as a result of its collection policies. Or, you may be inclined to feel that the lower the total debt to total assets ratio, the healthier the business. Low debt leverage does indicate low financial risk, but many companies can and should avail themselves of the benefits of debt leverage; for them, very low debt ratios may point up financial policies that are unnecessarily timid.

In short, judging the appropriateness of a particular ratio is not easy. An analyst's judgment is aided by two techniques, however: reviewing trends over time, and comparing the company's ratios with those of similar companies in the same industry.

Trends

Suppose that in, the year 2005, companies A and B both have current ratios of 2.0; they apparently have equivalent liquidity. Now, suppose the current ratios for these two companies over the past three years have been as follows:

	2003	2004	2005
Company A	1.4	1.8	2.0
Company B	2.6	2.2	2.0

Now, company A seems to be building somewhat more comfortable liquidity (its current ratio has strengthened over recent years), while company B's liquidity has deteriorated. Trend information shapes your conclusion.

Table 9-7 shows the trend in selected Merck ratios over a recent five-year period. Note the declining trend in ROS as the mix of Merck's sales move toward lower-margin resale products and away from products manufactured by the company. (Unsurprisingly, R&D expenditures as a percentage of sales has also fallen, as the denominator of this ratio has grown rapidly.) Yet the ROE has improved to a handsome 45.4 percent. This increase has occurred, in part, because of greater debt leverage, but also because asset turnover has accelerated. The huge increase in working capital turnover in 2001 results from a sharp decrease in working capital: a decrease in current assets (despite substantial growth) and an increase in current liabilities somewhat in line with its sales growth.

TABLE 9-7. Trends in Financial Ratios, Merck & Co., Inc. 1997–2001

	1997	1998	1999	2000	2001
Profitability:					
Return on sales (ROS)	19.5	19.5	18.0	16.9	15.3
Return on equity (ROE)	36.6	41.0	44.5	46.0	45.4
R&D expenditures ÷ Sales (%)	7.1	6.8	6.3	5.8	5.1
Capital Structure:					
Long-term debt to total capitalization (%)	9.6	20.1	19.2	19.5	23.0
Asset utilization:					
Total asset turnover	0.92	0.84	0.91	1.01	1.08
Working capital turnover	8.9	6.5	13.1	11.1	33.7

Comparison with Similar Companies

A company's ratios should be interpreted in light of its particular industry. Comparisons of Merck to an electric power utility or a retailer are irrelevant. Utilities typically have high fixed assets and little working capital; their fixed-asset and total-asset turnover ratios are very low, but they employ substantial debt leverage to earn competitive returns on equity. Large retailers have few fixed assets or accounts receivable, and thus their asset turnovers are high. With relatively little investment, and despite low gross margins, a well-run retailer can earn a very competitive return on equity.

But comparing Merck to other large pharmaceutical companies might be revealing. Comparisons between direct competitors may be particularly useful, but comparisons with industry averages may be even more revealing. Industry (trade) associations often facilitate these comparisons by collecting from their members, typically on a confidential basis, detailed information on financial ratios and then compiling and publishing these data to their members. Certain government agencies, banks, auditing companies, and security analysts provide similar industry financial ratio data that serve as useful benchmarks for industry participants.

LINKAGE AMONG FINANCIAL RATIOS

By now, you probably surmise that some interesting linkages exist between ratios, particularly between asset utilization, capital structure, and profitability. For example, a company that can achieve higher total-asset turnover can reduce its total investment in assets, thus requiring less capital; if this condition results in lower total owners' equity, then the return on equity (ROE) is enhanced. A capital structure employing higher debt leverage can result in higher ROEs. Also, a company that can improve its return on sales without changing total-asset turnover or capital structure earns higher ROEs.

The product of three ratios, which equals the ROE, illustrates these interrelationships or linkages:

$$\text{ROE} = \frac{\text{net income}}{\text{total sales}} \times \frac{\text{total sales}}{\text{total assets}} \times \frac{\text{total assets}}{\text{total equity}}$$

The first ratio is return on sales (ROS); the second is total-asset turnover; and the third is a debt leverage ratio. (Since the difference between the numerator and denominator of this third ratio equals total liabilities, the higher this ratio, the greater the level of debt, and therefore the higher the company's debt leverage.) This entire expression must be true, since, by the rules of algebra, total sales and total assets cancel out of the equation and the ratio of net income to total equity—the definition of return on equity—remains.

I focus on this expression for two reasons: (1) the ROE is the fundamental measure of how well the company earns on shareholders' funds; (2) the rate at which a company can grow without seeking additional equity financing is a function of its ROE, and (3) the three individual ratios, or fractions, suggest three distinct approaches for improving the return on equity.

This third point deserves further elaboration. The first fraction indicates that the ROE is improved if the ROS increases, that is, if the company earns more profit per dollar of sales. The second fraction focuses on asset management: if the company achieves the same sales with lower asset investment, the total of the right-hand side of the balance sheet is reduced (in other words, less debt and equity capital are required); this phenomenon also improves the ROE. Finally, if the company substitutes debt for equity in its capital structure—that is, if it takes on greater debt leverage—the ROE is increased. (Remember, however, that this last action also increases the company's risk of failure.)

Merck's ROE for 2001, based on year-end equity, is 45%. This value is the product of the three fractions just described:

$$\text{ROE} = \frac{7{,}282}{47{,}716} \times \frac{47{,}716}{44{,}007} \times \frac{44{,}007}{16{,}050} = 0.15 \times 1.1 \times 2.7$$

$$= .45 \text{ or } 45\%$$

Suppose Merck wished to decrease its debt leverage from the 2.7 shown here (assets at 2.7 times owners' equity) to, say 2.0; this action would reduce Merck's ROE from 45% to 33%. Or, suppose the company could, by reducing its investment in various securities, improve its asset turnover from 1.1 to 1.4; if this reduction in asset investment were reflected by proportionate decreases in both debt and equity, so that the company's debt leverage remained unchanged, its ROE would be improved from 45 to 59 percent. Or, these two offsetting actions might be taken together; assuming a leverage ratio of 2.0 and no change in the ROS, Merck would need to increase its total-asset turnover to about 1.5 in order to maintain its current handsome ROE of 45%. For a variety of reasons, these changes may be impractical or undesirable, but the linkage equation focuses attention on the importance of

Return on sales

Asset utilization

Debt leverage

in the company's fundamental measure of profitability, namely, the return on equity.

Why does the ROE relate to the company's ability to grow sustainably without seeking external equity financing or increasing its debt leverage? As annual sales volume grows, total investment in assets grows—recall the total asset turnover ratio discussed earlier in this chapter. How will these additional asset investments be financed? One or a combination of only three ways: additional borrowing, the sale of additional equity securities, or retained earnings. The debt to equity leverage ratio remains unchanged as long as the growth in borrowing matches the growth in equity. The growth in equity equals, by definition, the ROE if—and this is an important *if*—the company does not pay dividends but instead reinvests in the company its entire net income. One caution: my statement earlier was that a company's sustainable growth rate is a function of its ROE; the two are not equal.

SUMMARY

The fundamental tool used to extract useful information from financial statements is ratio analysis. It is comparisons of amounts appearing in the financial statements, rather than their absolute levels, that provide insight into the two key questions: How sound is the company's financial position, and how well is the company performing in earning returns (both reported profits and cash) on the capital employed?

Many different ratios are in active use. Ratios that are highly relevant to certain industries or companies may be of marginal or no interest in other circumstances. The most common ratios, grouped into the five categories discussed in this chapter, are as follows:

Liquidity
    Current ratio
    Acid-test ratio (or quick ratio)
Working capital utilization
    Accounts receivable collection period
    Inventory turnover
    Accounts payable payment period
    Working capital turnover
Capital structure
    Total debt to owners' equity
    Total debt to total assets
    Long-term debt to total capitalization
    Times interest earned
Profitability
    Percentage relationships on income statement
    Return on sales (ROS)
    Return on equity (ROE)
    Return on assets (ROA)

Cash adequacy

   Cash flow from operations to net income

   Dividends to net cash flow before financing

   Cash flow from operations to cash flow for investments

   Cash flow from operations to owners' equity

A number of caveats apply to the interpretation of ratios. Ratios that compare data from the income statement with data from the balance sheet (for example, the working capital utilization ratios) can be significantly distorted by seasonality and by high rates of growth (or decline). No absolute standards exist for ratios; indeed, with respect to most ratios, one cannot even conclude that the higher the value the better, or the lower the better. Judgment is required to ascertain when a particular ratio is providing a danger signal or revealing a particular company strength.

Trends in ratios typically reveal additional information about a company, particularly trends in the liquidity and capital structure ratios. Comparisons between similar companies in like industries are also helpful. However, bear in mind that the optimum level of a ratio depends very much on the nature of the particular company's business, as well as on, for example, the company's future plans, management policies, and access to additional capital.

NEW TERMS

Accounts payable payment period. The average number of days from the receipt of merchandise or service to the cash payment of the associated invoice, calculated as follows:

$$\frac{\text{Average accounts payable}}{\text{Average credit purchases per day}}$$

Accounts receivable collection period. The average number of days from the invoice date to the cash collection date, calculated as follows:

$$\frac{\text{Average accounts receivable}}{\text{Average sales per day}}$$

Acid-test ratio. The ratio of the sum of cash, cash equivalents (including readily marketable securities), and accounts receivable to current liabilities; a measure of liquidity. An alternative name is *quick ratio*.

Capital structure. The composition of the capital employed in the business.

Cash equivalents. Short-term investments that are highly marketable and can be quickly converted to cash.

Cash flow from operations to cash flow for investments. The ratio of these two values, which are drawn from the cash flow statement; an indication of the company's ability to finance internally (without external financing) its new investments.

Cash flow from operations to net income. The ratio of cash flow from operations (from the cash flow statement) to net income. Sometimes called simply *cash flow to net income.*

Cash flow from operations to owners' equity. The ratio indicating the cash (rather than reported earnings) returns on shareholders' investments.

Current ratio. The ratio of current assets to current liabilities. The ratio measures liquidity.

Debt leverage. The extent to which the company relies on borrowed funds (debt) for financing. Companies with a high debt leverage are exposed to greater financial risk but also enjoy the potential of greater returns to shareholders.

Dividends to net cash flow before financing. The ratio of dividends paid to the sum (net) of cash flow from operations and cash flow for investing (from the cash flow statement.)

Earnings before interest and taxes (EBIT). A company's earnings before the payment of (1) interest on borrowed funds, and (2) income taxes. EBIT is used to calculate the times interest earned ratio and the return on assets ratio.

Earnings per share (EPS). The net income attributable to each share of common stock, calculated as follows:

$$\frac{\text{Net income}}{\text{Average number of shares of common stock outstanding}}$$

Fixed-asset turnover. A measure of the efficiency with which a company is using its net investment in fixed assets, measured in times per year:

$$\frac{\text{Sales}}{\text{Average net fixed assets}}$$

Inventory flow period. The average number of days from receipt of inventory to its shipment to customers:

$$\frac{\text{Average inventory}}{\text{Average cost of goods sold per day}} \quad \text{or} \quad \frac{365}{\text{Inventory turnover}}$$

Inventory turnover. The number of times per year that the firm's inventory turns over, calculated as follows:

$$\frac{\text{Cost of goods sold}}{\text{Average inventory}} \quad \text{or} \quad \frac{365}{\text{inventory flow period}}$$

Long-term debt to total capitalization. A financial ratio that measures debt leverage, calculated as follows:

$$\frac{\text{Long-term debt}}{\text{Total capitalization}}$$

Marketable securities. Those investment securities for which a ready market exists. U.S. Treasury securities and certain other interest-bearing short-term investments are marketable securities that are cash equivalents.

Quick ratio. An alternative name for *acid-test ratio.*

Return on assets (ROA). A measure of the company's return on investment that is independent of the company's capital structure, and calculated as follows:

$$\frac{\text{Earnings before interest and taxes}}{\text{Average total assets}}$$

Return on equity (ROE). A measure of the company's return on shareholders' investment, and calculated as follows:

$$\frac{\text{Net income}}{\text{Average shareholders' equity}}$$

Return on sales (ROS). A measure of the company's rate of profitability on sales, and calculated as follows:

$$\frac{\text{Net income}}{\text{Total sales}}$$

Times interest earned. A financial ratio that measures a company's ability to meet its interest payment obligations, and calculated as follows:

$$\frac{\text{Earnings before interest and taxes}}{\text{Annual interest expense}}$$

Total-asset turnover. An assessment of the efficiency with which a company is using its total investment in assets. It is measured in times per year:

$$\frac{\text{Sales}}{\text{Average total assets}}$$

Total capitalization. The sum of long-term debt plus owners' equity.

Total debt to owners equity. A debt leverage ratio calculated as follows:

$$\frac{\text{Current liabilities} + \text{long-term debt}}{\text{Total owners' equity}}$$

Total debt to total assets. A debt leverage ratio calculated as follows:

$$\frac{\text{Current liabilities} + \text{long-term debt}}{\text{Total assets}}$$

Working capital turnover. A working capital utilization ratio (in times per year) calculated as follows:

$$\frac{\text{Annual sales}}{\text{Average working capital}}$$

EXERCISES

1. Why is the return on equity (ROE) comparable across different industries and types of companies, while most other ratios are not?

2. What ratio(s) seem to you most relevant in assessing the following?
   a. The company's ability to make interest payments when due
   b. The mix of capital sources utilized by the company
   c. The "safety" of the current rate of dividends paid by the company
   d. The time it takes customers to pay their bills
   e. The time it takes the company to pay its bills
   f. The liquidity of the company
   g. The company's risk of inventory obsolescence
   h. The probable incremental investment in assets that will be required to support growth in sales
   i. The amount of noncash expenses included among the company's total expenses
   j. The attractiveness (for possible investment) of the market price of the company's common stock
   k. The reasonableness of total marketing expenses at the company
   l. The financial riskiness of the company

3. Indicate whether each of the following statements is true or false.
   a. Seasonality in sales affects the accounts receivable collection period but should have little or no effect on the ratio of long-term debt to total capitalization.

b. Shareholders should encourage high debt leverage at the companies in which they invest in order to capture the potential of higher shareholder returns.

c. The times interest earned ratio uses EBIT rather than net income in its numerator because interest payments are deductible for income tax purposes.

d. While trends over time provide useful information for income statement ratios, trends are generally meaningless for balance sheet ratios.

e. Accurate assessment of trends requires that ratios be calculated with precision, and generally with four or five significant digits.

f. Cash equivalents are typically deposits in foreign banks denominated in the foreign currency.

g. No generally accepted guidelines exist as to what constitutes either "satisfactory" or "dangerous" ratio levels.

h. Total capitalization is the sum of long-term debt and retained earnings.

i. The return on assets (ROA) ratio is unaffected by the company's debt leverage, but the return on equity (ROE) is so affected.

4. What factors should a company consider in deciding the following?

a. What its accounts payable payment period should be

b. How high its debt leverage should be

c. How much of its earnings it should pay out in the form of dividends

d. How rapidly it should turn over its inventory

e. What an optimum accounts receivable collection period is for the company

5. If a company could get its customers to make substantial down payments (advance payments) with their orders, would the company thereby improve its current ratio? Explain.

6. If the ratio of net income to total assets for a company approximates its return on equity, what can you conclude about the composition of the company's balance sheet?

7. Niemz LLP achieved sales of $8.6 million during 2005 and had an accounts receivable balance at year-end of $2.1 million. If that A/R balance grew by 20 percent over the course of the year, what was Niemz's average accounts receivable collection period for 2005?

8. Fill in the missing word or phrase in the following statements:

a. Because total credit purchases are seldom reported by companies, a financial analyst calculating the A/P payment period frequently uses _____ as a proxy for credit purchases.

b. The primary difference between the current ratio and the acid-test ratio is _____.

c. Interest payments are _____ for income tax purposes, but dividends are not.

d. To properly account for seasonality in sales, the numerator of the accounts receivable collection period ratio should be _____.

e. In determining the ROE, net income is calculated as a percentage of invested capital plus _____.

f. To assist in analysis, the individual line items within the "current assets" and "current liabilities" sections of the balance sheet are listed in order of _____.

g. In assessing a company's ability to maintain its dividend payments at their current level, annual dividend payments are compared with _____ on the cash flow statement.

h. The liquidity ratios help assess the ability of the company to meet its _____.

i. Working capital is the difference between _____ and _____.

j. The unit of measure of the inventory turnover ratio is _____.

9. A grocery retailer generally earns a low margin on sales, but nevertheless can achieve quite competitive returns on equity. Explain.

10. An analyst at the Herbert Company calculates the following ratios for Herbert as of the end of fiscal year 2006:

Return on sales	6%
Asset turnover	1.5 times
Assets divided by owners' equity	1.3

a. What is Herbert's return on year-end equity?

b. If Herbert's return on sales fell to 5 percent, and its asset turnover remained unchanged, how much would its leverage have to increase to maintain the same ROE?

11. If the Raval Company has a current ratio of 1.7, total assets of $4.6 million, and current assets of $2.7, what is Raval's total capitalization?

12. If a financial analyst has the following information for Sauro Corporation for fiscal year 2004:

Pretax return on assets	10%
Total debt to total assets	65%
A/R collection period	65 days
Pretax return on sales	15%
Total assets	$3.6 million
Working capital	$800,000

Determine the following:

    a. Accounts receivable balance (in dollars)

    b. Pretax return on shareholders' equity

    c. Current ratio, assuming current assets equal noncurrent assets

    d. Long-term debt (in dollars)

13. Construct a balance sheet for the Nadim Company as of December 31, 2005, in as much detail as possible, using the following information. Make estimates as required, and state any assumptions you make:

Current ratio	2.2
Acid-test ratio	2.0
A/R year-end collection period	55 days
Return on year-end equity	17.6%
Total debt to total equity ratio	1.2
Net income to total sales	10%
Net income to total assets	8%
Sales for 2005	$24 million
Cash and cash equivalents at year-end	$1.4 million

14. If the Pinella Company has a ROE (based on year-end equity) of 15 percent, a ROS of 8 percent, total assets of $520,000, and a total asset turnover of 1.1, what is the value of Pinella's shareholders' equity?

15. The condensed, audited financial statements for the June 30, 2003, fiscal year for McNeil Corporation are as follows:

BALANCE SHEET, JUNE 30, 2003 ($000)

Assets	
Cash	$16,400
Accounts receivable, net	25,590
Inventory, net	36,930
Other current assets	16,110
Total current assets	95,030
Fixed assets, net	48,580
Total assets	$143,610
Liabilities and Owners' Equity	
Accounts payable	$21,370
Bank loan payable (8% interest rate)	15,000
Other current liabilities	21,690
Total current liabilities	58,060
Long-term debt (10% interest rate)	20,000
Capital stock	12,000
Retained earnings	53,550
Total liabilities & owners' equity	$143,610

INCOME STATEMENT, YEAR ENDED JUNE 30, 2003 ($000)

Sales	$163,700
Cost of goods sold	96,610
Gross margin	67,090
Operating expenses	38,220
Operating profit	28,870
less: Interest expense	3,810
Profit before taxes	25,060
less: Income taxes	9,770
Net income	$ 15,290

Calculate the following ratios for McNeil Corporation:

a. Times interest earned

b. A/R collection period, year-end

c. Inventory flow period, year-end

d. Current ratio

e. Acid-test ratio

f. ROA, year-end

g. ROS

h. ROE, year-end

i. Long-term debt to total capitalization

j. Total-asset turnover, year-end

k. Effective income tax rate

16. Refer to the financial statements for McNeil Corporation shown in exercise 15, and answer the following questions:

    a. If the company's retained earnings were $4,930 at June 30, 2002, estimate total dividends the company paid to its shareholders during fiscal year 2003.

    b. If McNeil had, just prior to year-end, used $10 million of its cash balance to pay down its bank loan, how would that have affected its year-end current ratio?

    c. If McNeil had, at the beginning of the fiscal year, sold $10 million of additional common stock and used the proceeds to reduce its long-term borrowing, what would be the value at June 30, 2003, of the following?

        i. ROS

        ii. ROE

        iii. Long-term debt to capitalization

    d. If, in the course of the year-end audit, inventory valued at $5 million had been determined obsolete and had been "written off", how much would this action have changed:

      i. The company's net income for fiscal 2003

     ii. The company's inventory turnover at June 30, 2003

    iii. The company's current ratio at June 30, 2003

  e. If McNeil had been permitted by its auditors to switch from the cost value to the market value method of valuing its fixed assets, its fixed-asset value at June 30, 2003, would have been $12 million higher and its accumulated depreciation would have been $4 million higher. What would be the effect of this switch on each of the following?

      i. The company's working capital at fiscal year end

     ii. The company's ratio of total debt to total equity at fiscal year end

    iii. The company's total-asset turnover for fiscal year 2003

17. Locate, on the web, the most recent annual report for a company of your choice. Find the audited financial statements within the annual report.

  a. Assess the company's liquidity, capital structure, and profitability using the ratios you think most appropriate.

  b. Skim the notes to the financial statements, and determine what additional information contained in those notes helps you assess the financial position and operating performance of the company.

CHAPTER 10

ACCOUNTING AND FINANCIAL MARKETS

An important "audience" for the financial statements that we have been constructing and analyzing is that set of people who themselves buy, sell, loan, borrow, and trade in the financial markets, or advise and assist those who do. That set is large. Moreover, it is quite demanding of accurate and reliable information. Financial statements are, of course, not the only source of information available to this audience; other information, much of it subjective or conjectural, may come from company executives, competitors, economists, industry observers, analysts, and personal observation. But, if financial statements are not entirely sufficient for making decisions about actions in the financial markets, they are certainly the fundamental starting point.

In this chapter, I give equal attention to the needs of those on each side of financial market transactions: the borrowing company, and the lending institution; the company obtaining funds through the issuance of new securities, and the underwriters and purchases of those securities; those who trade—buy and sell—in the financial markets.

In recent decades, financial markets have grown markedly in size, complexity, and sophistication. In general, the U.S. markets have led the way, but, increasingly, all the markets of the developed world are tied closely together and influence each other. With the revolution in electronic communication, monetary funds flow instantly, thus eliminating inconsistencies among markets through a process known as *arbitrage*. To a first approximation, we now have a single, worldwide capital market on which the sun never sets: transactions can be effected 24 hours per day—just like in Las Vegas.

TYPES OF SECURITIES

With the increased sophistication of the financial (capital) markets has come a baffling array of new securities, some quite difficult to describe and even more difficult to analyze. Each fills a perceived need both of capital providers and of those companies and institutions seeking capital. Here, we will focus our attention on the most widely used securities—if you will, the "plain vanilla" securities. The more esoteric securities are generally designed to combine some of the features of two or more of these basic types.

Common Stock

All for-profit corporations have, in the past, issued, and now have outstanding, shares of common stock. The common stock owners—the common share-holders or equity holders—own the company. Others may have future claims on common stock—for example, those who hold stock options or warrants or securities convertible into common stock, all of which are explained later in this chapter.

A company's common stock has no fixed maturity; it remains outstanding as long as the company is a going concern. The owner of common stock seeks returns on his or her investment in some combination of dividends paid by the corporation and future appreciation in the share price—that is, the opportunity to sell shares at a price higher than he or she paid for them.

Corporations are not obligated to pay dividends to their common share-holders; payment is wholly at the discretion of the corporation's board of directors. With no fixed obligation to pay dividends and no requirement that the invested capital ever be repaid, the sale of common stock represents the lowest-risk avenue for the corporation to acquire capital. On the flip side, common stock is a high-risk investment to the shareholder.

In general, dividends are higher, both in relation to earnings and to market price, for a company whose net cash inflow is higher than its need for cash to reinvest in the business. For companies with attractive opportunities to reinvest (including acquiring related companies), dividends are typically low or nonexistent. Some companies such as Microsoft Corporation and Berkshire Hathaway, Inc., go decades without paying any dividends on common stock. Owners of these shares must look solely to market price appreciation for their return.

One more important feature of common stock that distinguishes it from all other securities is the sole right to vote in the election of directors. The board of directors carries the fundamental responsibility for operating the company in the interests of the shareholders (and, arguably, other stakeholders as well.) The board hires and fires the corporate officers, sets their compensation, de-clares dividends, decides on capital transactions, decides on mergers and ac-quisitions, approves key policies, and so forth.

In the United States, paid dividends, unlike paid interest, are not deductible by the corporation for income tax purposes. Thus, ignoring risk for the moment, this condition contributes to the fact that common stock is a higher cost source of capital than is debt. On the other hand, shareholders must pay income taxes on dividends received. This condition leads some people to say that dividends are subject to double taxation: once to the corporation and again to the shareholder. Why this lack of symmetry? Simply because that is the way the U.S. Congress drafted the tax laws.

Repurchase of Common Stock. From time to time, some companies repurchase and "retire" a small portion of their outstanding common shares—by negotiating with individual shareholders or by simply placing "buy" orders on an organized stock exchange. Companies are motivated to buy back stock by one or a combination of the following reasons:

1. Management feels that the market price of the common stock is low—that is, the shares can be purchased at a favorable price.
2. The company has excess cash that it prefers not to pay out in the form of dividends.
3. Management wishes to buy out a dissident, disruptive, deceased, or retiring shareholder.
4. The company seeks to support the market price of its common stock through its own demand in the marketplace.

How are other shareholders advantaged by such a repurchase? They end up owning a slightly higher percentage of the outstanding, post-repurchase stock, and thus their equity position increases. Obviously, the repurchase results in each share that remains outstanding having claim to slightly more earnings, and within limits a rational stock market should, therefore, accord a slightly higher price per share to the remaining outstanding shares.

Share price appreciation, or **capital gain,** that is realized when a shareholder sells the stock is (in the United States) taxed at the so-called **capital gains rate,** which is generally lower than the ordinary income tax rate. In some other countries, capital gains are not taxed at all. Thus, many shareholders and corporate executives argue that companies should be managed so that most of the return to shareholders comes in the form of market price appreciation rather than dividends. This argument leads to generous share repurchase programs in preference to generous dividends. On the other hand, some investors, including particularly institutional investors, feel more comfortable owning common stock on which they regularly receive dividends.

Stock Options. Most medium- or large-sized companies issue **stock options** to key employees—and sometimes "key" is defined as a large fraction of all employees. Designed to motivate the employees to act in the best interests of the common shareholders, options have in many companies, particularly small and start-up companies, proven to be the major form of compensation for these key employees. For example, a very large number of Microsoft employees have become wealthy through exercising stock options.

While many forms of stock options have been invented over the years, some in response to changing income tax laws and accounting rules, they typically provide for a purchase price (often called a **strike price,** and typically equal to the market price at the date the option is granted to the employee), a number of shares, and a term of years (typically five to 10) during which the option can be exercised. An employee who holds an option enjoys the benefits of a one-way street. He or she exercises the option—that is, purchases part or all of the specified number of shares at the strike price—if the market price of the common stock has appreciated well above the strike price. Incidentally, stock options are typically not fully exercisable on the day they are granted. Often, the option "vests" over four or five years: that is, 25 or 20 percent of the number of shares become exercisable each year, but the optionee need not exercise the option until just before it expires—and then only if the option is "in the money" (when the stock market price exceeds the exercise price). Options will be discussed further in Chapter 11.

The tax consequences to both the issuing corporation and the employee exercising the option are complicated, and, over the years, the tax laws regarding options have changed considerably. In general, though, any person realizing a gain on the exercise of an option can count on that gain being taxed. In turn, the issuing company will benefit from some tax advantage.

Proper accounting for the "cost" to the company of providing its key employees with stock options is a controversial issue. Some executives argue that there is no cost to the company, as the company pays out no cash and, in fact, receives cash from the employee when he or she exercises the option. However, outstanding stock options represent a contingent claim on equity ownership and, when exercised, dilute the ownership position of all other shareholders. Without question, this dilution is a real cost.

Bonds: A Form of Debt

Companies can borrow from a wide assortment of lenders. In addition to friends and relatives, banks, insurance companies, and leasing companies (see below), large corporations can borrow through the sale of bonds to the "public." Unlike common stock securities, bonds have a maturity date, when the face value of the bond must be repaid; they also specify the interest that must be paid to the bondholder, often quarterly or semiannually, between the date of issuance and the maturity date. Bonds are typically sold at or near their face value. Some are secured by the assets of the issuing corporation; in the event of default, bondholders have first claim to the secured assets. Others are unsecured; bondholders are simply general creditors of the issuing corporation as are, for example, its trade creditors.

Some bonds are **callable** prior to the maturity date, giving the issuing corporation the opportunity to repurchase the bonds, typically at a small premium to the bond's face value. The option to call the bonds provides financing flexibility to the issuing corporation. This feature, however, may be a negative to the bondholder; bonds are frequently called when prevailing interest rates decline and the company has the opportunity to refinance the borrowing at a lower interest rate.

The interest rate at which a bond sells depends on three factors: prevailing market interest rates, the creditworthiness of the issuing corporation, and the term (time to maturity) of the bond. The credit markets in this country generally view bonds issued by the U. S. government as risk-free. All other bonds involve some greater risk, and, therefore, interest rates at which a corporation can sell its bonds are an increment (referred to as a *risk premium*) above rates on government bonds of comparable maturity. Corporations with poor credit ratings must offer very high interest rates in order to entice purchases of their bonds. In the 1980s, when these high-risk bonds first became popular as a source of corporate capital, they were referred to as *junk bonds;* now, more genteelly, they are referred to as simply **high-yield bonds.** Of course, bonds issued at rates close to Treasury bond rates can become junk bonds if the fortunes, and thus the creditworthiness, of the corporate issuer plummets.

After issuance, corporate bonds typically trade in organized markets. As we saw in Chapter 2, when prevailing rates decline (and assuming no change in the creditworthiness of the issuing corporation), bond prices increase—and vice versa. The primary owners of corporate bonds, which tend to trade in the market in very large dollar amounts, are institutions: banks, insurance companies, mutual funds, and sometimes other corporations with substantial excess cash.

If borrowed capital is cheaper than capital obtained through sales of common shares, why don't corporations always turn to the debt market for new capital? The answer is simple: risk. Failure to make interest payments when due, or to repay principal at maturity, is a corporate default event and can result in the corporation's creditors forcing it into bankruptcy. And, even if the creditors permit the corporation to continue to function while in default, these creditors will, in effect, be in control of the corporation. Moreover, in the event of liquidation, the bondholders have to be fully repaid before the common shareholders divide up whatever little may remain.

Again, the flip side of the risk coin is that, for the investor, holding bonds is less risky than holding common stock, precisely because of the promises of cash flows (interest and return of principal) to creditors and the absence of such promises to shareholders, together with the bonds' preferential position in liquidation.

Leasing

Many lease arrangements are close cousins of borrowing arrangements. When the term of a lease for a piece of equipment or a building, for example, is approximately equal to the useful life of that physical asset, the risks of ownership of that asset are in reality undertaken by the lessee, not the lessor. That is, if the lessee is obligated to pay on the lease throughout the asset's useful life and has the use of the asset throughout that useful life, then the lessee has essentially "acquired" the asset, even though legal title remains with the lessor. Therefore, the lessee's balance sheet should reflect this fact: the value of the physical asset is shown as a long-term asset and the equivalent present value (discussed in Chapter 12) of the future lease payments is recorded as a long-term liability. These long-term leases, noncancelable by either the lessor or the lessee, are referred to as **capital leases.**

Note that the previous paragraph refers to a particular type of lease, one that runs for the useful life of the asset. Short-term leases—**operating leases**—or rental agreement are different animals. A rental agreement (for example, for a copy machine) may have a term of one month, several months, one year, or several years, but in any case, the term is substantially less than the full useful life of the asset. Such rental arrangements are quite dissimilar from purchase arrangements, and the accounting for rentals is also very different: the rental payment due in April for use of the copy machine during that month is recorded as an expense in April, and the value of the asset does not appear on the lessee's balance sheet.

If a long-term lease is economically equivalent to a purchase, and both are accounted for similarly, what are the advantages, if any, to the lessor or lessee of the lease agreement? A lessor, in contrast to a lender, retains ownership of the leased asset and, therefore, has immediate access to—and can repossess—that asset in the event of the lessee's default. Of course, the asset can secure a loan, but, if the borrower defaults, gaining clear title to the asset can be an involved process for the lender. As a result, a company with limited credit-worthiness may be able to lease an asset even when lenders are unwilling to finance the company's acquisition of the asset—particularly if the asset in question has a ready resale or re-lease market.

Further, the tax deduction for depreciation remains with the lessor. If the lessor is in a high tax bracket (for example, a partnership of high-income individuals), or if the lessee is in a loss position and therefore is not paying current income tax, the *depreciation tax shield,* as it is called, is more valuable to the lessor than to the lessee. This tax advantage can be reflected in the lease rate charged to the lessee. Consider, for example, the airlines in the United States, which seem to operate almost perpetually at a loss. Aircraft are enormously expensive assets. The depreciation on these aircraft is far more valuable to leasing companies than to the airlines themselves. Accordingly, most airlines lease rather than purchase their aircraft. Start-up companies that expect to be in a loss position for a number of years also find leasing a cost-effective way of obtaining long-term assets.

Preferred Stock

Preferred stock falls between common stock and bonds in terms of its risk to both issuer and investor and its attractiveness as a source of capital. Preferred stock typically has no maturity date but carries a contingent promise to make scheduled payments, called *dividends* rather than interest. Failure to pay these dividends on schedule does not constitute a default event; rather, the dividends are "accumulated." Typically, all such accumulated dividends on the preferred stock must be paid before any dividend can be paid on the common stock. Also, in most cases, if dividends are not paid for a considerable period of time, preferred shareholders are given the right—sometimes the sole right—to vote in the election of directors.

Once again, preferred dividends are not tax deductible to the corporation, but they are taxable as ordinary income to the recipient. As a result, preferred stock is not widely used to finance corporations, although it can be very useful in certain circumstances, as we shall see.

Most preferred stock is convertible, a concept considered shortly. First, review Table 10-1, which summarizes and compares the characteristics of the three primary security types just discussed.

Hybrid Securities

Convertibles. When the words *bonds* or *preferred stock* are modified by the word **convertible,** the securities may, under certain conditions, be converted into common stock—that is, the bond or the preferred share can be traded in for a specified number of common shares.

For example, when a corporation's common stock is selling for $35 per share, the corporation might elect to sell convertible preferred shares at $100 per share, with each share convertible into two common shares. If all works out as the company hopes, the share stock price will, in time, increase to

TABLE 10-1. Comparison of Characteristics of Securities

	Common Stock	Preferred Stock	Bonds (Debt)
Stated face value	No	Yes	Yes
Specified return to holder	No	Contingent	Yes
Maturity date	No	Typically no	Yes
Vote (for directors)	Yes	Contingent	No
Relative position in event of liquidation[a]	3	2	1
Risk:			
To the issuing corporation	Low	Intermediate	High
To the investor	High	Intermediate	Low

[a] As a matter of law, the claims of employees for unpaid salaries and of government taxing authorities for unpaid taxes stand in preferential position even to bondholders.

more than $50 (the *indifference price,* where the value of the two common shares is equal to the value of the unconverted preferred share) and investors can be induced to convert their preferred shares into common shares; when that happens, the company will have, in effect, sold the common stock for $50 per share (1/2 of $100)—an attractive price compared to today's $35 price.

Why is this deal attractive to the individual investor? The investor has the protection on the "downside" of the dividend on the preferred share, plus the "preferred" position in event of liquidation, while at the same time participating in the "upside" from any substantial common stock appreciation. As hinted above, the issuing corporation needs to be able to induce this conversion when it is clearly attractive to the owners of the convertible security. This conversion can be induced if the convertible security is callable—that is, it can be repurchased by the issuing corporation at its original issuance price; when the issuing corporation gives notice that it intends to call, any sensible investor will convert to common shares, given their higher market value.

Warrants. Similar to convertibility, debt securities that have "warrants attached" provide the purchaser the opportunity to participate on the upside in stock price appreciation, while enjoying the protection on the downside that debt entails. A **warrant** is simply an option to purchase a certain number of common shares at a guaranteed price for a certain number of years. A bond with warrants attached (attached economically, that is, not physically attached) is thus a hybrid security, providing the creditor with the protections of a bond and some of the benefits of common stock.

Other. A plethora of other hybrid securities has reached the market in recent decades, limited only by the imagination of investment bankers on Wall Street and of corporate treasurers. As the needs and desires of investors—both individual and institutional—become more specialized and the complexity of global businesses increases, new securities are designed to fit niches in the capital markets. Some bonds may carry interest rates that vary with future market conditions. Bonds can be issued in any one of many international currencies. Some bonds issued by a corporation may be *senior* while others are *subordinated;* holders of **subordinated debt** will, in the event of liquidation, receive nothing until the senior bonds have been redeemed in full.

The key to both corporate issuers and investors is to assess accurately the trade-off between risk and potential return for each type of security under consideration.

MARKETS

As securities have become more sophisticated and complex, so also have the markets in which the securities are bought and sold. We can, however, make a couple of simple distinctions among markets.

Private Markets

A willing seller and a willing buyer constitute a *market*. Subject only to quite complex securities regulations promulgated by various governments, the buyer and the seller can strike a deal. You can approach your father-in-law to invest in your business. Similarly, you can approach a venture capitalist or another corporation to invest in your business. If I own shares in Ford Motor Company, I can sell them to you, my daughter-in-law, or my neighbor next door. All of these transactions occur in the *private market*. No organized exchange or securities dealer or investment banker is involved. We should not forget that the very large majority of new businesses in the United States and around the world obtain their start-up and growth capital in the private market. Some are financed through the entrepreneurs' own consumer credit cards— the ultimate in the private market!

The phrase used in the previous paragraph—"subject to securities regulations"—should not be lightly dismissed. The regulations prohibit you from standing on the street corner hawking to passersby new shares in your company. These regulations, by and large designed to protect the unsophisticated investor, place significant obligations on securities issuers to provide full and accurate disclosure of the facts relevant to the prospective investor. The regulations may also require issuers to assure that each investor can afford the investment and is sufficiently knowledgeable to make an informed investment decision. That said, however, investors are still free to make stupid decisions.

Public, or Organized, Markets

You know well that the daily newspapers are replete with price quotations and related information for many thousands of securities that are traded in *public,* or *organized, markets* or exchanges. The New York Stock Exchange (NYSE) may be the largest, but almost every developing or developed country in the world has one or more exchanges. Common stock, preferred stock, bonds, and many hybrid securities are traded on organized exchanges.

Table 10-2 shows the NYSE listing in the *Wall Street Journal* for Merck (pharmaceuticals), Ford Motor Company, and the Washington Post (publishing) for March 21, 2002. The footnotes explain each column. Note the very considerable differences in price per share and trading volume. Yield and price/earnings ratios will be explained later in this chapter.

While the trading mechanics vary somewhat from exchange to exchange, they are greatly affected by very rapid communications and the Worldwide Web. In turn, these developments have sharply reduced trading transaction costs.

Individuals and firms providing investment advice surround the public financial markets. As a result, some will argue that markets for securities of large corporations traded on major exchanges have become *efficient*—that is, all available relevant information on the corporation is reflected in the exchange price of its securities. The more efficient a market is, the harder it

TABLE 10-2. NYSE Reports for Selected Corporations (as of March 21, 2002)

%CHG[a] YTD	52 WKS[b]		SYM[c]	DIV[d]	YLD%[e]	P/E[f]	VOL[g]	LAST[h]	NET CHG[i]
	Hi	Lo							
−1.3	80.85	56.71	Merck MRK	1.40	2.4	18	64,275	58.05	−0.03
+6.2	31.42	13.90	Ford F	0.40	2.4	N/A	94,402	16.70	+0.20
+13.8	617.75	470	Washington Post WPO	5.60	0.9	25	61	602.95	−6.68

[a] Year-to-date percent change.
[b] High and low prices over the previous 52 weeks.
[c] Abbreviated corporate name and New York Stock Exchange ticker symbol.
[d] Annualized dividend rate.
[e] Yield percentage.
[f] Price/earnings ratio.
[g] Volume: the number of shares traded (in hundreds).
[h] Last price at which the shares traded.
[i] Net change in closing price from the previous day's trading.
Source: Wall Street Journal, March 22, 2002.

is for the individual investor to gain a competitive advantage over other investors. This realization has led many investors to *buy the market:* invest in special investment vehicles that replicate closely a broad market index such as the Standard & Poor's 500 or the Russell 1000.

Government agencies (in the United States, the Securities and Exchange Commission) extensively regulate public markets. Other nongovernmental entities, such as the National Association of Securities Dealers (NASD) and the NYSE itself, require that their members follow certain practices. Again, these are largely designed to protect the individual investor. For example, these regulations restrict **insiders**—individuals who have privileged knowledge about a particular corporation—in trading the corporation's securities for their own account. Individual brokers are forbidden from taking advantage of their investing clients, although just what falls into the category "taking advantage" is subject to wide interpretation.

I pointed out earlier the huge difference in price per share between the Washington Post ($603) and Ford Motor ($16.70). If the Washington Post wished its common stock to sell in the general range of other common stocks, say at $30 per share, it could easily achieve its objective by declaring a 20-for-1 **stock split.** To accomplish this split (which would have to be authorized by its board), it simply needs to send each shareholder 19 additional shares for every one share he or she now owns. The shareholder is neither better nor worse off following this split; he or she continues to own the same fraction of the Washington Post. Washington Post's market price should—assuming rationality, always a dangerous assumption—fall to 5 percent of its former value, about $30. Why would the Washington Post do this? To address the widespread feeling (again, not entirely rational) that stocks selling in the $20 to $50 range attract more buying attention; more buying attention may well translate into a higher price per share. **Stock dividends** are analogous to stock splits: a 10 percent stock dividend results in each shareholder getting one more share for every 10 shares that she or he now owns.

New Issues Versus Trading

The huge majority of transactions on public exchanges do not involve the corporation whose securities are being traded. That is, neither the buyer nor the seller is the corporation itself. Buyer and seller are, instead, passive investors in the corporation's securities. These investor-to-investor trades drive market price changes, but market prices are not reflected in the financial records of the corporation.

The **capitalized value** of the corporation is equal to the number of its common shares outstanding times the market price per share. Remember that this amount bears no necessary relationship to the value of owners' equity on the corporation's balance sheet. Typically, capitalized value will be higher if the investors are optimistic about the corporation's future, and lower if and when the company falls on hard times and investors shun its common stock.

Corporate executives are, nevertheless, hardly indifferent to the trading price of their company's common stock. If and when the company needs or wishes to sell additional common stock, the share price received will be a function of the then-market price. Moreover, corporate chieftains and their colleagues often own substantial numbers of shares themselves and stand also to gain from stock options they hold but haven't yet exercised. And, of course, shareholders who are happy with the value of their stock holdings cause management much less trouble than do unhappy shareholders!

How do corporations go about selling newly issued shares (or bonds) to the investing public? A detailed explanation is beyond the scope of this book, but a short-form explanation is not. Most sales of new shares are *underwritten* by an investment banking firm, the **underwriter.** Incidentally, the first public sale of common stock by the company is referred to as its **initial public offering (IPO);** prior to its IPO, the company's stock was not publicly traded. When an offering is underwritten, the offered shares are actually purchased from the company by the underwriter (quite likely, in fact, a large number of underwriters acting as partners in what is typically called a *syndicate*); the underwriters, in turn, sell the shares immediately, or as soon as possible, to the public, both individual and institutional investors. The costs associated with a public stock offering are very high, including accounting and legal fees; the costs of printing a lengthy *prospectus,* which provides great detail on the company's business, prospects, financial position, and recent performance; and the commission paid to the underwriter(s). Since most of these expenses, except underwriting commissions, are fixed regardless of the size of the offering, a small offering is uneconomic. The relationship between the lead underwriter and the issuing corporation is complex and conflicts of interest can arise; a high level of trust between the two parties is essential. While, theoretically, common stock can be sold to the public at any time, even if at a distress price, as a practical matter there are times when access to the public equity market is effectively closed to new issuers. This fact complicates considerably the decisions regarding financing: when to raise capital, how much capital to raise, and what form of security to use.

INVESTMENT RATIOS

Chapter 9 discussed operating ratios, the data for which were drawn directly from the company's income statement, balance sheet, and cash flow statement. We now need to consider a set of ratios that draw data both from the financial markets and from the company's financial statements. We'll call these "investment ratios" and illustrate them by reference again to Merck & Company, whose financial statements appeared in Tables 9-1, 9-2, and 9-3.

Earnings per Share (EPS)

The ratio introduced in Chapter 9—*earnings per share* (EPS)—shows the annual earnings attributable to each share of common stock, and is simply the net income divided by the total number of shares outstanding:

$$\text{Earnings per share} = \frac{\text{net income}}{\text{number of shares outstanding}}$$

Table 9-1 shows Merck's 2001 earnings per share was $3.18.

Many—indeed most—corporations have outstanding certain claims on common shares arising from employee stock options and perhaps also from warrants and convertible securities. Thus, the current shareholders' position will be diluted if, or when, these options are exercised or securities are converted. The Financial Accounting Standards Board (FASB) has set forth a fairly complicated manner of reflecting this potential dilution. When the strike price of the option or warrant is below the value of the securities to be received upon exercise, the option or warrant is "in the money." In this case, the FASB rules require the accountant to assume that the options or warrants are exercised and the cash realized is used to repurchase common shares at the current market price; the net difference between number of shares issued and number of shares repurchased is assumed to be additional shares outstanding for the purpose of calculating diluted earnings per share.

For 2001, Merck's fully diluted earnings per share was $3.14.

Price/Earnings (P/E) Ratio

Market prices for common shares, as they move day to day and are reported in the newspaper, do not permit us to draw meaningful comparisons between different corporations. For an example, see Table 10-2: the fact that a share of Washington Post stock sells for over 10 times the price of a share of Merck does not tell us that Washington Post is 10 times as "expensive" or "valuable" as Merck. And, neither does a comparison of earnings per share provide useful comparative information. It is the ratio of these two pieces of data— the **price/earnings** (P/E), ratio—that is useful for purposes of comparison:

$$\text{Price/earnings ratio} = \frac{\text{market price per share}}{\text{earnings per share}}$$

As shown in Table 10-2, Merck's P/E ratio was 18 at March 21, 2002.

Think of this ratio as indicating the multiple of earnings per share that investors are willing to pay to purchase a share of the stock. Indeed, the P/E ratio is often referred to as the company's **multiple.** Investors might be willing to pay a high multiple—say, 35—for a company with a particularly bright future (good growth in revenues, a high ROS), but only seven times earnings for a slow-growth, low-earning company. We can say that the first company is five times more expensive than the second company, regardless of what its price per share is. Washington Post's P/E ratio is 25, or nearly 40 percent higher than that of Merck. But a large caution: these P/E ratios do not tell you which of these two companies represents a better investment today. More on that in a moment.

Book Value per Share

As you may have already guessed, a company's **book value per share** is determined simply by dividing its total owners' equity (the book value of its equity as stated on the balance sheet) by the number of common shares outstanding. Thus,

$$\text{Book value per share} = \frac{\text{total owners' equity}}{\text{total number of shares outstanding}}.$$

Merck's book value at December 31, 2001 was = $16,050/2,273 = $7.06 per share, compared with a *market* value per share of $58.05 on March 21, 2002.

What relevance does this ratio have for investors? Not much. Again, companies with bright prospects will probably have a market price higher than book value, and vice versa. You might infer that book value per share indicates what the shareholder would receive in the event of liquidation. That conclusion is valid only when the value of assets on the balance sheet approximates their liquidation value, a quite unlikely situation.

Dividends per Share

It follows that the **dividends per share ratio** is simply the total dividends paid (not earnings, but dividends actually paid out) divided by the total number of shares outstanding. Actually, when boards of directors declare dividends, they normally do so in terms of an amount per share.

Merck's dividend per share in 2002 was $1.40 (see Table 10-2).

Payout Ratio

The **payout ratio** shows the portion of total earnings paid out in the form of dividends:

$$\text{Payout ratio} = \frac{\text{dividends per share}}{\text{earnings per share}} \quad \text{or} \quad \frac{\text{total dividends paid}}{\text{net income}}$$

Merck's payout ratio in 2001 was 3,145*/7,282) = 43 percent.

Let me anticipate a question you may have: yes, the payout ratio can exceed 100 percent. This may happen when a company paying a consistent dividend year after year has a bad earnings year but expects to return soon to traditional levels of profitability. The company's shareholders may be dependent on these dividends for their livelihood, and a reduction in or discontinuation of the dividend, even for a year, would cause real hardship. As long as the company has or can get the required cash, it is not prohibited from paying dividends in excess of earnings (but generally not in excess of accumulated retained earnings).

Yield

Finally, **yield** is the relationship between dividends paid and the market price (not book value) of the company:

$$\text{Yield} = \frac{\text{dividends per share}}{\text{market price per share}}.$$

For Merck, on March 21, 2002, the yield was 2.4% (see Table 10-2).

We spoke earlier of bond yields: annual interest paid divided by the current market price of the bond. Dividend yield on common stock is analogous.

INTERPRETING INVESTMENT RATIOS

Most investors who are new to the game look in vain for the one or perhaps two ratios that signal which stock or which bond they should buy, or when they should sell a stock they currently own. Remember that investors who trade in securities are matching wits with each other, and all potential investors in a common stock have—or can and should get—the same ratio information. Remember, too, that when you invest in a common stock, you are investing in the *future* of that company: the right to receive dividends and participate in the appreciation (or suffer the decline) in the market price of the stock. Ratios are historical.

Moreover, common stock prices are the *result* of collective buying and selling by investors. When more investors are attracted to buy a particular

---

*See Merck's cash flow statement (Table 9-3) for total dividends paid.

common stock, its price will increase. When current shareholders are running for the exits, the supply of shares (that is, shares offered for sale by their current owners) will greatly exceed the demand (the number of shares that investors seek to purchase) and the market price per share may plummet.

The key to common stock investing, then, is to compare your expectations about the future of the company and the future price of its shares to the prevailing expectations of all investors as revealed by such ratios as the P/E and the yield. Suppose a company's stock is selling at a P/E ratio of 60 and a zero yield when the average P/E and yield for all publicly-traded stocks are 17 and 2 percent, respectively. You should buy the stock if and only if you believe that the future for the company, and thus its future price per share, is even rosier than the average investor thinks it is. Of course, this investment may be a high risk: if the P/E ratio moves to the "mean" of the market, you will lose more than 2/3 of your investment. If you already own the stock, and your view of the company's future is more pessimistic than the generally prevailing view, you should sell your shares.*

On the other hand, a stock selling at a P/E ratio of 8 when the average P/E is 17—that is, when the stock's selling price is less than half that of the market—is not necessarily a good deal. There may be sound reasons (such as loss of market share or persistent operating losses) why current investors are selling and prospective investors are not attracted to the stock. But if you think present and prospective investors in this company are excessively pessimistic about its future, this may be a good time for you to buy shares of its common stock.

DESIGNING A CAPITAL STRUCTURE

Now let's return from the investor's viewpoint to that of the company managers who must decide how to finance the company. As discussed earlier, all companies have some common stock outstanding; and they all borrow some money—at least, from their trade vendors and probably from banks or leasing companies as well. The challenge the managers face is deciding what portion of their total required capital should be obtained by selling common stock to investors and what portion should be obtained by means of borrowing. That is, what should its **capital structure** be? Recall the capital structure ratios we reviewed in Chapter 9.

As mentioned earlier, from the issuing corporation's viewpoint, common stock is typically the least risky (with no risk of default) but the most expen-

---

*Even if you don't currently own shares, you could sell shares "short." *Selling short* involves borrowing (for a small fee) shares from another investor with the promise to return them at a fixed future date. You then sell these shares with the expectation that you will be able to buy back the same number of shares at a lower price prior to the expiration of the borrowing period.

sive source of capital. Debt is the opposite: the most risky and the least costly. Obviously, preferred stock and most hybrid securities have risks and costs somewhere between common equity and debt. Let's, then, confine our analysis solely to common stock and debt.

We should recognize that another source of capital is the net earnings retained by the company and not paid out to shareholders in the form of dividends. Note that capital obtained this way really represents additional capital invested by the shareholders.*

Servicing debt requires cash—for the payment of interest when due, and the repayment of principal at maturity. The risk of debt is that the company will not have the cash available at those critical times. Thus, the more confident the company's management is that the cash will be available when needed to service the debt, the more it can feel comfortable in capitalizing the company with debt. As we will see shortly, there are some real advantages—as well as risks—of obtaining capital through borrowing by comparison with selling additional common stock.

What types of companies have reason to feel comfortable with a relatively high proportion of debt in their capital structures, that is, a highly leveraged capital structure? Those companies where cash inflows and outflows are highly predictable. Take, for example, a water company that serves my home. It has a monopoly, granted in effect by the local municipality. I cannot choose to buy my water from another supplier. I can be counted on to take a bath regularly, wash my clothes and dishes, and water my lawn and flowerbeds. I'm unlikely to change my water use habits much from year to year (except that I will water my yard less in "wet" years). The water company has a convenient and foolproof way to force me to pay my water bill on time: threaten to turn off my household water. Thus, this company can be confident that customers will use water and pay on time. The company's cash expenses are also very predictable. This company, then, can afford to borrow a large portion of its required capital and refrain from selling additional common stock that would dilute the present common shareholders.

On the other hand, a small biotechnology company is likely to rely to a greater extent on common stock financing. The unpredictability of federal regulation of new drugs and of manufacturing processes, and the very competitive nature of the industry, make cash flows much less predictable. Perhaps the regulators will find cause to require the company to withdraw from the market a drug that is currently generating substantial cash. Or a competing company may suddenly get approval to introduce to the market a drug that threatens to erode the market position of one of the company's blockbuster drugs.

---

*Of course, the shareholders didn't choose to make this investment; the board of directors made the investment decision for them. However, if shareholders remain persistently unhappy with decisions made by the directors, including dividend decisions, they can vote to replace them with directors more to their liking.

While many analytical techniques are available to help determine an appropriate capital structure, in the final analysis, the choice is one of judgment based largely on the risk tolerances of the board of directors and senior managers. Some have described the choice as being between "sleeping well" (low debt and low financial risk) and "eating well" (higher debt and higher risk, but greater opportunity for positive leverage effects). Bill Hewlett and David Packard, founders of the Hewlett-Packard Corporation, eschewed long-term debt, consciously limiting the firm's growth to the rate that could be financed from internally generated earnings. Directors and executives of other companies have high risk tolerances and are willing to borrow all that the creditors will lend at reasonable rates; the capital structure of such companies is really set by the capital markets.

UPSIDE AND DOWNSIDE OF DEBT LEVERAGE

High debt leverage has both an upside and a downside. Comparing two companies, one with high debt leverage and one with low debt leverage, we see that the earnings per share of the high-debt company will be leveraged up in good times, and leveraged down in bad times, by comparison with the low-debt company.

Suppose company A decides on a capital structure consisting of 80 percent common stock capital and 20 percent borrowed funds, while company B obtains its capital in equal proportions of equity and debt. Assume the two companies are the same size, in the same business, and achieve identical EBIT (earnings before interest and taxes), in both good years and bad years. Tables 10-3 through 10-5 illustrate this situation. This illustration assumes that both companies incur the same interest rate on their debt, but in practice company B might incur a higher interest rate because of the added risk associated with its higher debt leverage.

In the normal year shown in Table 10-3, company A earns higher net income but its earnings per share (and EPS is what counts to shareholders!) is virtually identical to that of company B. Table 10-4 illustrates a boom year: both companies have increased their EBIT by 50 percent from the normal year. Company A still has higher net earnings than company B, but its earnings per share are lower. Company B's interest cost remains the same in this boom year, and thus more earnings are left over for its shareholders. The reverse is true during the recession year illustrated in Table 10-5: company A now has higher total earnings and higher EPS than company B. Company B must still pay the high interest cost associated with its debt even in this recession year, and thus its EPS has been "leveraged down."

Even the third scenario, illustrated in Table 10-5, hardly represents disaster for either company. But if the two companies encounter a string of loss years, company B is going to get in trouble—that is, be unable to service its debt and, thus, face the risk of bankruptcy—long before company A. Such is the two-edged sword of debt leverage!

TABLE 10-3. Neutral Effect of Leverage in Normal Times

	Company A	Company B
	Capital Structure	
Debt[a]	$ 4.4 million (20%)	$11.0 million (50%)
Company equity[b]	17.6 million (80%)	11.0 million (50%)
Total	$22.0 million	$22.0 million
	Earnings	
EBIT[c]	$1.760 million	$1.760 million
Interest expense	0.352	0.880
Profit before tax	1.408	0.880
Income tax (40%)	0.563	0.352
Net income	$0.845 million	$0.528 million
Earnings per share	$0.423	$0.422

[a] At 8 percent interest rate.
[b] 2.0 million shares for company A and 1.25 million shares for company B.
[c] Earnings before interest and taxes is 8 percent, equal to the pretax cost of debt.

TABLE 10-4. Positive Effect of Debt Leverage in Boom Times

	Company A	Company B
	Capital Structure—see Table 10-3	
	Earnings	
EBIT[a]	$2.640 million	$2.640 million
Interest expense	0.352	0.880
Profit before tax	2.288	1.760
Income tax (40%)	0.915	0.704
Net income	$1.373 million	$1.056 million
Earnings per share	$0.687	$0.845

[a] Earnings before interest and taxes increases to 12 percent of total capital.

TABLE 10-5. Negative Effect of Debt Leverage in Recessionary Times

	Company A	Company B
	Capital Structure—see Table 10-3	
	Earnings	
EBIT[a]	$1.100 million	$1.100 million
Interest expense	0.352	0.880
Profit before tax	0.748	0.220
Income tax (40%)	0.299	0.088
Net income	$0.449 million	$0.132 million
Earnings per share	$0.225	$0.106

[a] Earnings before income and taxes at 5% is well below the pretax cost of debt.

INTRODUCTION TO THE COST OF CAPITAL

How much does capital cost the company issuing new securities? This may sound like a simple question, but it is not, particularly with respect to equity capital.

The cost of debt capital is relatively straightforward: The cost to the company is essentially the interest rate. When the borrowing company is profitable, the interest paid is deductible and thus the after-tax cost is the relevant figure: the interest rate $\times$ (1 − income tax rate).

At first glance, the cost of equity capital might appear to be the yield on the common stock or the inverse of the P/E ratio. The yield on common stock seems to be analogous to the interest rate on debt, but it is not. Remember that earnings retained by the company (not paid out in dividends) represent (involuntary) reinvestment by the shareholder; they are not "free," no-cost funds to the corporation. Managers who view retained earnings as free capital will soon be in trouble with their shareholders.

Then, how about earnings as a percentage of the price of the common stock—the inverse of the P/E ratio? The problem with this measure of the cost of common stock is that it leads to absurd results. For companies with high P/E ratios, shareholder expectations are also very high; that is, the shareholders expect the company to be able to invest its new cash at very high returns, thereby growing the company's earnings rapidly and justifying its high P/E ratio. Using the inverse of a high P/E would imply a low cost of equity capital for this very promising firm, a result that is inconsistent with shareholder expectations. Similarly, low P/E ratios would imply a high cost of equity capital for the company with mediocre or poor prospects, which is also inconsistent with shareholder expectations.

Recall from Chapter 2 that we found the market price of a bond to be equal to the present worth of the cash flows of interest and principal payments from the present date to the bond's maturity date, all discounted at the prevailing market interest rate. Thus, we found that, when prevailing interest rates increase, future cash flows are worth less, when measured in today's dollar, and the present worth of the future cash flows—and the bond's market price—decrease. The same valuation process applies to common stock: its market value is equal to the present value of future cash flows to be derived from owning the stock. The difference is that the bond's cash flows are fixed by contract, while the cash flow of common stock dividends out into the distant future is anything but certain.

Common stock valuation can also be thought of as the present worth of the dividends to be received during the period the stock is owned, plus the present worth of the cash receipts from the sale of the stock at a certain date in the future; this formulation may sound better, but in fact it doesn't help much, since the future selling price is equal to the present worth of cash flows that will accrue to all future owners of the common stock.

Just as risk affects the valuation of bonds (as risk increases, the price declines, so the effective yield increases), similarly, risk has a major impact on the cost of equity capital. A shareholder must be compensated for the risk which he or she is accepting. The higher the risk, the higher the required compensation. A venture capitalist who invests in start-up or early-stage companies expects to earn a substantially higher return, on average, than does the investor in Blue Chip, mature, stable companies. This idea is illustrated conceptually in the **capital market line,** shown in Figure 10-1. The horizontal axis shows increasing investment risk, while the vertical axis shows (1) the increasing inherent discount rate, and (2) the increasing return required by the investor—and (1) and (2) are the same thing! The point where the market line intersects the vertical axis represents the risk-free rate; only investments in U.S. Treasury short-term debt instruments (**T-bills,** as they are called) are considered risk-free. Returns on all other investments, debt or equity, must be higher in order to compensate investors for the greater risk they are assuming in comparison with the alternative of investing in T-bills. (Note that this intersection point moves up and down with prevailing interest rates, which are affected by inflation and by demand and supply of investment capital.)

Note points A through E on the capital market line in Figure 10-1. Conceptually, point A might represent the risk and cost of long-term government bonds (slightly above the risk-free rate). B might represent corporate bonds of intermediate term (somewhat above the cost of government bonds). C might represent high-yield (junk) corporate bonds. D might represent common stock for a mature, stable company. And, E might represent venture capital.

Once we know the cost of each component of the capital structure, we can weight these costs to reflect the long-term capital structure sought by the company. The result is a **weighted-average cost of capital (WACC).** Suppose the Selix Company seeks a long-term capital structure consisting of 65 percent

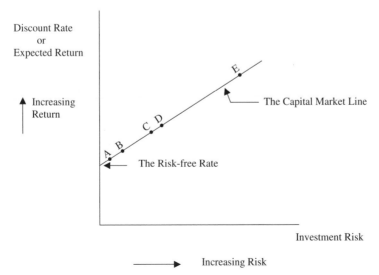

Figure 10-1. Capital market line.

equity and 35 percent debt. Further assume that the company's after-tax cost of debt is 4.5 percent, and its cost of equity is 9.7 percent. We arrive at its weighted-average after-tax cost of capital by the following simple calculation:

$$(0.35 \times 4.5 \text{ percent}) + (0.65 \times 9.7 \text{ percent}) = 7.88 \text{ percent}$$

We will see in the next chapter why we need to get a good handle on the weighted-average cost of capital.

SUMMARY

This chapter links our discussion of accounting and financial statement analysis with the all-important consideration of the financial (capital) markets, from the viewpoint both of investors (buyers and sellers) and of industrial and commercial firms seeking capital from these markets.

The three fundamental securities that trade in the financial markets are common stock, preferred stock, and bonds. The dominant contrasting features of these three types of securities are (1) a specified return to the holder; (2) a stated maturity; (3) the right to vote for the issuers' directors; (4) priority in event of liquidation; (5) risk; and (6) tax treatment to both the issuing corporation and the security owner. Stock options provide their holders with the opportunity, but not obligation, to purchase common shares at a predetermined price for a period of years into the future; options are used primarily as incentives for management. Corporations may, from time to time, elect to repurchase (and typically retire) their own common shares. Leases with a term generally equivalent to the useful life of the leased asset can be viewed as—and are accounted for similarly to—long-term debt.

Private and public (organized) financial markets operate considerably differently, but both are subject to extensive regulation. Remember that a market can be as simple as the coming together of a willing buyer and a willing seller. Organized markets are dominated by trading among investors rather than transactions that generate new capital for issuing corporations.

The primary investment ratios are earnings per share (EPS), price/earnings (P/E) ratio, book value per share, dividends per share, payout ratio, and yield. Investors' decisions are aided by these ratios, all of which are, of course, historical, but, since investors are either buying or selling participation in the company's future, the ratios do not themselves signal when to buy or sell a particular security.

The mix of sources of long-term capital, primarily debt and common stock equity, defines a company's capital structure. Capital structures are influenced by market conditions, the stability of the company, and the company's risk tolerance. Increasing debt leverage is a two-edged sword, enhancing shareholder returns in good times but lowering them when the company's operating results are poor; in either case, increasing debt leverage exposes the borrowing company and its shareholders to greater financial risk.

A company's cost of capital is a function of its capital structure and of the after-tax cost of each source of funds. The cost of capital plays an important role in assessing the attractiveness of long-term investment decisions, as will be discussed further in Chapter 12.

NEW TERMS

Book value per share. The ratio of the company's total owners' equity, as reported on its financial statements, to the number of common shares outstanding.

Callable. Refers to securities, typically bonds or convertible preferred stock, that may be repurchased (called) by the issuing corporation at or near their original selling price at some time, or during a specified period, in the future.

Capital gains. The difference between the price realized in the sale of an asset (common stock, for example) and the price paid at the time of its earlier acquisition.

Capital gains rate. The income tax rate applied to an individual's or a corporation's capital gains, typically less than the rates applied to earned (ordinary) income.

Capital leases. Leases having a term generally equivalent to the life of the leased asset, thereby transferring the risk of ownership from the lessor to the lessee, and accounted for in a manner similar to long-term debt.

Capital market line. A graphic representation of the trade-off between risk and expected (or demanded) return on securities.

Capital structure. The proportions of various sources of permanent capital used by the corporation.

Capitalized value. Refers to the aggregate market valuation of the company's common stock: the market price per share × the number of shares outstanding.

Convertible. Refers to a security that may be exchanged, at the election of the holder, for a specified amount of other of the issuer's securities at a specified time, or during a specified period, in the future.

Dividends per share. The ratio of total dividend payments over the course of a year, to the company's number of shares outstanding; dividends are declared by the company's board of directors in terms of an amount per share.

High-yield bonds. Bonds that carry high interest rates because of the issuers' relatively poor creditworthiness; sometimes referred to as *junk bonds*.

Initial public offering (IPO). The first sale of the issuer's common stock into the public (as opposed to private) financial markets.

Insiders. Existing or potential investors who have privileged information about a company's operations and are therefore restricted by law in their freedom to trade in the company's securities; not limited to individuals employed by the issuing corporation.

Multiple. Another name for *price/earnings (P/E) ratio.*

Operating leases. Lease or rental arrangements that provide for payments over a time period shorter than the useful life of the asset; contrasts with *capital leases.*

Payout ratio. The ratio of a company's dividends per share to its earnings per share, expressed as a percent.

Preferred stock. A form of security providing a specified dividend payment amount and schedule, and further providing that the failure to pay dividends does not constitute default on the issuer's part.

Price/earnings (P/E) ratio. The ratio of a company's market price per share to its earning per share. An alternative name is *multiple.*

Stock dividends. A dividend declared by the corporation's board of directors, paid in additional common shares rather than cash. A stock dividend can be viewed as a fractional stock split.

Stock options. A form of security that provides the holder the opportunity, but not the obligation, to purchase (exercise the option for) a specified number of shares of common stock at a predetermined price (most often the market price on the day the option was granted) for a specified period of time into the future.

Stock split. Action by a corporate board of directors to issue to shareholders a certain number of additional common shares for every common share owned.

Strike price. The predetermined price stated in the stock option or warrant.

Subordinated debt. Debt that, in the event of liquidation or default, is entitled to delinquent interest payments and return of principal only after the issuing company's obligation to so-called senior debt has been satisfied.

T-bills. A short-hand name for *U.S. Treasury bills,* a form of short-term borrowing by the U.S. government; generally viewed as risk-free to the investor.

Underwriter. A securities firm that purchases newly issued securities from the issuing corporation and assumes the responsibility and risk of reselling those securities to individual and institutional investors.

Warrant. A security, sometimes packaged with another of the issuer's securities, permitting, but not requiring, the holder to purchase (exercise the warrant for) a specified number of shares at a predetermined price and at a specified time or for a specified period in the future.

Weighted-average cost of capital. The after-tax cost of acquiring capital by means of each of the company's sources of permanent capital, each source weighted by its amount in the company's capital structure.

Yield. The ratio of a company's dividends per share to its market price per share, expressed as a percent.

EXERCISES

1. What are the key differences between common and preferred stock?
2. For what reasons might a common stock with a high yield not represent an attractive investment?
3. Why is a price comparison between two common stocks better based on their price/earnings ratios than on their market prices or earnings per share?
4. What are the primary reasons why a company might repurchase its own common stock?
5. Why is a stock with a low price/earnings ratio not necessarily a more attractive investment that another stock with a high price/earnings ratio?
6. What is meant by the "risk-free rate of interest"? Does any security enjoy such a rate?
7. Show, in T-account format, the entries to record Marer Corporation's repurchase of 1,000 of its common shares at $37 per share. Indicate whether each account is an asset, a liability, a revenue, an expense, or an owners' equity account.
8. Why is the cost of obtaining equity capital from a venture capitalist typically higher than the cost of selling common stock in the public market? And, if it is more costly, why do corporations seek equity capital from venture capitalists?
9. Indicate whether each of the following statements is true or false:
   a. Stock options are a zero-cost method of compensating a corporation's key executives.
   b. Raising capital through the sale of preferred stock is more risky for a corporation than through the sale of common stock.
   c. All corporations have shares of common stock outstanding.
   d. Securities underwriters assume little financial risk since they serve simply as brokers between the corporation issuing securities and the purchasers of those securities.
   e. Many investors avoid acquiring high-yield (or junk) bonds because they involve excessive risk.
   f. With respect to convertible securities, the issuing corporation has the option of demanding, at any time, that the owners convert the security.

g. Corporations have little interest in the day-to-day market prices of their common stock, since trading between investors has no effect on the corporations.

h. The capitalized value of a corporation is the product of the market price of its equity securities times the number of outstanding shares of those securities.

i. So-called fully diluted earnings per share are always at or below the primary earnings per share.

j. When a corporation repurchases its common shares, one effect is to increase the corporation's earnings per share.

k. Dividends per share in any year may not exceed earnings per share for that year.

l. A corporation's shareholders often do not welcome a high payout ratio on the common stock.

m. High debt leverage signals that the corporation has encountered financial difficulties.

n. Because of its higher risk, the cost of debt is higher than the cost of common stock to the issuing corporation.

o. The phrase "efficient market" is used to indicate that a particular exchange effects trades rapidly and at low cost.

10. Fill in the word that best completes the following statements:

a. The capital market line shows the relationship between the cost of securities and the _____ of the issuing corporation.

b. The market price of a share of common stock can be thought of as the present value of all future _____.

c. Regulations promulgated by the Securities and Exchange Commission are largely meant to protect _____.

d. The best way to compare prices among common stocks is to look at their _____.

e. The book value of a share of common stock equals the _____ divided by the number of common shares outstanding.

f. To determine the weighted average cost of capital for a corporation, the cost of the various sources of capital are weighted by _____.

g. IPO stands for initial _____.

h. The payout ratio is the percentage that dividends paid during a year are of _____.

i. Assuming operating profit remains unchanged, when a corporation increases its debt leverage, it will _____ its net income but may increase or decrease its _____.

j. Returns to shareholders come in the form of dividends and price appreciation; a shareholder's tax rate on dividends is _____ than the tax rate on capital gains (that is, price appreciation).

11. Why might a company elect not to use the lowest-cost source of capital when raising additional funds to invest?

12. For what reasons might a corporation with strong earnings and cash flow decide not to pay dividends on its common stock?

13. What factors or forces cause the day-to-day fluctuation in market prices of common stocks?

14. Explain why a convertible bond might be more attractive to an investor than would a normal bond, while at the same time the convertible bond might also be more attractive than a normal bond to the corporation seeking additional capital.

15. If a company's common stock has a price/earnings ratio of 15 and a yield of 4 percent, what is the company's dividend payout ratio?

16. If a company's common stock is trading at about $23 per share when the company is preparing to sell convertible bonds at $100 per bond with a competitive interest rate, what conversion ratio would you suggest that the company offer? Explain.

17. If a particular stock has a yield of 3.5 percent, a payout ratio of 50 percent, and a market price of $21 per share, what is its price/earnings ratio?

18. Assume that LeRoy Corporation had a net income in 2005 of $4.7 million and earnings per share of $6.25, and that the current market price of its common stock is $73:
    a. What is LeRoy's total capitalized value?
    b. What is the yield on its common stock if its payout ratio in 2005 was 30 percent?
    c. What is the ratio of book value per share, to market value per share if LeRoy's return on equity was 15 percent in 2005?

19. Companies X and Y have EBITs of $140,000 in 2004. In addition, both have outstanding borrowings on which they pay 8 percent annual interest, both have a 40 percent income tax rate, and they have the following capital structures:

	X	Y
Borrowing (loans)	$200,000	$400,000
Owners' equity	$600,000	$400,000
Shares outstanding	600	400

By how much will the two companies differ in the following?
a. Net income for 2004
b. Earnings per share for 2004

20. The Franzia Company can borrow on a long-term basis at 9 percent interest, and its risk profile and current market price indicate a cost of equity of 16 percent. If the company seeks a long-term debt to equity ratio of 1:3 and has a 30 percent income tax rate, what is Franzia's weighted average cost of capital?

CHAPTER 11

DIGGING DEEPER . . .

Let's take a deep breath now, and dig a bit deeper into just what each of us, depending on our particular perspective, should be looking for in the financial statements. I mentioned in Chapter 1 that different audiences have different questions they seek to answer by studying financial reports. We can now get more specific about these questions. And, in the Preface, I stated that one of my objectives in this book is to develop in you a healthy skepticism about valuations appearing in these statements. The accounting scandals that surfaced in the early part of the new century heighten the need for readers of financial statements to be alert to ways in which financial statements can be "dressed up." We'll look at both of these issues in this chapter.

FRAMEWORK FOR FINANCIAL STATEMENT ANALYSIS

To repeat, because different audiences have different decisions to make, they need to focus on different information within the financial statements and the accompanying notes. But this difference is only one of degree: to some extent, all audiences are interested in all perspectives—and that is particularly so for the company's managers.

That said, we can develop an analytical framework for each different audience. Let's consider creditors, investors (actual and prospective), employees (including prospective employees), and tax collectors.

Creditor, Short-Term

A short-term creditor (such as a trade supplier or a bank considering, say, a 90-day revolving credit) is primarily concerned about getting repaid, preferably on schedule, over the next several months. Accordingly, the creditor is less interested in the long-term financial position or performance of the borrowing company, and more interested in liquidity. The ratios of primary interest, then, are these two:

- The current ratio
- The acid-test (or quick) ratio

But this short-term creditor is also interested in the accounts payable payment period, because this ratio speaks directly to the timeliness of payments to vendors, and the accounts receivable collection period, because the most immediate source of additional cash for debt repayment is the accounts receivable. Of course, trends in all of these ratios are informative.

Creditor, Long-Term

Long-term creditors include the following:

- Banks and finance or insurance companies providing credit with maturity greater than a year
- An investor, or a prospective investor, in a bond issued by the company
- A lessor of a building or of a major piece of equipment
- A vendor selling equipment on an installment basis

Obviously, the long-term creditor takes a longer view than the short-term creditor. Nevertheless, the long-term creditor is also interested in the borrower's liquidity, for, if the company is driven to bankruptcy by failure to meet its short-term obligations, the long term is quite irrelevant. And, the longer the term of the credit, the more this creditor's perspective begins to resemble that of an owner (such as a prospective purchaser of common stock.)

In addition to the liquidity measures, then, the long-term creditor is concerned with how well the company uses its assets. Reducing asset investments by, for example, shortening accounts receivable collection periods or increasing inventory turnover, is a potential source of cash for repaying loans. Or, the reverse: if the company either (1) becomes more inefficient in the use of its assets, or (2) grows but does not increase its efficiency in asset use, the company will have to raise additional cash to invest in working capital, thus jeopardizing its ability to repay borrowings. The long-term lender is also interested in the borrower's debt leverage, that is, how reliant on debt the company is to finance its activities, including growth.

The long-term creditor focuses, then, on the following:

- The accounts receivable collection period
- The inventory flow period (or inventory turnover)

- The total-asset turnover
- One or more of the many leverage ratios: total debt to total assets; total debt to total owners' equity; times interest earned; and long-term debt to total equity

The longer the credit's maturity, the more the lender looks to corporate earnings as the primary source of cash for debt repayment. Accordingly, the lender is concerned with the financial performance (profitability and growth) of the borrower as well as with balance sheet ratios. Once again, trends are key, as is *benchmarking* the borrower against other companies in the same industry.

Investor, Long-Term

A present or prospective investor in the company's common stock has a fundamentally different perspective than does a creditor. Nevertheless, the investor certainly cannot be indifferent to the company's liquidity or efficiency of asset utilization; if the demands of creditors are not met, and the company is forced into bankruptcy or into an unfavorable merger, the shareholder's return on his or her investment will be jeopardized.

The primary focus of the present or potential shareholder is on profitability, growth, and risk. Recall that profitability is measured in two primary ways:

1. Profit as related to revenue (percentages on the income statement)
2. Profit as related to investments in the company

Some of the specifics the investor should look at are these:

- Return on sales (ROS)
- Gross margin
- Contribution margin—to be discussed in Chapter 14
- Return on assets: EBIT divided by total assets, where EBIT refers to earnings before interest and taxes
- Return on equity (ROE)
- Growth rate, in both sales and profits
- Sustainable growth rate, as an indication of whether the company's long-term growth will require additional capital to be raised externally
- Potential equity dilution from several source: the exercise of stock options and warrants, the conversion of outstanding convertible securities, and newly issued common shares sold to the public or used to acquire other companies

The matter of risk is harder to get a handle on. Remember that risk is neither desirable nor undesirable. Often the flip side of risk is opportunity.

Some investors seek risky investments in hopes of capturing the "upside," while other investors eschew risk for fear of being hurt by the "downside." The company's financial statements can tell the investor something about the company's financial riskiness (look at debt leverage ratios) but not much about such operational risks as technology changes, competitive threats, marketing and pricing risks, and employee turnover.

Suppose the company under analysis is not yet earning revenue and/or not yet profitable, a frequent condition of start-up biotech and e-commerce companies, for example. For these companies, the investor must focus primarily on information not revealed in the financial statements. Nevertheless, the financials reveal some information regarding (1) **burn rates** (that is, the rate at which the company is chewing up cash received from creditors or shareholders); (2) the rate of growth of expenses, particularly on R&D and marketing; and (3) the rate of revenue growth and the gross margin earned.

Of course, the present or potential investor also needs to heed what the market for the company's shares is telling him or her. Some key ratios for the investor to pay attention to:

- Price/earnings ratio (remember, the absolute per-share price is irrelevant)
- Yield (dividends divided by market price)

A conservative investor, one more interested in preservation, than in appreciation, of capital, is content with lower-growth, lower-risk companies, and probably is interested in higher yields, that is, current cash returns. An aggressive investor, in contrast, seeks out growth companies; is willing to accept higher technology, competitive, or financial risk; and is generally content with little or no yield.

Short-Term Investor or Speculator

The common stock speculator isn't much concerned about the *fundamentals* of companies—all those issues just discussed—except to the extent that changes in them may signal buying or selling pressure in the market for the stock.

The ultimate stock speculator is the *day trader,* who was much in evidence in the late 1990s. The speculator is matching his or her wits and expectations against those of both investors and other speculators trading in the company's shares. Remember that stock prices go up when the demand to buy shares exceeds the supply of shares available for sale; the reverse holds true for falling prices. If the speculator's confidence that the share price will increase exceeds the prevailing view in the marketplace, the speculator should buy. If the reverse is the case, the speculator should sell. If this sounds a bit like gambling, it is!

Note that the speculator doesn't have to own the shares in order to sell them; he or she can **sell short:** borrow the shares (from another investor, say an endowment fund) for a small fee, sell them, and expect to "repay" the borrowing by buying shares back later at a lower price. Selling short is risky business, best left to hedge funds and speculators.

Employee, Present or Prospective

The viewpoint of the employee, or prospective employee, is much the same as that of the long-term shareholder—concern for stability, growth, profitability, and risk—but with much less concern for the machinations of the market where the common shares trade. Employees invest their careers in the company, just as shareholders invest their financial resources.

Income Tax Collector

The income tax collector doesn't give a darn about the financial position or performance of the company. The tax collector is interested only in assuring that the company pays its fair share of taxes, in accordance with current tax law. Thus, the tax collector looks for inappropriate deferrals of the recognition of revenue, or inappropriate acceleration of expenses—both of which result in understating the company's current profits and thus its current income tax liability. A *private company*—one that doesn't have to report to public shareholders—may be particularly inclined to understate taxable profits.

"COOKING THE BOOKS"

I hope by now you realize that, while the "recording, classifying, and summarizing" part of accounting may be a science (or at least a straightforward process), "observing and valuing" is very much an art. A good deal of discretion can and must be exercised to decide just what needs to be observed and, once observed, how it should be valued.

Given human failings, some accountants simply cheat, producing fraudulent financial statements. Every company should have in place good office procedures to facilitate early detection of embezzlement or other fraudulent activities. External and internal auditors are always on the lookout for fraud, though some illegal activities, particularly if undertaken or sanctioned by top management, are extremely difficult to uncover.

However, "cooking the books"—that is, dressing up financial reporting—need not be fraudulent nor necessarily wrong. Accordingly, you should be generally familiar with—and therefore appropriately skeptical of—the common methods used to "cook" or "dress" the financial statements. That is the purpose of the following discussion.

Some accountants and their management colleagues are aggressive: they seek out opportunities to account in ways that will increase reported profits, or, the same thing, accelerate in time the reporting of profits. They also look for ways to improve the appearance of the balance sheet. Other accountants and their management colleagues are conservative: they are eager to avoid overstating profits and to provide full balance sheet disclosure to their audiences. And, of course, a whole spectrum of accountant behavior exists between being so aggressive that you stay just one step short of illegal activity, and being so conservative that you are just one step short of willfully understating the company's profits and its balance sheet position.

Recall the accounting rules outlined in Chapter 6. While providing useful guidance, they still leave the accountant with a good deal of discretion. For example, what is *material* to one person may be immaterial to another. Indeed, one way to "cook the books" is to set a high materiality threshold. One accountant may feel, for instance, that inventory that has not turned over for three months should be written off as obsolete, while another may want a time threshold of a year, not three months. One accountant, reviewing legal proceedings against the company, may decide that no liability for possible adverse judgments should be recorded until court action is complete; another may feel it prudent to establish early-on a reserve for adverse legal judgments.

Let's consider five categories of aggressive accounting behavior. Each of the examples discussed in these five categories has led recently to one or more companies having to restate their financial statements—typically decreasing reported profits.

Overstate Revenue or Accelerate Its Recognition

Sell to Related Entities. To be recorded as revenue, a sale must be to a party that is at "arm's length" from the seller. What constitutes "arm's length"? How "unrelated" must the purchaser be? Clearly, if the purchaser is a subsidiary of the seller, the two are very much related; exchange of goods or services between them does not constitute a bona fide sale. Now, suppose the seller owns 5 percent of the purchaser and the managements of the two companies are entirely separate. Here the seller is almost surely at arm's length from the buyer.

"Stuff" Channels of Distribution or Ship in Anticipation of Orders.
Channel stuffing is, I'm afraid, quite prevalent. In an effort to increase revenue and profits in a particular accounting period, the selling company ships excessive quantities of product to its distributors and retailers, generally providing extended credit terms on these extra purchases. Assuming the shipments were made in response to valid purchase orders, this channel stuffing is not illegal. And, this action does accelerate, into the current accounting period, revenues (and thus profits) that would otherwise be recognized in a future accounting period. Of course, it is a short—but illegal—step from shipping product against valid purchase orders to shipping against anticipated orders not yet received. And from there to the clearly fraudulent practice of shipping to just the company's own warehouse rather than to a customer.

Account for Installment Sales at Low Interest Rates. Recall earlier that revenues associated with installment sales or so-called finance leases are properly recognized at the time of shipment as the time-adjusted value of the stream of future payments to which the purchaser is obligated. What interest rate should be used to calculate the time-adjusted value? The rate should appropriately reflect both current conditions in the debt markets and the customer's credit rating. At times, this rate should be quite high, resulting in correspondingly low revenue. The lower the *discount rate,* the higher the revenue that the seller can recognize currently. Over time, of course, the seller is going to receive cash payments totaling neither more nor less than the aggregate of those specified in the agreement. However, discounting at a low interest rate accelerates into the current period a larger portion of those aggregate payments.

Adjust Pension Plan Reserves. From time to time, particularly following a run-up in the stock market, a company finds that its employee pension plan(s) is overfunded: the asset value of the fund is more than sufficient to meet pension payout obligations. How should this overfunding be accounted for? One possibility is to take no action, assuming that the value of the assets of the pension plans will fluctuate as securities markets fluctuate, and thus the overfunding may be only temporary. That's the conservative approach. Some companies, however, will remove assets from their pension funds, adding them to current profit. If one is searching for ways to improve today's reported profits, such action is very tempting—but it is certainly aggressive! Permitting pension plans to remain underfunded is also aggressive.

Treat Nonrecurring Dispositions as Ordinary Income. When fixed assets, product lines, or small divisions are sold, any realized gain over book value is treated as "other income" and appears on the income statement below the operating profit line. Such **below-the-line gain** is assumed to be nonrecurring, and thus may be viewed by readers as largely irrelevant to an analysis of how well the company is performing in its normal business operations. Now, suppose such sales are relatively small and/or occur quite often. The temptation will be to view them as recurring rather than nonrecurring and thus to record them "above the line," in turn "dressing up" the company's income statement.

Record Income for Future Services or Returns. The price of many products includes (is *bundled with*, into a single price) the fee for future services: warranty repairs, periodic maintenance, future updates, and so forth. The temptation here is to underestimate the requirement for such future servicing that will be required, thus justifying taking into current income—and hence boosting current profits—a disproportionately large portion of the **bundled price.**

Another example: Franchisers of certain fast food and similar operations charge their franchisees a substantial startup fee. What portion of that fee should be recognized as current income, and what portion should be recognized over future accounting periods as the franchiser services the franchisee?

And another example: How much of a magazine subscription price received by the publisher should be recognized currently (matched against the marketing costs associated with obtaining the subscription), and how much should be deferred and spread over the subscription's duration (matched against the cost of "fulfillment," as it is called in the publishing world)?

And yet another: Suppose a publisher of books or computer software launches a new product that is expected to have high demand but also high customer returns. The publisher may be tempted to record revenue currently for every book or software package shipped. This accounting is pretty aggressive. A more conservative approach would be to reduce current revenue to create a sales return reserve reflective of the publisher's best estimate.

Understate Expenses or Delay Their Recognition

Amortize or Depreciate Over Unrealistically Long Lives. Patents and fixed assets have limited lives, and investments in them must be spread over those lives through the process of amortization (for intangibles such as patents) and depreciation (for fixed assets). The U.S. Internal Revenue Service defines, for the purpose of calculating income tax expenses, minimum depreciable lives by class of asset; tax accountants, who typically seek to minimize tax liabilities, will adopt these guidelines. But for public reporting purposes, accountants are free to ignore these guidelines and choose useful lives that are longer (more aggressive) or shorter (more conservative).

A strong argument can be made for the conservative approach of using accelerated depreciation, rather than straight-line depreciation, for fixed assets. A fixed asset, particularly a computer, machine tool, or other technology-based device, is typically of most value to its owner early in its life. As years pass, the device obsolesces, requires more repairs, and thus its value declines. The best expense matching, then, may suggest higher depreciation expense early in the asset's life.

Capitalize Questionable Expenses. Throughout this book, we have encountered examples where a choice must be made between capitalizing an expenditure and expensing it. I hope it is now obvious that capitalizing is aggressive and expensing is conservative. Capitalizing the expenditure defers the expense to future periods (when it will be amortized or depreciated), thus accelerating the recognition of profit.

Here are some examples. Suppose a company engages in an extensive advertising and marketing effort before a product is released for sale; think, for example, of a high-budget movie, or a new auto model, or a piece of software to be released next year. Should these advertising expenditures be capitalized and then written off in future accounting periods when revenues are being earned from the new product? A good argument can be made for this approach, but the practice can be easily abused.

An excellent theoretical argument, again based on the matching concept, can be made for capitalizing product development expenditures and amortizing them over the product's future life. Nevertheless, this accounting practice is seldom followed. The arguments against the practice are several: Companies, in order to stay competitive, typically must continue or even increase their product development expenditures, and, thus, the effect on profits of capitalizing and then amortizing is minimal. Also, if a product development project fails, the company will have to take a large "hit" to profits in the year that it writes off the accumulated capitalized R&D.

In 2002, WorldCom, the second largest U.S. telecommunications and cable operator, was driven into bankruptcy as a result of having capitalized billions of dollars of expenditures that should have been expensed. While the cost of constructing new networks should, of course, be capitalized (a fixed asset), these WorldCom expenditures were apparently for network maintenance, and thus properly should have been treated as a current expense. While some gray area may exist between new construction and maintenance, many billion dollars is hard for anyone—inside or outside the company—to overlook. Perhaps this action was just aggressive; in the final analysis, it was declared fraudulent!

Among the "cooking the books" activities of some telecom companies in recent years has been this ploy: Two companies swap equal amounts of telephone capacity. Each treats the income from this swap as current income, but treats the capacity purchased as an asset to be written off against future revenue earned when the purchased capacity is used. Is such action fraudulent? Perhaps not, if it was thoroughly and carefully disclosed, but one would hope that an independent outside auditor would blow the whistle on such deception, if a company employee didn't do so first.

Ignore the Cost of Stock Options. Perhaps no accounting issue has been more hotly debated in recent years than the appropriate method to account for management incentive stock options at the time they are issued. In the past, stock options were assumed not to give rise to any expense for the corporation at the time they are issued, assuming they are issued with a strike price equal to or above the stock's market price. As these options become **in the money**—that is, the market price climbs above the strike price—the unexercised options begin to count (in a complex manner) in the denominator (number of outstanding shares) of the earnings-per-share ratio.

Options, when first granted to one of the company's managers, must have no value if they are to give rise to no expense for the issuing corporation. Is this reasonable? Certainly the managers to whom the options are granted think the grant has value, as they are keen to receive it. If the option recipients are better off with the grant than without it, then surely there must be some cost to the issuing corporation. Something of value has changed hands.

Senior executives in high-tech companies argue that stock options have been the "magic" ingredient in the United States over the past several decades for achieving rapid technology development, new company formation, and job creation. They go on to argue that to require issuing corporations to expense options at the time of granting them will eliminate this magic. Even if this argument were valid (a questionable assumption), the argument is an economic one, not an accounting one.

But how does one calculate the value of an option at the time of the grant, when it is not "in the money"? Good question. Valuation methods are available, but they are esoteric, require several assumptions, and are not easily understood by the layperson. Still, that fact does not argue for the typical—and very aggressive accounting approach—of assuming that options are costless.

Incur Nonrecurring Expenses Repeatedly. I argued earlier that below-the-line expenses—those treated as nonoperating—are less painful to company managers because they do not affect operating margins and because they are labeled nonrecurring. In recent years, some companies have recorded so-called nonrecurring expenses nearly every year! One has to question just how "nonrecurring" these expenses are.

Various actions give rise to these below-the-line expenses, for example:

- Discontinuing a product line may give rise to the lump sum write-off of associated inventory.
- Laying off a sizable percentage of employees, perhaps during an economic recession, can give rise to substantial, one-time severance expenditures.
- Merging with another company may give rise to the one-time write-off of redundant assets and any other expenses (including severance payments to redundant employees) associated with effecting the merger.

Treating these as nonrecurring expenses may be quite legitimate, but the temptation is to sweep extra expenses into the nonrecurring category; after all, if earnings are going to take the hit of a large nonrecurring charge, the hit might as well be big. Security analysts may be persuaded to ignore these nonrecurring expenses as they assess the company's future prospects.

And, the company that *over* allows for nonrecurring expenses this year can in future years reverse the action, thus boosting operating earnings in those future years. This somewhat insidious practice of recognizing excess nonrecurring expenses in one year and then reversing that action in above-the-line income in future years is anything but conservative—but, I regret to say, this maneuver has been all too common in recent years.

Delay the Accrual of Certain Expenses. The temptation exists to delay the accrual of certain expenses—and the corresponding creation of associated reserves or liabilities—when the cash outflows will occur long into the future. A couple of examples are pension expenses, medical expenses of retired employees, and pollution cleanup expenditures. From time to time, companies have found that their allowances for future pensions are woefully inadequate; this condition is typically described as *underfunded pensions.* As medical costs have accelerated in recent years and the U.S. Medicare program proves inadequate to cover these expenses, a company's long-term obligation to retirees can become very substantial. How accurate is today's balance sheet if these contingent, long-term liabilities continue to be unaccounted-for liabilities that overhang the company?

Overstate Assets or Understate Liabilities

You surely recognize that all of the examples of overstating revenues and understating expenses just discussed also have impacts on the balance sheet: over- or understating assets or liabilities. But here are a few other examples of dressing up the balance sheet.

Delay Recognizing Declining Asset Values. Every time an accountant develops a balance sheet, he or she is charged with the responsibility of assessing the value of all assets, and adjusting them appropriately. Conservatism requires the accountant to lean in the direction of understating, rather than overstating, the following assets:

- Accounts Receivable—by providing for a fully adequate Allowance for Doubtful Accounts. If the company's sales are relatively infrequent but very large (as with commercial aircraft, for example), the probabilistic approach to determining bad-debt allowances may be inappropriate.
- Loans Receivable—by reassessing frequently the creditworthiness of the borrower. Beginning in the 1990s and continuing into the new century, Japanese banks refused to recognize the volume of bad loans (euphemistically called "nonperforming loans") carried on their books. Had they done so, many of them would have had to report substantial negative owners' equity.
- Inventory—by the diligent and timely recognition of obsolete inventory, or by the creation of an allowance for obsolete inventory.
- Fixed Assets—by writing off, to zero, machinery, equipment, instruments, facilities, and leasehold improvements no longer being actively used by the company.

- Investments—by **marking to market** (reflecting current market values) when valuing securities owned by the company as passive investments. The Japanese banks also resisted this action as the Japanese stock market plummeted.

Fail to Disclose Encumbrances on Assets

Companies occasionally pledge a portion of their cash or other assets to secure an obligation of the company itself or perhaps of an affiliated company. Failure to disclose this pledge results, in effect, in overstating the liquidity of the company. Readers of financial statements are easily misled unless such asset encumbrances are clearly disclosed.

Fail to Disclose All Liabilities

Judgment is always required to decide what contingent liabilities should be valued and recorded on the balance sheet. Valuation is often very subjective, but not impossible. For example:

- Every large company must defend itself against a steady stream of lawsuits, some of merit, many not. While pending lawsuits must be discussed in the footnotes to the financial statement, when does the probability of an adverse judgment become great enough that the company should record a specific liability? Consider tobacco and pharmaceutical companies.
- Mentioned earlier were pension liabilities; obligations for retiree medical benefits; and toxic cleanup obligations, which in some instances can be enormous and difficult to value.
- As executive pay packages have ballooned in recent years, some provide that the company will provide very substantial deferred compensation and related benefits to top executives following their retirement. Conservative accounting would suggest that these growing liabilities be accumulated and reflected on the balance sheet during the period when the executive is actively working.

Use Unconsolidated Debt. Balance sheets can be dressed up by off-loading debt (or other obligations) to affiliated companies whose balance sheets are not consolidated with the parent company. For example, the major auto companies in this country have substantial financing subsidiaries that lend or provide lease financing to car buyers. These subsidiaries are, appropriately, very highly debt-leveraged. Typically, the financial statements of the finance subsidiaries, because they are in a distinctly different business from the parent company, are not consolidated with the parent, and thus these large debt obligations do not appear on the reported balance sheet of the auto manufacturer. The condition of the finance subsidiary must, however, be disclosed in the footnotes to the financial statements.

The now-famous collapse of Enron Corporation in late 2001 was occasioned by disclosure that Enron was benefiting from loans made to corporations that were purported to be at arm's length from Enron—"special-purpose entities." Enron did not disclose these loans, either on its balance sheet or in the accompanying notes. Upon investigation, it turned out that Enron, in fact, wholly controlled these special-purpose entities, and thus their debts were, in truth, the obligations of Enron itself. This revelation soon led to Enron's bankruptcy. Although in the Enron case, fraud was involved, special-purpose entities can be legitimately used under certain circumstances; nevertheless, their use can hardly be considered conservative!

Adjust Reserves to Meet Earnings Estimates. In recent years, the market price of a company's common stock has been punished sharply when a company's earnings per share for a particular financial period fails, even by one or two cents, to reach the level that Wall Street expects. But, perhaps surprisingly, the stock price is not moved upward by much if earnings estimates are exceeded by a penny or two. For most companies, quarterly EPS can be shifted by one or two cents by simply adding to, or reducing, one or more reserves (adjustments to asset or liability accounts such as Reserve for Doubtful Accounts, Reserve for Inventory Obsolescence, Accrued Pension Liability, Reserve for Lawsuit Settlement, Deferred Taxes, and so forth). These adjustments are, of course, also reflected on the income statement and thus in earnings per share.

Remember, all these reserves are estimates, and evidence can always be found to justify some alteration in the estimates. Accordingly, financial officers often "squirrel away" (by increasing expenses) a few pennies of EPS in a quarter that comes in over the estimate. These few pennies are available, then, in future quarters if actual results come in slightly below the then-prevailing estimate. To my amazement, many analysts and investors fail to understand this flexibility that is available to corporate financial officers and therefore continue to attribute unwarranted importance to having a company meet the EPS projections of "the street.".

Some Final Thoughts

By now you have surely concluded that no "bright line" exists between appropriate accounting for a particular event or condition and inappropriate (even if not illegal) accounting, that is, between aggressive and conservative accounting treatments. The line is a great deal brighter between aggressive accounting and fraudulent accounting, that is, go-to-jail accounting. The decision as to just how aggressive or conservative the company's financial accounting should be is for senior executives and boards of directors to make.

They abrogate their duty if they leave the decision solely to the senior accounting personnel. When excessively aggressive or fraudulent accounting is exposed, the lament by senior executives or directors that "we didn't know" is no defense in either a court of law or the court of public opinion; they have the responsibility to know.

Consensus regarding where on the conservative-aggressive spectrum a company should operate can be difficult to reach. Too frequently the chief executive puts heavy pressure on the financial officers to "make the numbers": achieve a certain earnings level in a fiscal quarter or year, perhaps an earnings level to which the chief executive has publicly committed. A financial officer often needs a backbone of steel—and perhaps an updated resume!—to stand his or her ground under such pressure. But standing one's ground is wholly worthwhile, and not just to avoid spending time in jail. A manager's integrity—and certainly a financial manager's integrity—is his or her most important asset.

Moreover, when a company gives in to the temptation to do a little "cooking the books" in order to report favorable earnings in a particular fiscal period, this is very frequently just the first step down a slippery slope: the pressure often becomes even greater to "fudge" earnings for subsequent periods. With each period, the "cooking" becomes more unreasonable until, all too soon, "fudge" turns into fraud. This situation is analogous to the bookkeeper who "borrows" a few dollars from the petty cash box to bet at the racetrack; then, when his horses don't win, he takes a few more dollars with full intent to repay all he has taken from his future winnings—winnings that somehow never materialize. "Borrowing" soon becomes embezzling.

Please note that the purpose of this chapter is not to teach you how—and certainly not to encourage you—to cook the books! Rather, I hope it alerts you to the kinds of "cooking" you should beware of—both as an operating manager and as an investor. It also should remind you again that accounting is not a precise science; judgment or "art" is also required, and over the past decade we have encountered some "art" that certainly is not pretty.

SUMMARY

This chapter recognizes explicitly that (1) different audiences of financial statements have different financial requirements, depending on the particular decisions they must make; and (2) financial statements can be "dressed up"—appropriately, deceptively, or fraudulently—in many different ways.

The audiences and their interests can be categorized as follows:

- *Short-Term Creditor:* the liquidity of the borrower
- *Long-Term Creditor:* in addition to liquidity, asset utilization and debt leverage of the borrower

- *Long-Term Investor:* the above plus profitability, market-based ratios, risk profiles, growth rates, and overhanging equity dilution
- *Speculative Investor:* none of the above; rather, short-term changes in equity market dynamics
- *Employee, Present or Prospective:* interests that generally parallel those of the long-term investor
- *Income Tax Collector:* none of the above; solely the company's compliance with applicable tax laws

Financial statements can be "cooked"—made to appear more attractive than perhaps they should be—by one or more of the following:

- Overstating revenue, or accelerating its recognition
- Understating expenses, or delaying their recognition
- Overstating assets, or understating liabilities
- Failing to disclose encumbrances on assets
- Failing to disclose all liabilities

Readers of financial statements must strive both to appreciate the useful information the statements reveal and to remain skeptical of ways the financials may be misleading.

NEW TERMS

Below-the-line gain (loss). A revenue (or expense) considered nonoperating and, thus, recorded on the income statement "below" operating income (as a nonoperating revenue or expense).

Bundled price. A price that entitles the purchaser to more than a single product or service, often a product *and* a related service.

Burn rate. The annual expenditure of invested capital (equity and loans) that startup companies incur before they are able to generate revenues and operating profits.

In the money. Refers to the condition of an option or warrant when the market value of the underlying security or securities is above its strike-price value.

Marking-to-market. Valuing an asset (typically a security) at its current market price rather than at its acquisition cost.

Selling short. Selling borrowed (nonowned) equity or debit securities into the market with the expectation that the seller can reacquire them at a later date at lower cost and return them to the original owner.

EXERCISES

1. Indicate whether each of the following statements is true or false.
   a. A short-term lender can focus solely on liquidity measures and ignore other ratios in judging the creditworthiness of a prospective borrower.
   b. A prospective employee has a perspective similar to a stockholder in evaluating whether to join the company.
   c. A common stock speculator (such as a day trader) has little incentive to evaluate the financial statements of a company whose shares he or she is considering buying or selling.
   d. An income tax auditor need pay little attention to balance sheet ratios.
   e. A common stock investor is concerned about past equity dilution but need pay little attention to probable future dilution.
   f. A low price/earnings ratio signals a buying opportunity for potential investors in the common stock of the company.
   g. The burn rate indicates the rate at which a new company is consuming capital supplied by its early investors.
   h. Both the accounts receivable collection period and the accounts payable payment period are ratios closely related to measuring liquidity.

2. Why should prospective common stock investors be at all concerned with short-term issues such as the company's liquidity?

3. If your company were considering providing substantial trade credit to a new customer, on what would you focus your analysis of this customer's financial statements?

4. If your company were considering signing a long-term supply contract with a particular company, on what would you focus your analysis of this supplier's financial statements?

5. Your company purchases a new generator system and relegates the old system to standby status. What accounting entries, if any, should you make with respect to the old generator system? Discuss.

6. On what matters does the common stock speculator focus particular attention? Why do companies' published financial statements shed little light on these matters?

7. Suppose a senior executive's five-year employment contract provides for a $2 million bonus if the senior executive successfully completes her employment obligation. When should this $2 million be recorded as an expense?

8. Suppose Mansfield Corporation anticipates receiving a major purchase order from customer X right at the end of Mansfield's fiscal year. The merchandise being ordered is ready to ship, and its shipment would have

a significant positive effect on Mansfield's financial results for the year. Unfortunately, the purchase order does not arrive by the last day of the fiscal year. Under what circumstances, if any, do you think Mansfield would be justified in making the shipment and recording the associated revenue within the current fiscal year?

9. Suppose the Kid Company, a manufacturer of toys, urges its distributors just prior to year-end holidays to load up on inventory of a just-introduced toy that Kid expects to be very popular. To assist the distributors, Kid agrees to extended payment terms (payment due in 90 days rather than the customary 30 days) and to unlimited merchandise return privileges. Nor surprisingly, distributors accommodate Kid with very large orders for the new toy. What issues does Kid face in deciding how to account for the revenue from these shipments?

10. The Leonard Company borrows heavily from its bank and pledges its accounts receivable to the bank as collateral for the borrowing. The bank agrees to relieve Leonard of the pledge of receivables from December 20 to January 10 so that the pledge need not be disclosed on Leonard's year-end financial statements. If you were Leonard's auditor, would you insist on disclosure of this arrangement?

11. Company X, a distributor, is wholly owned by Ms. Patton, and company Y, a retailer, is wholly owned by her sister. Might one reasonably argue that when X ships merchandise to Y, X should recognize revenue only after Y sells the merchandise to the end customer? Discuss.

12. The Internal Revenue Service assigns a guideline depreciation life of five years to a certain machine tool. The Waterman Company decides to depreciate this tool over eight years for public reporting purposes and five years for tax reporting. Is this action appropriate? Is it conservative? Discuss.

13. The Travers Company acquires, for $55,000, a patent with a remaining life of 14 years. What arguments might be made in favor of amortizing the patent over a period of substantially less than 14 years?

14. The Burger Buster Company franchises outlets throughout the Midwest, charging each franchisee a startup fee of $100,000. How would you decide the portion of the fee that should be recognized as revenue at the time the franchise agreement is signed and the portion to be deferred to future years?

15. The Rosso Company decides to discontinue operating one of its small plants. Rosso estimates that severance obligations, moving costs, and expenses associated with subleasing the plant will total about $2.2 million. Rosso's chief accountant suggests that, to be conservative, Rosso should record a nonrecurring, nonoperating expense of $2.4 million this year just

in case the closing costs exceed the estimate. If you were Rosso's auditor, would you agree with the chief accountant?

16. The Smith Company incurs an expenditure of $400,000 to reinsulate and repaint its facilities. Would you treat this expenditure as an expense of the current period, or would you capitalize it? Discuss.

17. You have, for three years, held a stock option that permits you to purchase 1,000 shares of Willets Corporation stock at $5 per share over a five-year period. Two years remain on this option period, and Willets stock now trades at $8.12 per share. Would you exercise this option now? Explain your reasoning.

18. Your company offers its customers a 30-day period during which they can return merchandise for full credit. What argument might you make in favor of recognizing revenue from sales at the time of shipment, rather than 30 days later when the return period has expired?

19. The Preston Company manufactures champagne corks that it sells to wineries. As you know, these corks are under substantial pressure, and occasionally they cause eye injuries during removal from bottles. Should Preston recognize the potential liability of customers' legal suits against it? If so, how, when, and in what amount?

20. Security analysts who follow your stock estimate that your company will earn 73 cents per share in the quarter that is about to end. This estimate has been circulated widely among brokerage firms. At the close of the quarter, your chief financial officer tells you that actual earnings will probably be 71 cents per share, but, by reducing the company's reserve for inventory obsolescence by half, an additional 2 cents per share of earnings could be reported. What would you do?

CHAPTER 12

FINANCIAL MARKETS AND CAPITAL INVESTMENT DECISIONS

A key set of decisions for company executives and their boards of directors involves capital investment projects: whether and how much to invest in machinery, instruments, new facilities, research and development projects, the acquisition of other companies or products, and other long-term projects or assets. Most such investments are motivated by an opportunity that extends over several years: the opportunity either to capture a new cash inflow stream or to reduce future expenditures and thus avoid a cash outflow stream.

Not all capital investment opportunities, of course, provide sufficient cash inflow (or, the general equivalent, avoid sufficient cash outflow) to justify the expenditure of the required funds. Such investments should be avoided. The challenge is to separate the profitable investments from the unprofitable ones.

Still other investments—for example, those required for safety or to meet environmental regulations—are really not discretionary and must be pursued regardless of the cash flow consequences. If, in the absence of such an investment, regulators close the manufacturing facility, the potential loss of cash flow is so great as to make obvious that the investment must be undertaken.

CONCEPT OF A HURDLE RATE

We'll focus here on discretionary investments: opportunities for new cash inflows or cost-saving opportunities to avoid cash outflows. Capital devoted to such investments should earn a rate of return that exceeds the cost of that capital to the investing company. If the return on the investment falls below that threshold, the investment will dilute the company's overall returns, and the shareholders will suffer. These statements imply that investment opportunities should be tested against some **minimum required rate of return,** or, as it is often called, the **"hurdle rate."** This simple proposition is the essence of capital investment decision making, often called **capital budgeting.**

That quick summary leaves us with three questions: (1) How do we determine a company's cost of capital? (2) How do we calculate the investment's expected rate of return? (3) Are there situations where the relevant hurdle rate is different from the weighted-average cost of capital? We'll deal with these questions in order.

WEIGHTED-AVERAGE COST OF CAPITAL

Chapter 10 introduced the concept of weighted-average cost of capital (WACC). Once a company's optimum capital structure is determined, each source of capital within the structure can be costed. The cost of borrowing is straightforward: the interest rate i. The borrower pays the lender interest in return for the use of the funds during the term of the loan. Because interest payments are tax deductible, the after-tax cost of borrowing is $i \times (1 - \text{tax rate})$.

The cost of equity capital is anything but straightforward. As outlined in Chapter 10, the cost of equity is a function of risk and expected returns. Think of the cost as the fraction of ownership in the company that must be offered to the investor to induce the investor to purchase common stock. A detailed explanation of the methods to "cost" equity is beyond the scope of this book. But, two approaches are worth a quick mention:

1. *Dividend Growth:* $K_e = d/p + g$, where K_e is the cost of equity, d/p is the stock's yield (d = dividend per share, and p = price per share), and g is the growth rate of the company (and therefore, presumably, the rate of growth in dividends). Note the use of dividends, not earnings, in the equation. The dividend is the cash return to the owner of common stock; the difference between earnings and dividends is reinvested in the company and fuels the final term in the equation, g (growth).

2. *Capital Market Line:* The capital market line was shown in Figure 10-1. Recall that the horizontal axis is a measure of risk, and the vertical axis a measure of return. The intersection of the capital market line on the vertical axis is the risk-free rate of return, as discussed in Chapter 10. The line says, quite simply, the higher the perceived risk of the investment, the higher the return (in whatever form—interest, dividends, price appreciation) required to induce the investor to provide his or her funds to the company.

To repeat an admonition from Chapter 10, a common error in "costing" is to assume that the cost of equity is E//P, the inverse of the price/earnings ratio. This equation leads to a nonsense cost of equity: a company with low earnings (high P/E ratio) would have a low cost of equity, and a company with no earnings would have a zero cost of equity. To the contrary, a very high P/E ratio implies, rather than a low cost of capital, that investors expect the company to generate very high future returns; thus, the "cost" to the other shareholders of giving up fractional ownership is high, not low as implied by the inverse of the P/E ratio.

Calculation of the WACC for the Martin Company is shown in Table 12-1. The WACC depends, of course, on the capital structure and lies somewhere between the after-tax cost of debt and the cost of equity (for which pre- and after-tax costs are the same, since dividends are not tax deductible). Had the company decided on a more leveraged capital structure—say, an equal mix of debt and equity—its WACC would decrease, in this case to about 10.6 percent. However, such a shift to a riskier capital structure (a higher proportion of debt) might well increase the interest rate that lenders would demand and/or decrease the P/E ratio acceptable to investors. Either of these shifts will increase the value of the WACC.

CALCULATING RETURNS

Now consider how to calculate the return earned by the investment. Two methods are popular: net present value and internal rate of return. Both are examined in the discussion that follows.

TABLE 12-1. The Martin Company: Weighted Average Cost of Capital

Components of Capital Structure	Market Value[a] of Components	Pre-tax Cost[b]	After-Tax Cost[c]	Weighted Average
Debt	$30 million (30%)	8%	5.2%	1.56%
Equity	$70 million (70%)	16%	16.0%	11.20
	$100 million	After-tax WACC =		12.76%

[a] Note the use of market values rather than book values:

- *Debt:* today's market value, not the value at issuance. The market value can be higher or lower than issuance value, depending on whether prevailing interest rates are lower or higher than they were on issuance date and on whether the company's risk profile has shifted during this interval.
- *Equity (Common Stock):* the number of common shares outstanding × today's market price per share.

[b] An estimate of today's cost.

[c] Assumes a 35% income tax rate.

For the moment, let's assume we know that the minimum required rate of return for a particular investment is 15 percent. That is, in order to be judged a profitable investment that the company should undertake, the investment must produce sufficient cash flows over the life of the investment to return to the company the value of the original invested funds plus 15 percent per year on the portion of the investment not yet recovered. You might think of the company "lending" capital to the investment and requiring both a 15 percent "interest" return on the investment and the payback of the principal (or capital) invested.

This calculation requires that the analyst estimate the cash flows year by year over the useful life of the investment. Remember that avoiding cash *out*flows is, for these purposes, equivalent to generating cash *in*flows. Note, also, that we are focused on cash flows, not earnings flows. As emphasized in Chapter 5, cash is the lifeblood of a business. The business must have cash to pay salaries and accounts payable, to repay debt or make dividend payments to shareholders, and to make new capital investments.

Net Present Value (Discounted Cash Flow)

Recall the discussion in Chapter 2 of the time-adjusted method to value loans, common stock securities, and other assets. In the capital budgeting context, this same method is referred to as the **discounted cash flow** or **net present value** method. Future cash flows are "discounted"—that is, time adjusted— to determine their equivalent value today, their net present value. Tables 2A-1 and 2A-2 in Appendix 2A provide the factors that we need to make this time adjustment, that is, to discount to the present.

Using the Interest Tables

Calculations using these interest factors are clarified by a few symbols:

$$P = \text{present value}$$
$$F = \text{future value}$$
$$A = \text{annual cash flow}$$
$$i = \text{discount, or hurdle, rate}$$
$$n = \text{number of years}$$

The columns and rows in Table 2A-1 are i and n: the factors permit you to calculate the present value equivalent P of a given future value F occurring n years in the future at a given interest rate i. Or, if you know P, F, and n (for example, for a real estate investment of P that you expect to sell for F in n years), you can determine (by trial and error using Table 2A-1) the inherent value of i. Or, if you know P, i (your required rate of return), and n

(the number of years you plan to hold this real estate investment), you can calculate the minimum price, F, you must receive upon resale in order to make this a good investment for you.

The factors in Table 2A-2 permit you to calculate the present value P over n years of uniform annual cash flows A discounted at interest rate i. Suppose you have today a savings account of P, earning at the rate of i. With Table 2A-2, you can calculate how much you can withdraw each year for n years and fully deplete the account at the end of n years. Or, if you know P, A, and n for an investment, you can determine (by trial and error) its inherent value of i.

A few simple examples will illustrate this:

1. You make a loan to a friend; she promises to repay the loan at the rate of $500 per year for five years. If your personal hurdle rate (required rate of return) is 8 percent, what is the minimum amount you would be willing to accept now in full repayment of your loan? The equation is:

$$P/A(i = 8\%, n = 5 \text{ yrs}) \times \$500 = 3.993 \times \$500 = \$1,996.50$$

2. Now turn example 1 on its head: calculate the annual installment you will be required to pay if you negotiate a 25-year, $100,000 mortgage from a lender charging 7 percent interest. You know P, i, and n and must calculate A:

$$A \times P/A(i = 7\%, n = 25 \text{ yrs}) = \$100,000$$

$$A = \frac{\$100,000}{11.654} = \$8,581$$

3. How much will you need to invest today at 10 percent annual earnings rate in order to have $120,000 in 15 years?

$$\$120,000 \times P/F(i = 10\%, n = 15 \text{ yrs}) = P$$

$$P = 120,000 \times (0.239)$$

$$= \$28,680$$

4. If you are a venture capitalist investing in young, high-potential, high-risk companies, you search for investment opportunities that can earn high returns commensurate with the high risk you are accepting, probably a compound rate of 25 percent or more. If you invest $1 million now in the Texit Company and expect that company to be acquired in five years, what will you need to receive from the acquiring company

in order for your investment in Texit to achieve your target rate of return?

$$\$1,000,000 = P/F(i = 25\%, n = 5 \text{ yrs}) \times F$$

$$F = \frac{\$1,000,000}{0.328} = \$3,048,780$$

From this calculation, it's clear that earning three times on your money in five years is about equivalent to achieving a rate of return of 25 percent per year.

An Industrial Example. Now for a slightly more complex example. Suppose a company is evaluating whether to purchase a fully automated packaging line to replace its semiautomated line. The new line costs $450,000 but is expected to produce annual cost savings of $80,000 over its 10-year useful life and to have a salvage value at the end of its 10-year life of $40,000. A widely used but inappropriate measure of investment attractiveness is the so-called **payback period.** Here the annual cash flows of $80,000 "repay" the original investment, ignoring interest, in 5.6 years. But this 5.6-year payback takes no account of either the overall life of the packaging line or its salvage value. In fact, here the salvage value is small and occurs a long time (10 years) in the future; accordingly, it has little effect on the attractiveness of the investment. A 5.6-year payback might signal an excellent return on an investment that has a 15-year life, and a quite inadequate return on an investment that has only a seven-year life. Thus, we cannot ignore the life of the investment, its salvage value, or the investor's required rate of return.

If the company requires a rate of return on this type of cost-saving investment of 12 percent, should the company make the investment? That is, does this investment clear the "hurdle" of generating at least a 12 percent annualized return? For simplicity, assume that the $450,000 investment is made at time zero and that the $80,000 annual savings occur at the end of each of the next 10 years. The $40,000 salvage is another relevant cash inflow that occurs 10 years from now. We can diagram the cash flows as shown in Figure 12-1.

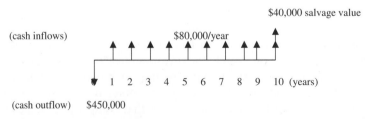

Figure 12-1. Purchasing a new packaging line: cash flow diagram.

Table 12-2 shows the calculation of the net present value of these cash flows at an interest rate of 12 percent. Since this investment has uniform $80,000 annual cash flows for 10 years, we can use the factor P/A (for $i = 12\%$, $n = 10$ yrs) from Table 2A-2 to calculate their present value, and the factor P/F (for $i = 12\%$, $n = 10$ yrs) to calculate the present value of the salvage amount. Since the net present value of this investment is +$14,880, it comfortably clears the hurdle of providing a 12 percent annualized return. Note that the statement would be true even if the salvage value were ignored; because the n (10 years) is relatively long and the i (12%) relatively high, the present value of the salvage is only 32 percent of the amount received 10 years from now.

Internal Rate of Return

We now know that the rate of return projected for this new automated packaging line exceeds 12 percent. By how much? That is, what is its actual return percentage—more precisely, what is its **internal rate of return?** Most financial calculators permit the quick calculation of the internal rate of return. Table 12-3 calculates net present values of the packaging line investment at both 12 and 15 percent and then interpolates to find the approximate internal rate of return, in this case 12.8 percent.

Thus, both the net present value method and the internal rate of return method lead to a "go" decision on the investment. One caution, however: when a project's cash flows are complicated, with positive cash flows in some years and negative flows in others, the project can have multiple internal rates

TABLE 12-2. Present Value Analysis of Packaging Line

Year	Amount	Factor	Value
0	−450,000	1.0	−450,000
1	+80,000		
2	+80,000		
3	+80,000		
4	+80,000		
5	+80,000		
6	+80,000	$P/A(12\%, 10 \text{ yrs}) = 5.650$	+452,000
7	+80,000		
7	+80,000		
8	+80,000		
9	+80,000		
10	+80,000		
10 (salvage value)	40,000	$P/F (12\%, 10 \text{ yrs}) = 0.322$	+12,880
		Total =	+14,880

TABLE 12-3. Internal Rate of Return Analysis of Packaging Line

Time (years)	Amount	Factor	Present Value
		Present Value at 12%	
0	−$450,000	1.0	−450,000
1–10	+ 80,000	P/A (12%, 10 yrs) = 5.650	+452,000
10	+ 40,000	P/F (12%, 10 yrs) = 0.322	+ 12,880
		Total	$14,880
		Present Value at 15%	
0	−$450,000	1.0	−$450,000
1–10	+ 80,000	P/A (15%, 10 yrs) = 5.019	+ 401,526
10	+ 40,000	P/F (15%, 10 yrs) = 0.247	+ 9,880
		Total	−$ 38,600

$$\text{Interpolation} = 3\% \times \frac{14,880}{14,880 + 38,600} = 0.8\%$$

$$+12.0\%$$

$$\text{Rate of return} = 12.8\%$$

of return. In such cases, the net present value is more reliable (and simpler to calculate) than the internal rate of return.

Factors Requiring Estimation

Note carefully that the relevant input data for these calculations are *differences* in *future cash flows* over the *life* of the project. As emphasized above, we are interested in cash flows, not changes in reported earnings.* We are interested only in future, not past, cash flows. A decision today cannot change previous cash flows. For example, suppose that, in the earlier example, the semiauto-mated packaging line that the company is now considering replacing was installed just two years ago. Presumably, at that time the semiautomated line passed the test of merit—that is, projected cash flows resulted in a positive net present value. Viewed from today, that earlier investment is now a **sunk cost.** We cannot now change that earlier decision. If the new, fully automated

---

*This statement needs to be tempered by the realization that managers must, of course, be cog-nizant of impacts on reported earnings of very large investment projects. For example, the Boeing Company, an airplane manufacturer, will worry about effects on near-term reported earnings of a very large investment in a new airplane model, even if the long-term demand for the model yields cash flow estimates that result in a handsome projected return on that investment. Managers must be sure they have, or can get, the cash to sustain the investments in the new model, and they may need to warn shareholders and security analysts about the negative impact on near-term reported earnings associated with seizing market opportunities that require very large investments.

packaging line offers sufficient *incremental* (or differential) positive cash flows (compared with the semiautomated line) to achieve a positive net present value for the fully automated line, then investment in the new line is advantageous regardless of how recently the semiautomated line was installed.

Often, an investment under consideration does not involve the replacement of another asset. The relevant alternative to which the new investment should be compared is a "do nothing" alternative. Every investment has a default alternative, and that is "do nothing." Be aware, however, that the "do nothing" alternative may involve consequential cash flows that must be included in the analysis.

Take, for example, an investment in research and development to design and develop a new product, model P. The alternative is to not develop the new product. We can estimate the amount and timing of the cash outflows required to complete the development. We can also estimate the timing and amount of cash inflows from the sale of the new model P, as well as the corresponding incremental cash outflows associated with manufacturing and selling the new product, to arrive at the net cash inflows over the product's life. Note the emphasis on incremental cash flows; if model P is to be sold by the company's present sales force, the incremental selling costs may be minor. Similarly, if model P is to be manufactured in the existing plant with existing tools, incremental manufacturing costs may be considerably less than so-called full cost as reflected in the company's cost accounting records (more on this in future chapters.) You can appreciate that estimating the number of years that model P will be a viable product, and its sales volumes, particularly in the later years, is a formidable challenge. Actions by competitors and changing requirements of customers may be major influences on these estimates. And, perhaps, additional engineering development expenses will be incurred in the "out" years, as the company strives to maintain the competitive position of model P.

Now suppose model P is expected to cannibalize, to a small extent, model N, another of the company's products first introduced three years ago. In this case, loss of sales of model N (offset by appropriate reductions in manufacturing and selling costs) must be taken into account when deriving the net incremental cash inflows associated with the investment in model P.

These examples illustrate that defining the relevant alternatives to be analyzed, and then assessing the incremental cash flows associated with each, are often both difficult and subject to considerable error. Even estimating the original investment can, in certain circumstances, be difficult; two examples are software development, and when product testing is subject to complex regulation, as in the case of pharmaceuticals. Sometimes the cost of installing or debugging new equipment or instruments, all of which is considered part of the initial cost, is difficult to estimate.

Finally, investment lives and salvage values—that is, cash inflows at the end of the asset's life—are particularly difficult to estimate. Fortunately, the process of discounting causes those cash flows that occurr a long time in the future to have little effect on the final decision; the higher the discount rate used, the more this is the case. For example, Table 12-3 shows that the present value of the salvage to be realized 10 years from now, when discounted at 15 percent, is only $9,880, which is less than 25 percent of its $40,000 nominal value. If the life of the investment were 20 years and the discount rate 20 percent instead of 15, the same $40,000 salvage value would have a present value of less than 3 percent of its nominal value.

Working Capital Changes. Not infrequently, capital investments require ancillary investments in additional working capital, for example, in accounts receivable and inventory. Suppose the investment in the automated packaging line requires the company to carry additional inventory of either packaging materials or finished product. This additional inventory investment must also be assessed in the project. Incidentally, it is typically a reasonable assumption that this additional inventory investment can be liquidated (giving rise to a cash inflow) at the end of the investment's life.

Introducing model P to the market will probably necessitate additional investment in both accounts receivable and inventory (less offsetting increases in accounts payable). Again it is the *net* additional working capital that is relevant; if sales of model P have a negative impact on the sales of other company models, the incremental investment in accounts receivable will be less than the gross receivable arising just from model P sales.

Tax Shield from Depreciation. Companies are well advised to make investment decisions based on *after-tax* cash flows. As noted earlier, because interest paid on debt is tax deductible, the weighted-average cost of capital—on which the hurdle rate is typically based—is best calculated on an after-tax basis. Moreover, large capital investments in equipment or facilities give rise to significant depreciation charges. As emphasized in Chapter 8, depreciation expenses do not represent cash outflows. However, because depreciation is tax deductible for corporate income tax purposes, it is a *tax shield:* depreciation expenses "shield" the company from some income tax payments that it would have to make in absence of the associated investment decision.

Table 12-4 derives the after-tax flows in connection with a $100,000 investment in a capital asset that can be depreciated to zero over 8 years while the cash inflows from the investment continue through 10 years. Note that accelerated depreciation (in this case, double declining balance) is used to maximize the depreciation, and therefore maximize the tax deduction in the early years of the asset's life. Remember that the earlier cash returns occur, the greater their present value, thus boosting the return on the investment. This example assumes a 40 percent corporate income tax rate. The pretax internal rate of return is 12.4 percent, but, due to the depreciation income tax shield, the after-tax return is almost 9 percent, equivalent to about three-quarters of the pretax return. The importance of cash flow timing is also illustrated in Table 12-5. The initial investment and the sum of the cash flows are exactly the same, but investment A's returns are skewed to the early years while investment B's are skewed to the later years. Note that investment A has a small positive net present value when discounted at 20 percent, while investment B fails to clear this hurdle: its net present value is substantially negative.

DECISION CRITERIA

Put simply, a company should proceed with those investments for which the anticipated return is above the company's WACC, and forgo investment opportunities for which anticipated returns are below its WACC. Unfortunately, the world of capital investments is not quite that simple.

Decisions When Capital Is Abundant

Suppose a company has more capital available for investment than it has projects with anticipated returns greater than the weighted-average cost of capital? It certainly should not invest in projects with anticipated returns lower than WACC. Instead, it should reduce its excess capital by paying down its debt, paying dividends, repurchasing its common stock, or some combination of these three actions.

Decisions When Capital Is Rationed

In situations where the company has insufficient capital (that is, it has more attractive investment opportunities than available capital) and cannot (or is unwilling to) obtain additional capital to make all of the investments that meet its criteria, the company must ration its capital. It wants to be certain that it invests where the returns are highest and forgoes lower-return projects. Ideally, the company would list all of its potential projects in decreasing order of returns. The company would then approve projects starting at the top of the list, and move down the list until it has used up its available investment capital or has reached projects with returns below the company's weighted-average cost of capital. Under this set of conditions, the company's hurdle rate would be the return on the first project below the cutoff, again assuming that the return available on the first forgone project exceeds the company's estimated WACC.

Implementing this exact procedure to determine the hurdle rate when capital is rationed is complicated by the fact that not all future projects can be known at one time. Still, the concept is valid. Managers must do their best to estimate the rate of return at that cutoff point. The company's past experience in capital budgeting serves as a good guide.

TABLE 12-4. After-Tax Return Analysis: 10-Year Asset

Assumptions

Original investment: $100,000
Annual cash inflow (or savings): $18,000
Life of project: 10 years
Depreciation method: double declining balance

Depreciation life: 8 years
Income tax rate: 40%
Salvage value: 0

Cash Flow Calculations

1	2	3	4	5	6	7
Year	Investment	Cash Inflow	Depreciation	Taxable Income	Income Tax	After-Tax Cash Flow
0	−100,000	—	—	—	—	—
1		+$18,000	$25,000	−$ 7,000	+$2,800	$20,800
2		+18,000	18,750	− 750	+ 300	18,300
3		+18,000	14,062	+ 3,938	− 1,575	16,425
4		+18,000	10,572	+ 7,428	− 2,970	15,030
5		+18,000	7,904	+ 10,096	− 4,038	13,962
6		+18,000	7,904[a]	+ 10,096	− 4,038	13,962
7		+18,000	7,904	+ 10,096	− 4,038	13,962
8		+18,000	7,904	+ 10,096	− 4,038	13,962
9		+18,000	0	+ 18,000	− 7,200	10,800
10		+18,000	0	+ 18,000	− 7,200	10,800

Pre-tax Return

$18,000 P/A (i, 10 yrs) = $100,000

Therefore, P/A (i, 10 yrs) = 5.555

At i = 12%: P/A (10 yrs) = 5.650

At i = 14%: P/A 10 yrs) = 5.216

By interpolation: i = 12.2%

After-Tax Return

Year	Cash Flow	P/F (i = 9%)	Present Value (i = 9%)
0	−$100,000	1.0	−$100,000
1	+ 20,800	0.917	+ 19,074
2	+ 18,300	0.842	+ 15,409
3	+ 16,425	0.772	+ 12,680
4	+ 15,030	0.708	+ 10,641
5	+ 13,962	0.650	+ 9,075
6	+ 13,962	0.596	+ 8,321
7	+ 13,962	0.547	+ 7,637
8	+ 13,962	0.502	+ 7,009
9	+ 10,800	0.460	+ 4,968
10	+ 10,800	0.422	+ 4,558
		Total	−$629

Therefore, the return is slightly less than 9%.

[a]Switch to straight-line depreciation over the remaining life.

Note:

Column 5 = volume 3 − column 4

Column 6 = Column 5 × 0.4

Column 7 = column 3 − Column 6

TABLE 12-5. Effect of Cash Flow Timing

Year	Cash Flows from Investment	
	A	B
0	−$32,000	−$32,000
1	+ 14,000	+ 6,000
2	+ 12,000	+ 8,000
3	+ 10,000	+ 10,000
4	+ 8,000	+ 12,000
5	+ 6,000	+ 14,000
Total	+$18,000	+$18,000
Net Present value @ 20%:	$48	−$4,248

Decisions in Cash-Rich Companies

Some companies, such as Microsoft and General Electric, among others, operate with very large cash balances. A number of conditions may drive a company toward maintaining cash balances that many would consider excessive. For example, a biotech company facing a very long product development cycle may carry high cash balances as a hedge against the possibility that it will be unable to access equity markets on a regular basis. Other companies may simply be extraordinarily cautious and therefore wish to maintain a balance that would see them through a severe recession, during which the company operates at a loss. Still others may be holding large cash balances for use in the acquisition of other companies; note that such acquisitions are essentially the same as very large capital investments and can be analyzed in the same manner. A company with very high cash balances, then, is constrained by neither its ability to access capital markets nor its current cash balance. One might make the incorrect assumption that any capital investment that yields a return higher than the nominal return available on cash-equivalent securities would be acceptable. Not so—for reasons considered in a moment. The company still should avoid all discretionary investments that promise returns below the company's weighted-average cost of capital.

Deciding Among Mutually Exclusive Alternatives

Suppose that Chico Corporation is considering installing a new processing line in its Cleveland factory. Chico has been presented with five alternative designs by the engineering firm it hired to design and install the processing line. Each of the five alternatives meets Chico's operating requirements. Some of the alternatives require greater investment, but in those cases the resulting cost savings, when compared with the current processing line, are also greater. Obviously, Chico is not going to choose more than one of these alternatives, as they all have the same purpose; that is, the five alternatives are mutually exclusive. How, then, does the company choose the most appropriate alternative?

The top half of Table 12-6 provides return information on each of these five alternatives. Assuming Chico has a 15 percent hurdle rate, all five alternatives are acceptable in that they all have returns in excess of 15 percent. You might be tempted to select alternative 1, since it involves the least capital outlay and its return exceeds the hurdle rate. But minimizing capital outlay is not the objective and, therefore, not the appropriate decision criterion. To the contrary, the company wants to undertake all investments that earn a satisfactory return. Alternatively, you might be tempted to select alternative 5 since it involves the maximum capital outlay and its return still exceeds the hurdle rate. That, too, would be an incorrect decision.

Remember that these alternatives are mutually exclusive. The company won't choose more than one. As we move down from alternatives 1 to 5, both the amount of the initial investment and the amount of the cash savings increase. The analytical process is this: move systematically down this list, comparing each alternative with the previous one that proved acceptable. Chico wants to invest all it can in this processing line provided that *each* increment of the investment earns a return in excess of the hurdle rate.

We can see in Table 12-6 that alternative 1 has an acceptable return. Alternative 2 involves $30,000 of additional capital investment to capture a

TABLE 12-6. Deciding Among Mutually Exclusive Alternatives

Design Alternatives[a]	Investment Outflow (year 0)	Annual After-Tax Inflows (years 1–12)[b]	Present Value at 15% Hurdle Rate[c]
1	$140,000	$26,000	+$ 946
2	170,000	33,000	+ 8,893
3	185,000	35,000	+ 4,735
4	205,000	40,000	+11,840
5	230,000	44,000	+ 8,524

Comparison of Alternatives	Incremental Increase in Investment	Incremental Increase in Cash Flow	Present Value at 15%	Accept/Reject
2 with 1	$30,000	$7,000	+7,947	Accept #2
3 with 2	15,000	2,000	−4,158	Reject #3
4 with 2	35,000	7,000	+2,947	Accept #4
5 with 4	25,000	4,000	−3,316	Reject #5

Therefore, invest in alternative 4

[a] All five design alternatives meet Chico's design criteria for the processing line. Thus they are mutually exclusive.
[b] Savings compared to costs of operating the current processing line.
[c] Factor used : P/A ($i = 15\%$, $n = 12$ yrs) = 5.421.

$7,000 incremental annual cash flow for the 12-year life of the project. Does the incremental cash flow of $7,000 provide a return on the incremental capital investment of $30,000 that is above the hurdle rate? The bottom half of Table 12-6 shows that it does. Thus, we can accept alternative 2 and compare alternative 3 with it: $15,000 more invested capital to capture $2,000 of additional annual cash inflow. As seen in Table 12-6, that rate of return is below the hurdle rate, and thus Chico should discard alternative 3. Now we compare alternative 4 with alternative 2. (Alternative 4 may have a lower internal rate of return than alternative 2, but that fact doesn't disqualify it.) Alternative 4 requires $35,000 more investment to capture $7,000 more in annual cash flow. The return on this incremental investment is above Chico's hurdle rate; accordingly, the incremental investment in alternative 4 is justified. The incremental investment in alternative 5 over alternative 4 is not justified (its return is below the hurdle rate). Therefore, the right decision for Chico is to invest in alternative 4.

SOME EVERYDAY APPLICATIONS OF THESE TECHNIQUES

Lest you think that these present value and return on investment analytical techniques are arcane and applicable only to large companies, consider the following personal situations where these concepts can help you make sound personal decisions.

Suppose you are presented with an exciting opportunity to invest $5,000 in the initial public offering of the common stock of a technology company founded by a friend. Currently, you have no excess cash reserves, and your only source of funds is your consumer credit card. Since the interest rate charged on this card is approximately 18 percent per annum, that rate becomes your hurdle rate. You want to make the investment in your friend's company stock only if you are pretty confident that its rate of return will substantially exceed that hurdle rate. "Substantially exceed" because the risk of investing in this IPO is probably high, while your obligation to pay your credit card installments is absolute.*

Suppose you have a one-year-old daughter and are already concerned about having sufficient funds to pay for her college education. You have 20 years until she completes college, and you want to start investing now, so that you can withdraw $40,000 each year she is in college. You plan to invest in a mix of stocks and bonds (perhaps by means of a mutual fund) that you believe will deliver an annualized return over these next 20 years of 8 percent. Figure

---

*I am continually amazed that people invest in the stock market while incurring interest charges on their consumer credit cards. In effect, they are betting that stock market returns will exceed something like 18 percent. Occasionally, markets do provide returns at this high level, but the historical returns on U.S. stocks are substantially below 18 percent. "Investing" by paying off credit card debt provides an absolutely assured return equal to the rate of interest charged its holder! But enough of my preaching.

12-2 shows both the relevant cash flow diagram and the appropriate calculations to determine how much you need to invest each year for 20 years to achieve your objective.

Note that the present value of four outflows of $40,000 in each of years 17 through 20 is $39,611. Thus, the present worth of your 20 annual investments must also be $39,611. The interest factors from Tables 2A-1 and 2A-2 permit you to determine that you need to invest $4,035 each year. By the way, if you decide you want to make these college fund investments over only the next 10 years, the present value of those 10 payments must also be $39,611, and you will need to invest annually $5,903.

Suppose your father recently retired and most of his net worth is wrapped up in a family farm that you now operate. The $2 million market value of the farm will be included in his estate and thus subject to *inheritance taxes* (sometimes called *death taxes*) of, say, $1 million when he dies. You estimate that his remaining life expectancy is 20 years, and he would like to build up a fund sufficient to discharge those death taxes at that future date. He is debating how "conservatively" to invest. If he invests in government bonds, returns will be low, say 5 percent, but his risk of not having sufficient accumulated funds in 20 years will also be low. Conversely, if he invests in growth stocks that he projects will return, on average, 10 percent per year, he will need to invest less each year but will incur a higher risk that his accumulated investment will be insufficient to pay the inheritance taxes in 20 years. Use the equation

$$\$1,000,000 \times P/F\ (x\%,\ 20\ \text{yrs}) = \text{annual investment} \times P/A\ (x\%,\ 20\ \text{yrs})$$

to determine that he will need to set aside $30,252 each year if he chooses the conservative route of investing in government bonds, but only $17,501 if he accepts the risk associated with investing in the higher-return growth stock.

And another scenario: Suppose you are offered an opportunity to purchase a lifetime subscription to your favorite magazine for a one-time payment now of $336. Your alternative is to subscribe annually for $56. (Incidentally, you

Annual investment

CC

$40,000/year

Present worth of payments required during college years = 40,000 × (P/A, 8%, 4 years) × (P/F, 8%, 16 years) = $39,611
Annual investments for 20 years = $39,611 ÷ (P/A, 8%, 20 years) = $4,035

Figure 12-2. Investing for a child's education: cash flow diagram and calculations.

must be confident that you will wish, and be able, to read this magazine for the rest of your life!) Is this offer a good deal for you? That all depends on how long you expect to live and the rate you can earn on investments that represent alternatives to investing in a lifetime subscription. Suppose your hurdle rate is 7 percent. You can easily calculate the minimum number of years you must live in order to make the lifetime subscription a sound investment. You compute the present value at 7 percent of the annual subscription fees for a range of years, and compare that amount to the lifetime subscription fee (which is already at present value). Or, the equivalent:

$$\$56 \times P/A \ (7\%, \ n \ \text{yrs}) = \$336$$

yields a factor that happens to be exactly 6.0; Table 2A-2, under the "7%" column, shows that this factor occurs between the eighth and ninth year. Therefore, the lifetime subscription is a good deal if you expect to live longer than about eight years.

And a final example: Suppose you are considering purchasing a new $22,000 car. The auto dealer offers you "zero interest" financing: a $2,000 down payment and four annual installments of $5,000. Using your personal hurdle rate (probably based on returns you expect to earn on other investments), you can calculate the present value of the five required payments. This present value, compared with the $22,000 price, tells you the effective price discount the dealer is offering you through this special financing arrangement.

We could go on with other illustrations. Should you invest in rental property? Should you purchase or lease your automobile? What financial plans should you make for retirement?

ADJUSTING FOR RISK

Obviously, an investment analyst has a host of methods for adjusting for above-average risk that he or she may perceive in a capital investment opportunity:

- Shade down the anticipated positive returns—that is, future positive cash inflows—or shade up the anticipated future expenses (cash outflows).
- Shorten the assumed life of the project for the purposes of the study.
- Increase the required hurdle rate for the project.

An ever-present danger is that the analyst will overadjust for these future risks and, thereby, reject too many promising investment opportunities. Unfortunately, the analyst, probably a middle-level manager in the corporation, is unlikely to be criticized or punished for *not* taking prudent risks but, on the other hand, is likely to receive plenty of criticism for approving a project that turns out to fail. Not surprisingly, then, middle managers tend to be quite risk-averse. Remember that companies should not try to minimize their capital expenditure budgets (except when cash is severely tight, and additional capital is unavailable); rather, they should search out and make all investments that promise a return in excess of the corporation's weighted-average cost of capital. Failing to do so shortchanges the stockholders.

Multiple Hurdle Rates

Companies can adjust hurdle rates—that is, required rates of return—for different categories of risk. For example, they may set relatively low hurdle rates for low-risk cost reduction investments and higher hurdle rates for higher-risk investments, for example in new product development.

Some multidivision companies are, in effect, in several different businesses. Suppose division A is a mature business with strong market share and a slowly evolving product line; division B, on the other hand, participates in a high-growth business with short product life cycles, rapidly developing technology, and many competitors. The overall company, of course, has a single WACC. If the two divisions were stand-alone companies, A would have a lower WACC than B. In general, A's investment opportunities will be more modest and offer less handsome returns than will B's. Accordingly, if the same hurdle rate is used to analyze investments in both divisions, B will capture most of the capital available for new investments. The company will tend to overinvest in division B and underinvest in division A. In time, the riskiness of the overall company will creep up, and with it the company-wide WACC. The increasing WACC will then amplify the over- and underinvestment phenomenon. This company should use different hurdle rates in the two divisions.

A Probabilistic Approach

A promising way to adjust for risk is to assign probabilities to the range of expected cash inflows and outflows associated with the project and then utilize the probability-adjusted expected value of each in determining the net present value or internal rate of return for the project. Suppose that a $1 million, four-year investment has the range of expected future cash inflows shown in Table 12-7. If the corporation's hurdle rate for such projects is 15 percent, the net present values of these future cash inflows are as follows:

Alternative A	$ 893,000
Alternative B	$ 982,000
Alternative C	$1,142,000

TABLE 12-7. Alternative Expected Cash Inflows, by Year

Alternative	Year 1	Year 2	Year 3	Year 4
A	$200,000	$300,000	$400,000	$400,000
B	150,000	400,000	400,000	500,000
C	400,000	400,000	400,000	400,000

The analyst could simply accept the alternative that he or she considers most likely and ignore the other two. Alternative B describes the most likely outcome. Using those estimates, the analyst sees that the project does not pass the test: the net present value of the future cash inflows falls short of the present $1 million investment.

However, the analyst can deepen the analysis by assigning probabilities to each alternative—say, a 20 percent likelihood that alternative A best describes the future cash inflows, a 50 percent probability that alternative B will come to pass, and a 30 percent likelihood that alternative C best describes the future. (Note that these probabilities must sum to 100 percent!) Thus, there is a not-insignificant chance—indeed, a 30 percent probability—that the attractive alternative C will come to pass. If we apply probabilities to these three alternatives, we determine that the "expected present value" of the future cash inflows does meet the test (just barely, to be sure) of exceeding the current $1 million investment:

Alternative	Probability		Net Present Value		
A	20%	×	$893,000	=	$179,000
B	50	×	982,000	=	491,000
C	30	×	1,142,000	=	343,000
Total	100%				$1,013,000

Multiphase Projects

Incorporating risk adjustments in the analysis of a project that involves several phases is particularly challenging. At the outset the project may be assessed as very risky: for example, researching a new pharmaceutical drug. The total cost of developing, testing, and bringing to market a new drug is exceedingly expensive, and relatively few new drugs successfully pass through all the required steps and then prove viable in the medical marketplace. However, as the drug successfully completes each step or phase of the development process, the risk of ultimate failure decreases. For example, a new drug that successfully completes Phase II clinical testing has a substantially higher probability of succeeding in Phase III trials, then manufacturing, and then marketing than does another drug that is just entering preclinical trials.

Importantly, the pharmaceutical company has many options: it can expand, contract, extend, abandon, or defer a project as it gains additional information about both its own likelihood of successful development and the actions of its competitors. This flexibility is valuable and must be factored into the analysis.

For example, if the drug fails in preclinical trials—that is, relatively early in the overall multiphased process of bringing a new drug to market—the company will exercise its option to cease further investment in it. The company will never put at risk the amount of investment capital that was anticipated at the time the research commenced.

And, on the other hand, this drug may ultimately come to have an even brighter future than was originally anticipated. Suppose that clinical trials indicate greater efficacy than originally thought possible and, moreover, potentially competitive products failed to gain regulatory approval. The drug has the potential of being a "blockbuster," as highly successful drugs are called in this industry. If so, the company has the option to increase its investments—in the capacity of manufacturing facilities and additionally in marketing expenses—so as to exploit, to the maximum, this new blockbuster drug. That is, as prospects for the new drug brighten, the company can pour on additional capital investment to realize its maximum potential.

In concept, the hurdle rate should decrease for each subsequent phase of this long drug development process, because the risk of failure decreases with each step. Sophisticated techniques are available to analyze these multiphased investment projects (typically referred to as the *real-options approach*), but these techniques are beyond the scope of this book. Just bear in mind that the high hurdle rate for phase I of the project—when the risk of failure is typically the highest—should be reduced for subsequent phases because risk is squeezed out of the project as each step is successfully completed.

ADJUSTING FOR INFLATION

The investment project examples studied thus far have ignored inflation. "Assuming away" inflation is justified only when the inflation rate is near zero, as was the case in Japan and the United States just after the turn of the 21st century. In all other cases, we need to adjust for inflation in one of two ways: (1) Eliminate the inflation factor in the hurdle rate (required rate of return), and then perform the analysis in "constant dollars;" or (2) build in assumed inflation rates in all future cash flows.

It may not be obvious to you that the cost of capital has a built-in inflation factor. But, in setting interest rates on debt, lenders make explicit assumptions regarding the pattern of inflation over the term of the loan. Similarly, expectations of present and potential shareholders—expectations that are reflected in the market price of common shares—also have built-in inflation assumptions, even though shareholders are far less likely than are professional lenders to be explicit about just what those inflation assumptions are. Thus, when the required rate of return is a function of the weighted-average cost of capital, the estimates of future cash flows should include inflation.

SUMMARY

Capital budgeting, the subject of this chapter, links the financial (capital) markets with a key set of management decisions: selecting appropriate capital investment projects and discarding the others.

Acceptable capital investments are those that promise returns, factoring in the time value of money, that exceed the company or individual investor's minimum required rate of return or hurdle rate. When the investor has or can obtain abundant capital, the hurdle rate is equal to the weighted-average cost of capital (WACC). Alternatively, when capital is limited, it must be rationed and invested in projects with the most rewarding prospects; here the hurdle rate is the promised rate of return of the most promising project that has to be forgone because of the investor's capital shortage.

The WACC weights the after-tax cost of each source of capital by the proportion (figured at today's market values) that it represents in the company's capital structure. Assessing the cost of equity capital is particularly challenging.

Potential investments can be evaluated by either the discounted cash flow (net present value) method or the internal rate of return method, utilizing readily available interest factor tables or computer programs.

Defining the relevant alternatives to be analyzed and then determining the differences in future cash flows over the lives of the alternatives—and ignoring sunk costs—are the critical first steps. Working capital changes, tax shields from noncash expenses, income taxes, and inflation must also be worked into the analysis. Deciding among mutually exclusive alternatives— several ways to achieve a similar result—requires that each increment of the potential investment be tested for acceptability.

Adjusting for differences in risk among projects (or among phases within a single project) is essential but difficult. Some analysts shade estimates to be conservative, but when this personal risk aversion is overdone, some promising investments get discarded. Some companies use multiple hurdle rates, with higher hurdles for the more risky projects. Probability analyses can also help analyze projects having a range of potential future payoffs.

The analytical techniques discussed here have broad applicability for for-profit companies, governmental enterprises, and personal (private) investors.

NEW TERMS

Capital budgeting. The process of selecting, from among all possible capital investments, those that should be undertaken because they offer sufficient financial return.

Discounted cash flow. The equivalent value, at time zero, of a series of future cash flows (inflows and outflows) evaluated at a given interest rate. An alternative name is *net present value*.

Hurdle rate. Another name for *minimum required rate of return.*

Internal rate of return. The particular interest rate that results in zero net present value of a set of both current and future cash flows prescribed as to both amount and future timing.

Minimum required rate of return. The threshold return required of all discretionary capital investments, generally determined by a capital rationing process or by reference to the weighted-average cost of capital (WACC). Often referred to as the *hurdle rate.*

Net present value. Another name for *discounted cash flow.*

Payback period. A widely used, but inappropriate, figure of merit for potential capital investment projects. The investment today divided by the estimated annual cash inflow (or savings) yields the payback period in years.

Sunk cost. An investment made or cost incurred in the past that cannot be recovered by a decision today.

EXERCISES

1. What are the two ways in which an investing corporation's hurdle rate may be arrived at?

2. If a company is able to borrow additional capital, why is the cost of that borrowing not the appropriate hurdle rate to use in evaluating investments?

3. Since the before- and after-tax cost of common stock equity is identical, why must we calculate the after-tax cost of borrowing?

4. Why is the inverse of the price/earnings ratio not a good indicator of the cost of equity?

5. What effect does depreciation have on the cash flow associated with an investment in a fixed asset?

6. Why is an accurate estimate of salvage value less critical to a capital investment analysis than accurate estimates of annual cash flows?

7. When analyzing a multiphase investment opportunity, why may it be appropriate to reduce the required rate of return on later phases of the project?

8. Are inflation assumptions built into the costs of borrowing and of equity capital? Explain.

9. Capital investment decisions require comparisons between alternatives. If a company is considering an investment in an R&D project, what is the likely alternative to which it is being compared?

10. If you have two personal investment opportunities, projected to have returns of 25 and 12 percent, respectively, and insufficient cash to invest in both, under what circumstances might the choice of the lower-return investment be a rational decision for you?

11. What's wrong with using the payback period as a measure in deciding what investment opportunities to undertake?

12. What does the word *net* mean in the phrase "net present value"?

13. Tables 2A-1 and 2A-2 assume the *end-of-year convention*. What does this convention assume?

14. Indicate whether each of the following statements is true or false:
    a. Capital budgeting processes seek to minimize a company's investments in long-term projects.
    b. The farther in the future a cash flow occurs, the lower its present value.
    c. The higher the hurdle rate, the lower the present value of a future cash flow.
    d. In choosing among mutually exclusive investment alternatives, we select the alternative with the highest internal rate of return.
    e. Hurdle rates to evaluate risky investments are typically higher than those used to evaluate relatively "safe" investments.
    f. One cannot calculate the present value of a stream of identical annual cash flows that continue in perpetuity.
    g. The future value of a stream of annual cash flows is always higher than its present value.
    h. One method of adjusting for risk is to subject a range of possible outcomes (of cash flows) to a probabilistic analysis.
    i. All past expenditures should be considered irrelevant sunk costs when evaluating current investment opportunities.
    j. When the WACC (the weighted average cost of capital) is calculated, the various sources are weighted by their current book value on the company's balance sheet.
    k. Cash and cash equivalents currently available for capital investments are assumed to have zero cost.
    l. Most company mid-level managers are inclined to underestimate the riskiness of investment projects they analyze.

15. How much would you have to invest at 12 percent at the end of each year for the next 10 years in order to have $120,000 available at the end of that time?

16. How much could you withdraw from a $120,000 account at the end of each year for the next 10 years, assuming the investment is drawn down to zero at the end of the period and earns 12 percent during the period?

17. If you invest $8,000 at the end of each year for 10 years and wish to have $120,000 available at the end of that period, what rate of interest will you need to earn on your investment?

18. Today, you purchase a home for $200,000 and expect it to appreciate at 5 percent per year while you own it. What price are you assuming you can get for your house when you sell it in eight years?

19. Your brother-in-law promises that he can invest $10,000 of your money so as to make it double in 10 years. What is the implied interest rate for this investment?

20. The Devlin Company has an opportunity to acquire a new product line for an investment of $3 million. Devlin estimates that the annual net cash inflow from this investment will be $340,000 for the next five years and $250,000 for the following five years.

    a. Assume these are all of the cash flows associated with this investment. What is its internal rate of return?

    b. By how much would cash flows in the second five-year period have to increase in order to boost the return to 15 percent?

    c. Alternatively, assume Devlin feels that, after the first 10 years, the new product line will produce a perpetual net cash flow of $180,000 per year. Would the investment then be justified if Devlin's required rate of return were 15 percent?

    d. Return to the original estimates of cash flows: what can Devlin afford to pay for this new product line if its required rate of return is 15 percent?

21. Galas Corporation has the following mutually exclusive alternative investments available to it, each having a project life of 10 years:

Alternative	Investment	Annual Net Cash Inflow
A	$320,000	$55,000
B	380,000	68,000
C	420,000	78,000
D	440,000	81,000

   Assume that Galas's required rate of return is 12 percent. Which alternative should it select?

22. The Van Ness Company's capital structure has the following characteristics:

Security	After-tax Cost	Book Value	Market Value

Bonds	5 percent	$23 million	$24 million
Preferred stock	7 percent	10 million	10 million
Common stock	12 percent	40 million	60 million

Calculate Van Ness's weighted-average cost of capital.

23. The net cash inflow from a $27,500 project with an $8,000 salvage value is expected to be as follows over its five-year life:

Year	Cash Flow
1	$4,000
2	5,000
3	6,000
4	7,000
5	8,000

a. If the investor requires a 10 percent return, is this investment acceptable?

b. If the annual returns were equal at $6,000 per year, would this investment be acceptable?

24. Cohen Corporation is considering a $250,000 investment in new equipment with a 10-year estimated useful life and an estimated salvage value of 10 percent. If Cohen assumes that the net pretax cash inflow from this project will be $45,000 for each of the first three years, has a 40 percent incremental income tax rate, and assumes double-declining balance depreciation in its capital investment analyses, what is the after-tax cash flow relevant to this analysis during the first three years of the asset's life?

25. My wife and I are eager to help our children educate our grandchildren. I estimate that college tuition for my granddaughter, who just turned five, will be $55,000 per year by the time she goes to college at 18. Thus, she will need $55,000 at the end of her 18th through 21st birthday. How much will we have to invest at 8 percent annually now and each year through her 21st birthday in order to fund all of her college education? (*Hint:* draw a cash flow diagram carefully.)

26. Assume you could buy a nice car for $28,000 cash or lease the same car for four years at $5,500, payable now and at the beginning of each of the next three years. You plan to acquire the leased car at the end of the lease period for $16,500 and, whether you lease or buy it, sell it at the end of six years.

a. If your *cost of money* is 12 percent, should you lease the car?

b. What is your *break-even cost of money*, such that you are indifferent whether you lease or buy?

c. If your cost of money is 20 percent, how much of a discount would you have to obtain on the purchase price to make buying a better alternative than leasing?

27. Your company is considering buying a large computer system for $40,000 in order to avoid outsourcing charges of $15,000 per year. Your company's after-tax hurdle rate on such technological investments is 15 percent. Your company plans to use double-declining-balance depreciation, assuming zero salvage value and a four-year life.

   a. What is the after-tax internal rate of return on this investment, if you assume a four-year useful life?

   b. How long would the life of the computer system have to be in order for the return to meet the company's required rate of return? (Assume there is no change in depreciation life.)

CHAPTER 13

ANALYZING MANUFACTURING COSTS: AN INTRODUCTION TO COST ACCOUNTING

The next two chapters are devoted to what is popularly called **cost accounting:** developing detailed information on the cost to produce a product or service. Although traditionally treated as a subject quite apart from financial accounting, cost accounting is really just a subset of it: cost accounting data are part of financial accounting data. The object of cost accounting is to amplify and clarify the values of cost of goods sold and inventory, and provide selected supplemental information to help managers make operating decisions.

Recall, from our earlier discussions, that many transactions can be recorded in a number of alternative ways—for example, LIFO inventory accounting can be used instead of FIFO, or accelerated depreciation instead of straight line. The resulting financial statements are, of course, affected by these choices, but either alternative is both accurate and acceptable. Similarly, in cost accounting, we face choices. We arrive at somewhat different answers to the question What does the product cost? depending on our choice among alternative costing procedures.

We have succeeded in developing financial statements for a variety of companies without the benefit (and complexity) of cost accounting. Why do we now need more accounting techniques, and where will we use them? The primary answer: in manufacturing firms.

A manufacturing company typically owns (holds) inventory at various stages of completion. For example, a manufacturer of safety helmets carries raw-material inventory (plastic resins and strapping material); in-process, or partially completed, inventory (for example, helmet shells that have been formed but not finished or assembled); and finished goods inventory (helmets awaiting shipment to customers).* We must account for the increasing value of the inventory as it moves from raw to more finished states. Remember, however, the realization concept: a manufacturing company recognizes profit only when the critical transaction occurs, the ultimate sale to the customer. The cost model requires that costs incurred in effecting the transformation from the raw-material state to progressively more complete states be reflected in increased inventory values. By the way, we shall see that cost accounting is useful in service, as well as manufacturing, firms.

WHAT QUESTIONS? WHY? FOR WHOM?

Early in our financial accounting discussion we asked: What decisions are going to be influenced by the financial information supplied? Who is going to make them? What financial information influences these decisions? To whom are these data supplied?

Consider these same questions for cost accounting information. The answer to the question For whom? is easier. Unlike the multiple audiences for financial accounting, there is one key audience for cost accounting information: internal management—that is, operating managers within the manufacturing company or service firm.

Two relevant questions to be addressed by cost accounting data are (1) What is the value of inventory? and (2) What manufacturing cost should be matched against sales revenues? (that is, What was the cost of goods sold for the period?) These data could be supplied by the simplest cost accounting system. More elaborate systems are justified only if they supply additional useful information. The cost accounting function exists primarily to serve operating managers by providing data and analyses helpful in making day-to-day operating decisions. Typically, no outside authorities—stockholders, banks, security analysts, or governmental regulators—require or even have access to these more elaborate data.

Think about key decisions marketing departments face: what price to charge for the product or service, which products to promote, whether to discontinue a certain product, how much extra to charge to modify (customize) a standard product. While product costs are by no means the sole criterion for determining product selling prices or the composition of the product line, they are certainly relevant to the decision. Marketing is inclined to promote those product lines that offer the greatest margin between selling price and manufacturing cost. Moreover, the company should consider discontinuing a product when its margin gets very small or turns negative.

---

*Material that is considered raw to this manufacturer—for example, the plastic resin—would be considered finished materials to the company's resin supplier. It is essential to bear in mind the entity for which we are accounting.

Certain companies sell their products or services at **cost plus**—that is, at a price equal to the cost of the product or service, plus profit either in a fixed amount or as a percentage of the cost. Many defense contractors and certain construction companies price on this basis. In these instances, both supplier and customer must agree on the definition of the *cost* of the product or service. In effect, law firms, accounting firms, consulting companies, and other personal service operations bill their clients (customers) on a cost-plus basis.

In a manufacturing environment, consider key decisions production managers must make. Shall the company perform a certain manufacturing process itself, or subcontract it to an outside firm (a **make-or-buy** decision)? Should the company invest in new production equipment? Should it upgrade existing equipment? Should it schedule one long production run of 100 pieces or two shorter production runs of 50 each? Answering these questions requires the manager to consider both current production costs and the way those costs will be affected by each alternative.

Other questions that haunt operating managers are: Is the company operating efficiently? more efficiently this year than last? this month than last month? in plant A than in plant B? on product X than on product Y? Are operations generally proceeding "on plan?" Cost accounting helps answer these questions by highlighting deviations from plan on an *exception* basis.

Cost accounting also assists development and design engineers. The costs of existing products provide a basis for evaluating the probable cost of a new or redesigned product. Working together, the cost accountant and the industrial engineer (or value engineer) spot opportunities for significant cost reductions through product redesign, changes in manufacturing method, or material substitutions. A comprehensive cost accounting database is invaluable to engineering departments seeking to lower costs.

Because cost accounting is not constrained by rules imposed by outside authorities, few generally accepted accounting principles exist for cost accounting. The cost accountant is free to arrange and rearrange cost data to be most useful to line managers. Of course, the consistency principle applies: cost accounting methods within a single manufacturing company should remain consistent in order to facilitate comparisons of cost data on different products, from different departments, and across different accounting periods.

A final introductory caveat: Be careful in answering the question What does Product X cost? Both you and the person asking the question must understand the use to which the cost information will be put. You both also need to understand just what is included and excluded from the cost of product X. We focus now on that issue: Of all the expenses of a firm, which are properly included as costs of manufacturing the firm's products?

REMINDER: PRODUCT AND PERIOD EXPENSES

Back in Chapter 4, when considering the income statement format, we differentiated between the following:

- *Product Expenses:* those expenses (cost of goods sold) that are matched to revenue (sales)
- *Period Expenses:* those expenses (operating expenses, which include research and engineering, marketing and sales, general and administrative) that are matched to the accounting period (month, quarter, year) rather than to revenue

This distinction dictates the timing of expense recognition. In this chapter and the next, the focus will be on valuing *product* expenses—that is, the cost of goods (or services) sold and its complement, inventory—in manufacturing (and some service) companies. The question now is What cost elements and how much of each element should be included in the valuation of a certain unit of output, physical goods or intangible services?

Once we define product expenses, we define by deduction the period expenses: all nonproduct expenses accrued in the accounting period. What is so important about this distinction, about the timing of expense recognition? Note that if, in an accounting period, inventory expands—that is, manufacturing volume (measured in units, tons, meters, kilograms, and so on) exceeds sales volume—certain manufacturing expenditures of this period are reflected in the inventory account (an asset); cost of goods sold is less than the total manufacturing expenditures. Similarly, when inventory declines, more goods are sold than produced; since goods sold must include some manufactured in previous periods, the cost of goods sold for this period will be greater than the period's aggregate manufacturing expenditures.

UNDERSTANDING THE COST OF GOODS SOLD

Why does this problem only now arise? Heretofore, virtually all of our examples have been drawn from the merchandising industries, retail and wholesale, where the distinction between product and period expenses is quite straightforward. Merchandise is purchased at a known price; that known price is the historical cost for purposes of valuing both inventory and the cost of goods sold when the merchandise is resold to customers. The merchandise is resold in the same condition or state in which it was purchased; the merchandising firm has not physically altered it (except perhaps by combining it with other merchandise, or packaging it).

Manufacturing firms, by contrast, *convert* the materials or merchandise that they purchase into different products or materials. This **conversion** is the fundamental difference between a manufacturing firm and a merchandising firm.

Does this conversion cause the material to become more valuable? Of course, and therefore in valuing later stages of inventory—that is, semifinished products (typically called *work-in process* or *work-in-progress*) or completed products (typically called *finished goods inventory*)—the historical cost of the raw materials alone is only one cost element. Including only material costs as product expenses would understate the value of semifinished or finished inventory and, thus, would understate the value of the cost of goods sold when the inventory is sold.

Surely, managers in a manufacturing firm want to know what it costs to produce a particular product. This apparently simple question is difficult to answer. How about wages paid to manufacturing labor? all labor? including supervisors and quality inspectors? How about the plant manager or the vice president of manufacturing? And then, beyond salaries and wages, how about maintenance, depreciation on the facilities and equipment, power, and the host of other cost elements inherent in a manufacturing operation?

The distinction between product and period expenses is not always clear-cut, so we return to our old standby guideline: consistency. The way we divide up this gray area should be consistent—across accounting periods, across products, and across departments—in order to maintain comparability.

Note that resolution of this gray area affects the timing of expenses and, therefore, of profit. To the extent that more expenditures are categorized as product costs, fewer are period costs, with the following consequences:

- In periods when inventory grows, more expenditures are reflected as an asset (inventory), and fewer are expenses of the period. Therefore, profit is greater.
- In periods when inventory levels decline, both the inventory reduction and the cost-of-goods-sold value are higher. Therefore, profit is lower.

So, we come to the question at the nub of cost accounting. What are the relevant product costs? Where do we draw this important line between product and period expenses? Here are some alternatives.

Direct Material

One possibility is just to treat manufacturing environments like merchandising environments. The historical costs of materials purchased provide a reliable, verifiable, free-from-bias, and readily available basis for valuing inventories—whether raw, in-process, or finished—and, when the sale occurs, the related cost-of-goods-sold value.

This alternative would deprive manufacturing managers of some potentially useful information, but the work of cost accountants would be much simpler.

Before discarding this alternative, consider a very highly automated manufacturing plant, one into which raw material is inserted and from which a finished product in due course appears with no human intervention. The plant employs no more than, say, a diligent but unskilled overseer whose sole task is to shut off the power to the plant if it begins to run amok. While such plants are not numerous, they do exist. How different is such a plant from a merchandising operation? In a merchandising operation, materials typically arrive in bulk and leave singly or in small bunches. They are not converted in a manufacturing sense, but they are probably handled even more than in the highly automated manufacturing plant. Both the merchandising and the highly automated manufacturing firms require substantial investments in facilities and equipment, and in both cases the work force consists of material handlers, sales personnel, and various administrators.

You are probably inclined to think that the cost of operating the highly automated manufacturing facility (power, maintenance, or depreciation, for example) should be included in the cost of the finished product. If so, why not, in a parallel manner, include the cost of operating the retail store or wholesale warehouse facilities in the cost of the product delivered from these merchandising facilities? In a merchandising operation, the product may be unchanged physically, but it is surely altered in the eyes of the purchaser.

To complicate matters further, the definition of **direct material**—material directly related to the final product—is not always clear-cut. Every manufacturing firm purchases various materials that are used up in manufacturing but don't end up physically as part of the final product; examples are lubricants, solder, cleaning fluids, and compounds. Are these considered direct material? Typically not, but as you can see, the line between direct and indirect materials is not always sharp.

Direct Material and Direct Labor

A second alternative is to include the material, as just discussed, but also the wages of the work force directly involved with manufacturing the product, so-called **direct labor.** This sounds like a step in the right direction, but it opens a new dilemma. What is direct labor? What part of the work force is directly involved in manufacturing? Only those that physically transform (convert) the product in the direction of finished goods? Most transformation is actually done by machines, of course. Then perhaps machine operators should be included as direct labor. How about material handlers, inspectors (since they surely handle the product), and forklift drivers? All represent important categories of labor, but are they direct labor or **indirect labor?** Moreover, what does *directly involved* mean? Surely, the department supervisor, production scheduler, maintenance mechanic, and stockroom attendant think they are directly involved. Are they not as important as the machine attendant who spends most of his or her time watching an automatic machine tool operate?

Thus, as with the direct and indirect material, the distinction between direct and indirect labor is fuzzy. Nevertheless, we could decide arbitrarily which personnel to consider direct labor and which indirect labor. By now we have encountered many other accounting situations where we had to make arbitrary calls, and this is no more or less perplexing than the others. Note that we also have to determine which salaries are in the indirect labor category and which are administrative expenses, that is, which are period, not product, expenses. Is the manufacturing vice president a production person or an administrative person? Once again, we retreat to the comfort of consistency.

To repeat, a second alternative is to define product expenses as comprised of solely direct material plus direct labor—commonly called **prime costs.**

A common argument for using prime costs to value inventory and the cost of goods sold is that these costs are *variable*—that is they increase or decrease proportionately with increases or decreases in manufacturing volume. This assumption as to variability of costs is typically acceptable with respect to material but decreasingly appropriate for direct labor. For the assumption to be valid, management must be both able and willing to add or remove direct labor proportionate with changes in manufacturing volume. This assumption implies both that direct production workers can be laid off when activities slow, and that additional workers can be hired (or brought back from layoff status) when the volume of activity increases.

At least two conditions tend to invalidate the assumption that direct labor is variable. First, sound human-relations practices no longer permit hiring or laying off with little concern for employee welfare. Increasingly, companies work hard to dampen swings in employment, and many have virtually guaranteed stable employment, sometimes in the face of union pressure or government regulations, but often simply in response to societal pressure or a sense of responsibility. Second, as required production skill levels have increased, training periods for new employees have lengthened and thus the ranks of production workers cannot be augmented on short notice.

Be forewarned: In our study of cost accounting, we will typically treat prime costs as fully variable costs, though that assumption is imperfect.

Direct Material, Direct Labor, and Overhead

So, let's take the next step, and consider including overhead as part of the definition of product expenses. What is *overhead?* It consists of all expenses incurred by manufacturing that are neither direct material nor direct labor. Surely, indirect material and indirect production labor are overhead expenses. So also are the costs of occupying the manufacturing space: rent expense, depreciation expense on buildings and improvements, heat and light expenses, maintenance and janitorial expenses, and insurance and taxes on the property. Overhead also includes depreciation, maintenance, repair, and other expenses associated with owning and maintaining manufacturing equipment.

At the same time, we *exclude* the expenses of the engineering, development, marketing, selling, servicing, and administrative departments, and all other functions that are not strictly manufacturing. But dilemmas persist: where would you include the manufacturing engineering department? Is this a manufacturing function or an engineering function? Once again, it doesn't much matter which one you choose, as long as you remain consistent.

Direct material and direct labor are, by definition, *directly* identified with the product: the material ends up in it, and the direct labor workforce works directly on it. Not so with overhead. By definition, overhead is indirect. To move forward in defining product expenses requires that we find some rational basis for allocating—we will call it **absorbing**—these indirect, overhead expenses into the product.

Note that overhead includes many expenses that are fixed (that is, they do not change in the short term as manufacturing volumes fluctuate modestly) and relatively few that are fully variable. This fact will lead to complications as we consider, in just a moment, how best to absorb these indirect expenses (overhead) into individual products.

Material, Labor, Overhead, Engineering, Selling, and Administrative Expenses

We can now reasonably ask why we are so concerned with limiting the cost of products to their manufacturing expenses. If we really want to know what a product costs, shouldn't we also include the expenses associated with designing it, selling it, and administering the entire organization?

Such an all-inclusive definition of product expenses would, of course, destroy our efforts to distinguish clearly between product and period expenses. Most of what we have been treating as period expenses would, under this definition, be absorbed into product expenses. But these *non*manufacturing expenses are no less real and no less important to the successful mission of the manufacturing firm than those expenses incurred in the manufacturing departments. We could simply eliminate the distinction between product and period costs.

As with manufacturing overhead, we would have to find some method of allocating these nonmanufacturing expenses to the products produced; but, if we can find a rational basis for allocating indirect manufacturing expenses to the products, we could surely invent a way to allocate the engineering, selling, and administrative expenses.

These are sound arguments, and, indeed, some aerospace suppliers to the U.S. government are required to absorb both sets of expenses. However, commercial manufacturers follow the widespread convention of treating only manufacturing expenses as product costs, while treating all others as period costs.

Despite this convention, recognize that all functions in the company are crucial to delivering satisfaction to the customers. Consider, for example, the engineering department. Hardware design engineers are responsible for a great deal of the value added in a high-tech product. And, in a software company, since the cost of reproducing the software on disk—the manufacturing, if you will—is trivial, you could argue that software engineers (programmers) carry the primary responsibility for adding value to software products. And, increasingly, sales and marketing personnel provide services (both before and after the sale of the product) that are key to customer satisfaction.

CHALLENGE: ACCOUNTING FOR INDIRECT COSTS

To include indirect costs—commonly called overhead*, defined as all manufacturing costs beyond direct material and direct labor—in product expenses requires that the cost accounting system adopt a method of attributing, allocating, assigning, or (in cost accounting parlance) absorbing these indirect costs among all that is manufactured by the firm. The traditional and quite straightforward method simply aggregates all categories of indirect manufacturing costs and allocates them to the products in some logical but arbitrary manner. While this method has its decided shortcomings, it continues to be the dominant method employed today in U.S. manufacturing operations.

Here is the challenge: While one can relatively easily determine the value of direct material and direct labor to be assigned to a particular unit of output—time reporting and materials requisition systems provide the necessary data—determining how much of, for example, the factory rent or the shop superintendent's salary should be assigned to a unit of output is both arbitrary and imprecise. Such is the nature of costs that cannot be identified as direct.

To allocate these indirect costs, then, cost accountants select some rational, consistent, and fair—but, inevitably, arbitrary—method of dividing overhead among all that is produced. Bear in mind that a different allocation method will result in a different product expense. This fact needs to be emphasized repeatedly, simply because managers overlook it so frequently as they interpret cost accounting reports.

Another complication here is that overhead allocations must be made before we know exactly how much was spent on overhead or exactly how much was produced. We need to determine in real time—that is, as the products are being produced or services rendered throughout the accounting period—how much things cost. Typically, we can't wait, or don't want to wait, until the accounting period ends. This impatience is particularly true for a company that performs custom work (manufacturing or service) and prices its output as a function of cost; it cannot wait until the end of the accounting period to total up and allocate actual overhead because it needs to bill its customers as soon as it has fulfilled its obligations.

This need for timely data demands, then, that overhead be allocated to the various portions of production *during* the accounting period, not simply at the end. To do so, we must predetermine the basis for allocating overhead

---

*The term *overhead* carries the unfortunate connotation that these indirect costs are undesirable, unwarranted, or less worthy than direct costs. They are, instead, essential. No one working on manufacturing support functions likes to be thought of as "merely overhead"!

and then use it throughout the accounting period. Of course, the amount assigned to (absorbed by) in-process inventory is likely to be slightly different from the actual sum of expenditures on all overhead cost elements. This small difference is treated as a **variance;** we do *not* recompute, after the end of the accounting period, the amount of overhead allocated to each unit of output.

Consider now the procedures for allocating overhead and the meaning of the Overhead Variance account.

Overhead Vehicle

The first step is to determine the **overhead vehicle***, which is the basis for assigning overhead from the general pool of indirect costs to the particular production segments. The most common vehicles are direct labor hours, direct labor dollars, material dollars, machine hours, and units of production. The vehicle chosen must be directly identifiable with each production segment, and must be common to all items produced. It should represent a fair and equitable basis for the allocation, or absorption, of overhead. In a shop where direct labor hours is the overhead vehicle, if item A requires twice as many direct labor hours as item B, item A will cost twice as much in terms of overhead—A will absorb twice as much overhead as B. This assumption, while arbitrary, represents a reasonable basis for the allocation. It may be the best assumption you can make, but you should recognize its limitations when you later analyze manufacturing performance.

Although direct labor hours and direct labor dollars are the most common overhead vehicles, particularly in labor-intensive manufacturing or servicing environments, other vehicles are also possible. For example, in a custom printing plant, printing press machine hours are probably most appropriate since much of the overhead is connected with the presses themselves; thus, the overhead that a particular printing job should absorb is assumed to be a function of the number of press hours the job requires. A producer of safety helmets might use units of production as the appropriate overhead vehicle, each helmet absorbing the same amount of overhead.

Overhead Rate

Next, the **overhead rate** must be predetermined. Before the accounting period begins, the cost accountant calculates the amount of overhead that each unit of the overhead vehicle will carry. This calculation requires an estimate of total production activity for the accounting period. This planned activity level is then used to determine both (1) the expected, or budgeted, overhead expenditures needed to accomplish the planned output, and (2) the quantity of the overhead vehicle—for example, the number of direct labor hours or ma-

---

*As we shall see, sometimes more than a single overhead vehicle is used.

chine hours—to be utilized during the accounting period. The overhead rate, then, is simply item 1 divided by item 2.

For example, assume that a particular manufacturing operation expects to require 1,500 direct labor hours per month to accomplish its planned production. To support this activity level, the managers expect to spend $38,000 per month on indirect production costs. This operation's resulting overhead rate, assuming that direct labor hours are the vehicle, will be $25.33 per direct labor hour ($38,000 divided by 1,500 hours). When costing a particular item produced in this manufacturing operation, $25.33 of overhead is allocated to an item for each direct labor hour consumed in producing it.

Extending the example, the cost of an item that contains $88.60 of direct material and requires 3.5 direct labor hours to produce (wage rate of $11.00 per hour) is as follows:

Direct material	$ 88.60
Direct labor (3.5 × $11)	38.50
Overhead (3.5 × $25.33)	88.66
Total product cost	$215.76

Remember that this overhead rate is used throughout the accounting period. In an actual month, the number of direct labor hours will undoubtedly differ somewhat from the 1,500-hour estimate, and the amount actually spent on indirect cost elements in manufacturing will not total exactly $38,000. These differences are typically small, and we want to avoid reallocating overhead at month-end. What happens to these small differences between actual operations and planned operations?

Overhead Variance

The differences are reconciled (or drawn off) in the **Overhead Variance** account. Since actual operations almost always differ somewhat from plan, some debit or credit balance in the Overhead Variance account should be expected. Consider the sources of the debit and credit entries in this account. The Overhead Variance account is debited as the superintendent's salary is earned, the rent is paid, supplies are used, and depreciation on production equipment is recognized. (The corresponding credits are to Wages Payable, Cash, Supplies Inventory, and Allowance for Depreciation.) As hours of direct labor are expended to build in-process inventory, overhead is absorbed by this in-process inventory: $25.33 of overhead for each hour of direct labor. Throughout the period, credit entries are made to the Overhead Variance account as overhead is absorbed into inventory (with corresponding debit entries to in-process inventory.) At the end of the accounting period, the credit total in Overhead Variance will be the total overhead absorbed by production during the period.

Suppose that in a particular month (1) the total amount spent on overhead cost elements is $38,800 (a bit over the estimate), (2) 1,570 direct labor hours were spent, rather than the 1,500 hours estimated, and (3) thus, $39,768 of overhead have been absorbed (1,570 hours at a rate of $25.33 per hour.) The Overhead Variance account at month-end shows the following:

Overhead Variance

38,800	39,768

The Overhead Variance account has a net $968 credit balance. Production during the period absorbed more overhead than was actually spent. This credit balance is frequently referred to as a *favorable variance* in the sense that it represents a negative expense; accordingly, a debit balance is an *unfavorable variance*.

To repeat: Overhead is not redistributed at the end of the accounting period to eliminate the overhead variance. Instead, the balance in the variance account is typically charged to Cost of Goods Sold for the period. In this case, the credit balance is a negative expense serving to reduce Cost of Goods Sold. In the example above, the overhead rate would have to have been a bit less than $25.33 in order for the overhead variance to be zero: $38,800 actually spent divided by 1,570 hours of actual direct labor equals $24.71. This 2 percent difference (or error, if you will) in the assignment of overhead is simply not material; remember that overhead accounting is, in any event, only an allocation process. The use of a predetermined overhead rate assures timely cost accounting data; that advantage is well worth the disadvantage of generating small balances in the Overhead Variance account.

But we return, now, to consider the $968 credit overhead variance. What are possible explanations of this variance? We see that the company has had higher manufacturing activity than expected: equivalent to 1,570 hours of direct labor rather than 1,500 hours. But also, the company may have spent on indirect manufacturing costs more or less than it anticipated. Thus, without further analysis (see Appendix 13A), we are uncertain whether the Overhead Variance account signals "good" or "weak" manufacturing performance.

Multiple Overhead Pools

To this point, we have assumed that all indirect production costs (overhead) would be absorbed by the use of a single overhead vehicle, typically direct labor hours or direct labor wages. In certain kinds of manufacturing or processing operations, other overhead vehicles, such as machine hours in a contract printing company or direct material costs, are appropriate, but the assumption continues that only a single overhead vehicle in used.

Consider the possibility of using more than one overhead pool, with each pool absorbed by the use of a different vehicle. This approach seems sensible given the nature of overhead in a modern manufacturing environment. For example, a recent study of overheads in the electronics industry suggests the following breakdown of cost elements:

Indirect labor	12%
Facilities and equipment	20
Materials overhead	33
Manufacturing engineering	15
Manufacturing administration	20
Total	100%

Perhaps the first and the fifth of these cost elements—indirect labor and manufacturing administration—should be distributed (absorbed) by the use of an overhead vehicle based on labor, while the second should be absorbed by machine hours, the third and fourth by value of direct materials. Three cost pools, three different vehicles, three overhead rates.

While this more elaborate distribution of overhead has a certain logic to it, the final result is still simply an allocation of overhead to products. Overhead vehicles do not *cause* the indirect cost elements. Furthermore, an increase or decrease in the use of an overhead vehicle (say, direct labor hours or machine hours) typically will not *cause* a proportional change in the overhead it is absorbing.

Therein lies the danger. Multiple overhead pools absorbed by multiple overhead rates lend an air of precision and rationality to overhead accounting that is simply not present. Evidence from actual manufacturing operations indicates that most of the overhead cost elements tabulated above are, in fact, largely unaffected by modest swings in the volume of activity (measured by direct labor hours or machine hours consumed). Yet, overhead vehicles that vary directly with volume of activity are absorbing these costs elements.

Remember that we cost-account manufactured products not solely to value inventory and the cost of goods sold. We also want cost accounting data to help make a myriad of operating decisions:

- How should we price a product? Should we increase or decrease the price?
- Should we redesign the product to achieve lower manufacturing cost? If a redesign increases the product's functionality, how much of a manufacturing cost increase can we tolerate?
- Should we subcontract certain manufacturing activities?
- Should we increase productive capacity?

Addressing these questions requires a thorough understanding of how the cost data were determined and how various operating decisions (1) drive actual overhead costs up or down and (2) are reflected by the cost accounting system. Unfortunately action (1) is not always consistent with cost changes (2). We will dig deeper into this conundrum in the next chapter.

HISTORICAL PERSPECTIVE

Part of the problem is that the cost accounting procedures and rules still in use today were established a century or more ago, during the Industrial Revolution, as manufacturing became a key economic activity. As manufacturing firms grew in size and scope, managers became distant from shop-floor activities. Only then did managers' questions of the type just cited become relevant: What does a product cost? How efficient is a section of the manufacturing department? Should a product be redesigned, or phased out, or promoted more aggressively than a related product?

Prior to the Industrial Revolution, virtually all commercial transactions occurred between an owner—a farmer or a craftsperson—and a customer. To the extent the owner employed others, he (almost always he, not she) paid them wages on a piece-rate basis. Indirect labor was limited to that portion of the owner's time not devoted directly to production. Overhead was minimal, particularly relative to material and direct labor inputs. Formal cost accounting was unnecessary.

The Industrial Revolution, however, brought increased scale of operations, investment of substantial amounts of capital, hierarchical organizations, and the employment of workers on a long-term rather than piecework basis. In turn, these owner-managers—and increasingly hired managers—needed more information on costs and efficiencies. The earliest large-scale manufacturing activities were in process-type industries, particularly textiles and steel. In these operations, raw materials were converted into a single finished product. Costs were collected for materials and labor devoted to the process, and these aggregate costs were divided by total output to determine cost per unit. Overhead was a relatively small proportion of total costs, as capital investment per employee was minor by today's standards, and indirect labor was minimal, since virtually all employees were devoted to the single production process.

With the advent of metal fabrication companies during the late nineteenth century, the process-industry cost accounting techniques were adapted to this somewhat more complex environment. These manufacturing activities consisted of many discrete steps rather than a single process, and the labor force consisted of individuals of widely varying skill levels. Supervisors, inspectors, and material handlers now spread their attention and activities across several of the discrete production steps, and thus they became indirect labor rather than direct labor. Purchased parts were combined with fabricated parts, and labor was required to assemble finished products.

Still, overhead remained a relatively small portion of total costs; direct material and direct labor costs dominated. Product expenses were defined to include overhead in addition to direct labor and material, but indirect labor comprised much of that overhead. Workers were hired when work was available and laid off when the work was completed. Thus, labor costs—both direct and indirect—varied with manufacturing volume.

Interestingly, most cost accounting procedures in use today were developed prior to 1925. These procedures define product expenses as direct material, direct labor, and overhead. What has happened since 1925? A great deal to manufacturing industries, and unfortunately not much to cost accounting procedures. The "direct material, direct labor, and overhead" definition of product expenses continues to be the accepted one for industry today. It has become a progressively less satisfactory definition over the last 80 years.

Trends in Manufacturing

Four trends are particularly noteworthy.

First, industry has become far more automated, a trend that will surely continue. Automation involves the substitution of investment capital for labor—or, in a cost accounting view, direct labor is replaced by overhead expenses, specifically the depreciation, maintenance, power, and related expenses of the automatic equipment. Industry in highly industrialized countries has moved to the point where direct labor, instead of comprising 50 or 60 percent of the final selling price to the customer, is now often only 3 or 4 percent. Cost accounting systems designed to track direct labor costs when those costs predominated now focus a great deal of attention on what has become a minor cost element.

Second, both because of changes in employee relations practices and because of the trend toward automation, manufacturing costs increasingly do not vary with the volume of activity. They are **fixed costs.** Rent and depreciation of capital equipment are obviously fixed costs. As mentioned earlier, direct labor is becoming increasingly fixed in nature, simply because managers are committed to a stable employment policy; the potential net benefits of such an enlightened policy are substantial when direct labor is a small part of the total manufacturing costs. And, as the indirect labor force has become increasingly skilled (maintenance engineers and sophisticated production planners), the indirect labor force is also essentially a fixed cost, as essential to the company's operations as are the corporate officers, in both boom times and recessions.

Third, manufacturing expenses are less discretionary than they once were, and more like marketing and engineering costs. Take the costs of manufacturing engineering for a product: We can estimate the engineering hours required to develop the process for a new product or a redesigned product, but engineering is not a repetitive task, and we cannot set reliable time standards for the task. Moreover, we are willing to exceed our manufacturing engineering hours estimate if the result is a process with substantially better reliability or lower costs than we originally envisioned. This same situation may apply to employee training or equipment maintenance expenses. We cannot scientifically determine how much we need to spend on training in order to achieve a certain level of competence. Decisions as to how much to spend on these expenditures are in the realm of art, not science.

Now suppose that, in a particular accounting period, we spend more than we planned on manufacturing engineering salaries to improve a process. Does this represent poor performance? Not necessarily. It does indicate expenditures in excess of plan, but exceeding the planned expense level may be highly desirable. Suppose that, mid-project, the manufacturing engineers discover an opportunity, with a 10 percent increase in engineering hours, to leapfrog the primary competitor's quality specifications. To miss that opportunity simply to meet the engineering hours budget would be foolish. Similarly, to continue an unsuccessful project merely because the engineering hours are available is also foolish. But terminating the project early only results in expenditures below the planned level; it does not spell good performance, since, after all, the project itself was unsuccessful.

Fourth, if a company shortchanges activities in manufacturing, the downstream consequences can be costly—in decreased customer satisfaction to be sure, but also in higher warranty service expenditures. Thus, a preoccupation with lowering product expenses, when product expenses are limited solely to manufacturing costs, may have the undesirable result of adding cost to the selling and field service departments. The *optimum* manufacturing cost is often not the *lowest*-possible manufacturing cost.

So, the cost accounting environment has become much more complex, and yet traditional cost accounting still assumes the following:

1. Manufacturing expenses are of dominant relevance: manufacturing is where conversion occurs (where value is added) so that is where management's cost control attention should be focused.
2. The key manufacturing cost is direct labor; overhead should be assigned to products based on the amount of direct labor expended on that product.
3. Product expenses comprised of direct labor, direct material, and production overhead provide the relevant data for most management decisions—this in spite of the fact that overhead is increasingly fixed in nature and is growing as a proportion of total manufacturing expenses.
4. Managers should concentrate primarily on lowering product expenses, despite the fact that these expenses are increasingly discretionary in nature.

These assumptions are simply unrealistic, and yet they underlie much of current practice. While Appendix 13A recommends an alternative to conventional practice, you should have no illusions that traditional practices will be soon altered.

Comparing 1904 and 2004 Factories

A short history lesson illustrates why we need new cost accounting methods. Traditional cost accounting procedures have led, over the past century, to an enormous escalation of overhead rates. A late 19th century manufacturing enterprise was highly labor intensive, while a contemporary one is substantially more complex and more automated. The former had relatively low indirect production expenditures, while the latter has high overhead because it has a great deal of indirect labor, high depreciation, and heavy expenditures for maintenance and power.

Suppose that cost of goods sold had the following composition in the two factories:

	1904 Factory	2004 Factory
Direct material	40%	40%
Direct labor	50	10
Overhead	10	50
Total	100%	100%

The overhead rate for the 1904 factory is 20 percent (10/50) of direct labor dollars, while the overhead rate for the modern factory is 500 percent (50/10). When overhead rates approach several hundred percent or more of direct labor dollars, as in this example, it is questionable whether labor is the most rational basis on which to incorporate indirect costs into product costs. Quite clearly, direct labor is not in any sense causing or driving the indirect costs today, although it may have done so in 1904.

What is causing or driving today's much higher indirect production costs? If it is the quantity or value of the direct material used in the product, then we could shift the overhead vehicle from direct labor to a physical or monetary measure of direct material. In a few cases, this is a sensible shift and solves the problem. In other instances (for example, the commercial printing operation mentioned earlier), a shift of the overhead vehicle to machine hours is sensible and solves the problem.

But in many of today's complex manufacturing operations, none of these traditional overhead vehicles is appropriate. Why? Because complexity and automation are the primary causes of high indirect costs. These days, indirect production costs are associated with activities such as machine setup, ordering parts, inspection, material tracking and handling, scheduling, technician support, depreciation of automated equipment, manufacturing engineering support, and parts expediting. We might argue, therefore, that a product should bear (absorb) overhead to the extent that it causes or exacerbates these costs.

For example, suppose one model of a piece of furniture is made in small lot sizes of, say, 20 units, while another model, because it enjoys higher customer demand, is appropriately made in lots of 200. Then the first model will be more expensive to manufacture than the second even if the values of direct material and direct labor are the same for the two models. Suppose the same costs are incurred to schedule a production lot of each model and to set up furniture-making machinery and assembly fixtures, regardless of lot size—not an unreasonable assumption; the per-unit cost of these activities will be 10 times as high for the unit produced in small lot sizes of 20 as for the one produced in lots of 200. An overhead vehicle of either direct labor or direct material will not capture this important difference.

Another example: Suppose a single plant manufactures both standard motor-generator sets and customized motor-generator sets. A few customers who operate the units in particularly demanding environments require the customization. Traditional overhead accounting results in the customized sets costing, say, 20 percent more than the standard sets; this would typically be the case if the prime costs (direct materials plus direct labor) of the customized sets are 20 percent greater than the prime costs of the standard sets, and overhead is absorbed on the basis of direct labor or direct material. Yet, the customized sets are, in effect, the cause of the complexity of the factory; it is the customized units that are difficult to schedule, that require specialized purchasing, that consume hours of final testing and manufacturing engineering time to be certain that the customer's particular specifications are met. Any analysis of this manufacturing plant would quickly convince us that the customized sets should absorb more of the overhead than the amount allocated to them by the traditional procedures.

Similarly, an electronic instrument comprised of 400 separate parts is typically more expensive to manufacture than another instrument with only 150 parts, even if the dollars of direct material and direct labor for the two instruments are identical. More parts have to be ordered for the first model, and then these parts have to be received, inspected, inventoried, and moved to the production floor, and all of these activities have to be tracked. The traditional overhead vehicles are unable to capture these additional costs.

Shortcomings of Traditional Cost Accounting Processes

Why are we concerned about this issue? As long as the factory absorbs all of the overhead, variances are not extraordinarily large, and the cost accounting system yields a product cost that can be used to value inventory and the cost of goods sold, isn't the cost accounting system doing its job? The problem is that managers must make day-to-day decisions based on the information they obtain from cost accounting reports.

Note the kind of erroneous decisions that might arise from the examples just cited. First, and perhaps most important, are pricing decisions. While the extent to which product *costs* influence product *prices* varies widely among industries and companies, in many instances costs play some role—often an important role (and sometimes a dominant but inappropriate role). Assume that the instrument with 400 parts has the same manufacturing cost as the 150-part instrument, according to traditional cost allocation procedures. If that information leads the marketing staff to price the two instruments identically, the price of the 150-part instrument will be uncompetitive in the marketplace, and the price of the 400-part instrument will prove so attractive to customers that the company's sales will be skewed in that direction. The instrument that has 400 parts should be priced higher, and the one with 150 parts should be priced lower.

Similar distorted pricing of the two furniture models will result, and customized motor-generator sets will sell at such a small premium over standard sets that customers will be induced to prefer customized sets, with the result that the manufacturer's costs will escalate and its profits will suffer.

Second, design engineers of these furniture models, motor-generator sets, and instruments look to cost accounting information to guide their decisions about product redesigns or designs of the next product generation. Traditional overhead allocation methods, where direct labor is the vehicle, will signal the design engineer to reduce the direct labor demands of the design; note that, if the overhead rate is 500 percent of direct labor wages, a $1 savings in direct labor appears to result in a $6 savings in total product costs. At least, that is what the cost accounting system tells the design engineer, although even a cursory look at the plant would convince you that a small reduction in direct labor is unlikely to have more than a trivial effect on actual overhead expenditures.

In fact, the manufacturing engineer in the furniture factory might best focus on reducing the setup times required so that the company can better afford to produce certain models of furniture in small lots. The engineer in the motor-generator firm might consider whether the introduction of a new standard motor-generator set might satisfy the demands of a group of customers who have been purchasing customized sets, thus improving the standard-customized mix of manufacturing activity. And, the engineer considering a redesign of the electronic instrument should probably ignore direct labor and focus on reducing the number of discrete parts that comprise the design, as it is material ordering and handling that is the major cause of overhead costs in that factory.

ACTIVITY-BASED COSTING

Given that modern manufacturing contrasts sharply with those manufacturing activities that were prevalent during and following the Industrial Revolution when the traditional method of cost accounting was invented, is there a better way?

Over the last few years a new method of attributing overhead to products has gained some popularity. This method, called **activity-based costing (ABC),** utilizes certain characteristics of products, as well as certain measures of activity, as bases for assigning overhead to individual products. Advocates of ABC argue that the result is better information for managers to use in making key day-to-day pricing, product design, and marketing strategy decisions.

Cost Drivers

Activity-based costing seeks to identify those activities or product or process characteristics within the manufacturing or service operation that cause, or drive, indirect production costs. As just noted, these **cost drivers** are often not related to direct labor or direct material but instead are related to the complexity or automation of the manufacturing process.

In recent years, industry has dramatically reduced manufacturing costs by the following means:

- Reducing the amount of in-process time—the amount of time required to complete an activity or build a product.
- Reducing the number of parts that must be purchased, and the number of vendors from which they are purchased; the benefits sought are often improved quality and better quantity discounts, but an important related benefit is reduced indirect production activity.
- Eliminating inventory by means of a *just-in-time (JIT)* inventory system; the resulting reductions in the amount of capital tied up in inventory are important, but often of more importance is the reduced indirect production activity that results when parts move directly from vendor to assembly line without having to be inventoried and subsequently issued.
- Targeting small market segments with standard products for which setup times are so low (because of the use of flexible manufacturing systems) that small-lot-size production is practical and customization of products is reduced or eliminated.

In each of these cases, management has focused on, and reduced overhead by, changing the design of the product or the design of the manufacturing process.

ABC requires that we define the characteristics of products and processes that give rise to overhead expenditures—that is, the cost drivers—and then

1. Collect the costs associated with each activity; and
2. Allocate the costs to the products or services on the basis of the demand that each product or service makes on the activity.

Many of these cost drivers are unrelated to the volume of a production run. In our example of the instrument manufacturer, some possible cost drivers are as follows:

- Number of setups
- Number of discrete moves of material (material handling)
- Parts ordering required (number of orders placed)
- Product updating required (number of engineering change orders)
- Product complexity, as measured by the number of hours required in final system checkout
- Space consumed (occupancy costs), as measured by the number of elapsed days in process

Indirect manufacturing costs are then compiled by activity: purchasing, scheduling, material handling, manufacturing engineering, maintenance, and so forth. Each of these is then assigned to a driver. Estimates of the amount of the cost driver to be incurred or consumed for a period across all of the firm's products or processes determine the overhead charged to the products making demands on the activity.

Recognize that the selection of cost drivers can have behavioral effects, some desirable, some not. If the number of parts is selected as a cost driver, design engineers will be motivated to go to great lengths to eliminate or combine parts. To the extent that this effort simplifies the tasks of the company's purchasing, receiving inspection, and material handling departments, the motivation is productive; if this preoccupation with reducing the number of parts leads to a degradation of the performance of the product, the motivation may be dysfunctional. Similarly, using number of setups as a cost driver will encourage longer production runs, but that benefit needs to be weighed against the interest and obsolescence costs of carrying additional inventory.

But, when all is said and done, we are still dealing with allocations—or, perhaps more accurately, the attribution—of costs to products or services. And, like all cost accounting methods, activity-based costing involves compromises, approximations, and assumptions about causality. The resulting information is useful, but not unassailable as to accuracy and relevance.

When Is Activity-Based Costing Important?

You may by now have the impression that an activity-based costing system is substantially more expensive to operate than a conventional system utilizing a single cost pool and a single overhead rate. You are correct. The manufacturing process must be studied in detail, cost drivers selected carefully, and the use of each driver by each product measured. Cost accounting managers must trade off the additional costs of operating an ABC system with the added benefits of improved decision making.

Activity-based costing has been enthusiastically pursued primarily in those situations where a diversity of products emanates from the same production facility, where direct labor, machines, supervisors, production schedulers, material handlers, and manufacturing engineers spread their attention across an array of product lines and product models. The key dimensions of product diversity appear to be these:

- Diverse in the number of parts that each comprises
- Diverse in product size, with some large products consuming large dollar values of direct material and direct labor, other products consuming substantially smaller dollar values
- Diverse in lot size, with some products built in very large lots and others built a very few at a time
- Diverse in performance specifications, with some products demanding greater precision, more elaborate final checkout, and more attention from manufacturing engineers than other products that readily achieve their less demanding specifications
- Diverse in the extent to which they are customized for particular customers

Traditional overhead cost allocation methods, including those that use multiple vehicles, in comparison with ABC, will *overcost* the following:

- Products with fewer parts
- Larger products
- Products built in larger lots
- Products with less demanding performance specifications
- Standard (as contrasted with customized) products

This overcosting results from the fact that a disproportionate share of indirect costs is assigned to these products when in fact these costs are being driven by complexity (products with high part counts), frequent setups (products built in smaller lots) and difficult-to-meet standards (products that consume extra time on the part of manufacturing engineers and final test personnel).

The result is a misunderstanding of true manufacturing costs that often leads to cost *cross-subsidies* (to use an economist's phrase) and thus to errors in pricing or in product design. Competitors may exploit this lack of understanding.

Traditional methods die slowly. Both managers and accountants are reluctant to jettison a system of accounting for product costs that has been in use for many years and has resulted in volumes of comparative cost data. The fact that activity-based costing often results in product cost information that is very much at odds with the information yielded by the traditional cost allocation method contributes to the anxiety.

For our purposes here, we should recognize the shortcomings of the traditional cost allocation methods and keep an open mind regarding the possible benefits of the somewhat more complex activity-based system. In these days of increased worldwide competition and additional focus on competitiveness, we need a cost accounting system that focuses the attention of managers on the key competitive product, pricing, promotion, and design decisions. The benefits of ABC can often be gained from a study (outside of the formal cost accounting system) of activity costs; the information thereby obtained can go a long way to redirect the company's pricing, design engineering, process engineering, and capacity planning.

COSTING JOINT PRODUCTS AND BY-PRODUCTS

This is a good place to explore briefly an interesting cost accounting dilemma that arises when two or more different end products emerge from a single process or a single job order. These are called **joint products.** Examples are the various distillates derived from a catalytic cracking tower for processing oil; the end products made from a steer, including the various cuts of meats and the hide; and the various grades of agricultural products that are realized from a sorting or classifying operation. When one of these end products is not specifically sought and is of substantially less value than the primary product or products—for example, a lesser mineral when gold is mined—it is called a **by-product,** a special kind of joint product.

The dilemma in costing joint products is to decide what portion of the total labor, material, and overhead incurred is properly assignable to each joint product, or to the by-product. Some logical but inevitably arbitrary allocation method is used—not the first time we have encountered the need to be arbitrary, even if logical. Sometimes this method is based on the relative sales values of the joint products, or their relative weights, or some other criterion. For by-products, any amounts realized from the sale of the by-product may be offset against the process's operating costs, or the by-product cost may be set at zero. Obviously, no precisely accurate method of assigning costs among joint products exists; moreover, while joint product costs are adequate for valuing inventory and cost of goods sold, they are typically useless for making other management decisions.

COST ACCOUNTING IN OTHER FUNCTIONS AND INDUSTRIES

Remember that these techniques of product costing in a manufacturing setting also apply to many nonmanufacturing activities. For example, analogous techniques are widely used in the construction industry. Direct material, direct labor, and indirect construction costs are incurred in excavating, carpentry, concrete work, and landscaping. Some combination of these activities is required to construct a building, a road, or a dam. The contractor is typically building a single unit, not multiple ones, and the inventory (the construction project) does not physically move toward its finished goods state; however, in most other respects, construction and manufacturing are very similar.

Service industries also must cost-account—that is, account for the cost of providing a service. A plumber expends direct materials, direct labor, and indirect costs in completing a repair job, and he or she accounts for these expenditures as a basis for billing the customer. A lawyer incurs direct labor (the lawyer's time and perhaps that of associates and aides), as well as many indirect costs (rent, telephones, and insurance, for example), in completing an assignment for a client; again, cost accounting data are needed for billing purposes, but they are also needed to assure sound management of any sizable professional service firm.

Hospitals and medical clinics cost account. Research universities are required by their external funding sources (primarily the federal government) to cost-account research projects; faculty, graduate students, and research associates are the direct laborers. Local governments cost-account various municipal services, including road repair, custodial activities, library operations, and refuse collection.

And, within a manufacturing company, cost accounting is useful for activities outside the factory. For example, how much does it cost to develop a new product? or to make a sale? These costs are not used in valuing inventory and cost of goods sold, but they are very useful in managing these functions, in making staffing decisions, and in a host of other ways. So, just as manufacturing managers are interested in production costs, sales managers and engineering managers need data on the costs of developing products or effecting sales.

However, a word of caution: It is far more difficult to cost these nonmanufacturing tasks; unlike most production tasks, engineering and selling are relatively unprogrammed. You can estimate quite accurately the materials and labor hours needed to complete a production task; the same is not true of selling or of creative engineering. How knotty is the technical problem? How many different approaches will be tried before a solution is found? How creative is the engineer assigned to the task? How enthusiastic are customers about the product? How tough is the competitive battle on a particular order?

Nevertheless, even for these unprogrammed engineering and selling functions, historical cost information is useful. Knowledge of the labor, material, and indirect expenses utilized on an engineering design project helps the company track the deployment of its technical resources and provides a basis for forecasting and budgeting the future. Cost accounting is unable to cost the closing of a sale, since one cannot predict whether a particular sales call will lead to a sale, or how many calls will occur before the sale can be closed. However, making a sales call is a definable task for which cost data can be collected, and these data help in planning future selling activity.

So, cost accounting is pervasive in our modern society. But it shouldn't be overdone. We have said repeatedly that a merchandising firm—wholesaler or retailer—does, and probably should, include only direct material costs (that is, the purchase cost of the merchandise being resold) as product expenses. All other expenses should be treated as period expenses. This structure readily provides the merchandising managers with terribly useful information: the margin between selling price and acquisition cost. Of course, some merchandise takes up more space, requires more handling, is more difficult to sell, and flows through inventory more slowly. All of these conditions add to the cost of carrying such merchandise; that is, they contribute to higher overhead (indirect) costs. We could distribute this overhead to the various merchandise items as a percentage of their selling prices, or based on the number of days they are in inventory, or on some other basis, but merchandising managers have traditionally not found much use for elaborate cost accounting reports that distribute overhead. They use various rules of thumb—departments having lower annual sales per square foot of occupied space need to price for higher gross margins, or inventory that turns over rapidly can tolerate lower gross margins—but raw gross margins, unaltered by the distribution of indirect costs, remain the key cost accounting data.

Finally, one other accounting principle introduced in Chapter 6 deserves special emphasis: materiality. The cost accountant, more even than the financial accountant, tends to violate this guideline. There is almost no limit to the detail and apparent precision that can be included in cost accounting reports. Thus, we must strike a balance at the point of diminishing returns from increased detail, bearing in mind the following:

1. Information costs money. Operating personnel must keep detailed records, and these records must be processed. Final reports must be prepared and reproduced. Users must digest the reports. The more detailed is the output, the more detailed must be the input, and the more time spent in record keeping, processing, and analysis.
2. Masses of data tend more to confuse than to enlighten. Too often, the intended user despairs when presented with massive cost accounting reports, either because they are difficult to understand or because time pressures inhibit thorough analysis.
3. Apparent precision is often spurious precision. You are well aware by now that, in all areas of accounting, assumptions must be made. The more these data are manipulated, allocated, and processed, the more the final cost accounting reports take on an aura of precision. In fact, of course, imprecise input data cannot be turned into precise output information.

SUMMARY

Cost accounting develops detailed information on the cost of a product or service. That is, it focuses on the valuation of product costs as distinct from period costs. The primary audience for cost accounting information and reports is internal management rather than external audiences. Cost accounting procedures are less constrained by rules than are financial accounting procedures. While cost accounting is useful for valuing the cost of goods sold and inventory, its primary purpose is to aid managers in making sound manufacturing, marketing, product, pricing, and other operating decisions—and that should be the primary test of its usefulness.

The fundamental difference between merchandising and manufacturing enterprises is that manufacturing converts materials from one state to another and thereby increases their value. A fundamental cost accounting dilemma is deciding just what categories of costs should be included as part of this cost of conversion. For example, the categories might be (1) direct material alone; or (2) direct material plus direct labor (prime costs); or (3) material plus labor plus manufacturing overhead (indirect production costs); or (4) all of these costs plus engineering, selling, and administrative expenses. Traditionally, the third of these four definitions of product costs is used by manufacturing companies and is required for valuation of the cost of goods sold and inventory.

Today's typical cost accounting framework has changed little from the framework developed during and following the Industrial Revolution, even as large-scale manufacturing displaced owner-managed craft activities. At that time, direct labor and material dominated the costs of products, with indirect costs (overhead) of substantially less significance. Modern manufacturing—now incorporating much greater complexity and automation, changed employee relations practices, and more discretionary expenses—has substantially altered the mix of direct and indirect expenses. As a result, some of the preoccupation with manufacturing costs—and particularly direct labor costs—to the exclusion of other categories of expenses is increasingly inappropriate; so, also, are certain of the traditional cost accounting procedures, particularly those used to absorb overhead into individual products and services. The principle of materiality is often violated by cost accounting systems that develop excessive detail and apparent but spurious precision.

Determining the appropriate value of direct materials and direct labor to be assigned to a particular unit or output is relatively straightforward. However, distributing the indirect production costs (overhead) is substantially more challenging. Moreover, the resulting allocation is inevitably arbitrary, since determining the precise amount of overhead consumed by each product or in each step of manufacturing is impossible; accordingly, overhead allocations are frequently misinterpreted.

Traditionally, indirect production costs are allocated by (1) selecting an overhead vehicle (often direct labor hours or direct labor dollars) that is common to all segments of output, and then (2) predetermining an overhead rate that is applied throughout the ensuing accounting period to cost products or services as they are produced. The fact that overhead rates are predetermined necessitates that an overhead variance be developed to draw off, in each accounting period, the difference between the amount of actual overhead expenditures and the aggregate overhead absorbed.

In an effort to produce more exact per-unit product costs, some companies use multiple overhead pools, each absorbed by a different vehicle. While such allocations are more elaborate and appear more precise, they are still merely allocations; they do not reveal why a company spends more or less on overhead.

Activity-based costing (ABC), though not widely adopted by industry, attempts to couple more closely the allocation of overhead with the product and manufacturing characteristics that drive, or cause, the overhead costs. Indirect production activities are the focus of ABC, and these activities are traced through to the products in order to estimate the demand that each product (or service) places on the activity. The ABC system requires that a cost driver be identified to serve as the basis for distributing expenditures on each activity.

ABC is particularly useful where products that are diverse (in such characteristics as product complexity, production lot size, in-process time, and extent of customization) are produced in the same manufacturing facility. In these environments, traditional (and simple) overhead allocation techniques overcost the simpler, standard products manufactured in large lots.

A special case for costing individual products occurs when joint products, including by-products, are derived from a single processing activity.

Cost accounting techniques are widely applicable outside of manufacturing environments, for example, in service, construction, professional, and consulting firms, and in many other firms and functions.

NEW TERMS

Absorption. The process by which indirect production costs (overhead) are allocated to, and thereby included in, the cost of individual or groups of products or services.

Activity-based costing (ABC). A cost accounting procedure that traces costs to products according to the activities for which they are responsible; overhead costs are assigned to products by identifying the product-by-product consumption of *cost drivers.*

By-products. A category of joint products consisting of products of substantially less commercial value than the other joint products derived from the same process.

Conversion. The process of transforming raw material into in-process and finished goods inventory. The process of conversion distinguishes manufacturing from merchandising and service activities.

Cost accounting. The set of accounting techniques that develops detailed information about the cost of a product or service.

Cost drivers. The basis on which overhead expenditures are allocated to products in an activity-based costing system; examples are number of set-ups, number of discrete parts, number of times handled.

Cost plus. An arrangement between customer and supplier that provides that the price charged be a function of the cost of performing the contract—that is, cost plus a fixed fee, or cost plus a certain percentage of cost.

Direct labor. Wages paid for time spent on production activities that can be identified directly with products manufactured or services performed. Direct labor is a product cost, generally (but often erroneously) assumed to be variable.

Direct material. The portion of product costs represented by the cost of materials that end up on, in, or otherwise part of the final product. Direct materials are a variable product cost.

Fixed costs. Costs that remain unchanged (do not vary) with modest changes in volume of activity.

Indirect labor. Wages paid within the manufacturing function for activities that cannot be identified directly with products manufactured or services performed (as can direct labor).

Joint products. Two or more different end products that are derived from a single process or job. The cost of each joint product is determined by allocation of the joint costs.

Make-or-buy. A management decision whether to purchase a particular part or service from an outside vendor or produce the part (perform the service) within the firm.

Overhead rate. The amount of overhead to be allocated to a product or process as each unit of the overhead vehicle is consumed by that product or process. Overhead rates are generally predetermined as follows:

Overhead rate

$$= \frac{\text{estimated overhead expenditures at planned production volume}}{\text{estimated quantity of the overhead vehicle at that planned volume}}$$

Overhead variance. The difference between actual overhead expenditures for the accounting period and the aggregate of all overhead absorbed during the same period. When actual overhead expenditures exceed the amount absorbed, the variance has a debit balance, and overhead is described as being underabsorbed.

Overhead vehicle. An activity or cost (such as direct labor hours or direct labor dollars) common to all production activity in the enterprise that is used to allocate overhead to those productive activities.

Prime costs. A product cost definition that includes only direct material and direct labor as product costs and excludes all overhead.

Variance. General term that referes to the monetary difference between actual results and the original plan or budget.

APPENDIX 13A: VARIABLE AND FULL-ABSORPTION COSTING

This appendix explains the difference between two alternative definitions of overhead inclusiveness: (1) **variable** (or **direct**) **costing** defines overhead to include only its variable cost elements, the fixed elements being treated as period costs. (2) **Full-absorption costing** defines overhead to include both fixed and variable overhead cost elements. Full-absorption costing is more prevalent than variable costing but results in an overhead variance that is challenging to analyze and potentially misleading.

Review of the Overhead Variance Account

Recall that the Overhead Variance account is debited as various indirect cost expenditures are made, for example, indirect labor wages, power, rent, and insurance. Credit entries occur as the overhead rate is applied, by means of the overhead vehicle, to absorb overhead into in-process inventory. Thus, if the vehicle is direct labor hours, each time an hour of direct labor is charged to a particular job or process, overhead is also charged, in the amount of the overhead rate per direct labor hour.

At the end of an accounting period, the aggregate debit in the Overhead Variance account equals the amount spent (or accrued) during the period on all overhead cost elements, and the aggregate credit is the total overhead absorbed into production during the period. Only by shear and infrequent coincidence are aggregate debits equal to aggregate credits. What conditions would lead to a zero balance? First, spending on overhead cost elements would have to be smack on budget, with no deviation. Second, the amount of production activity—as measured by the quantity of the overhead vehicle by which overhead was absorbed—would have to be smack on budget, not off by even a few direct labor hours (or direct labor dollars, or machine hours, or whatever the overhead vehicle.) Neither of these conditions is likely to be met frequently, to say nothing of both. Therefore, some remaining variance balance is typical.

So, what does the overhead variance mean? That answer depends, in part, on how *overhead* is defined.

Interpreting the Total Overhead Variance Balance. Earlier in the chapter (in the subsection "Overhead Variance"), this overhead variance account appears:

Overhead Variance

38,800	39,768

The account has a net $968 balance. Because it is a credit balance, we are tempted to say that it is a favorable variance; however, please withhold judgment, as we will see that the forces causing this credit balance may not be favorable.

Now to the interpretation: First, we might assume that all overhead cost elements are variable expenses—that is, they vary directly and proportionately with the volume of production activity, just as we assume direct material does. *Fixed* indirect expenses, then, are considered, under this assumption, to be period expenses and are not absorbed into the products. If these assumptions hold, the variance is entirely a spending variance; that is to say, it is unaffected by differences between planned and actual volume of production, and the variance simply means the company spent less than planned on overhead.

Figure 13A-1 illustrates this situation. Two dotted lines show the derivation of the overhead rate: the planned overhead expenditures for the anticipated production level (P_2) are divided by the volume of the overhead vehicle

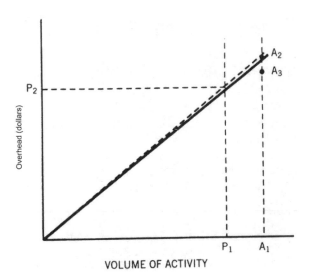

P_1 Planned activity level
P_2 Planned overhead at planned activity level
A_1 Actual activity level
A_2 Overhead actually absorbed = $39,768
A_3 Overhead actually incurred = $38,800

Figure 13A-1. Overhead accounting: variable costing

planned (P_1). The dashed diagonal line is the overhead budget line and, since all overhead cost elements are variable, the line starts at the origin and defines budgeted, or planned, overhead at all possible levels of production. As production occurs, the overhead rate is applied, and overhead is absorbed according to the solid line on the figure. Here, because all overhead is variable, the diagonal dashed and solid lines are coincident.

Figure 13A-1 illustrates a condition where actual production activity exceeds the planned level. Because all overhead costs are variable, the absorption of overhead—the credit to the Overhead Variance account—defines a volume-adjusted budget for overhead. Where is actual overhead (the debit to the Overhead Variance account) shown on this figure? It is, of course, unrelated to either of the two lines. It is point A_3, at the level of \$38,800 (in this example), below the overhead absorption line. The difference between that point and the absorption line defines the \$968 indirect production costs credit variance. Had the point occurred above the absorption line, the variance for the period would, of course, have been negative, or debit.

Now suppose that overhead is defined to comprise some cost elements that are fixed—costs (such as depreciation and supervision salaries) that don't vary with volume of production—and some that are variable (such as utilities and indirect materials). Here, two effects drive the overhead variance:

1. Spending above or below the planned budget level
2. Producing above or below the planned volume level

Figure 13A-2 illustrates this more complex variance. Once again, two dashed lines show the derivation of the overhead rate. Note that the dashed diagonal line (labeled "overhead budget line") does not begin at the origin but, rather, part way up the vertical axis, a distance equal to the value of planned expenditures on fixed overhead cost elements. Remember that these budgeted fixed costs are unaffected by the level of production activity, but also note that they will be exactly absorbed only if the production level exactly matches the planned level. If actual production exceeds or falls short of the plan, overhead will be, respectively, overabsorbed or underabsorbed.

To sum up, Figure 13A-1 illustrates a variable costing environment, while Figure 13A-2 illustrates a full-absorption costing environment.

Full-absorption costing is most prevalent in the United States today, but variable costing is gaining in popularity, and for very good reasons. Before discussing the advantages and disadvantages of these two costing schemes (schemes that differ only in the definition of cost elements included in overhead), we need to consider how to interpret the overhead variance in full-absorption costing. Because interpretation is not simple, misinterpretation of overhead variances is widespread, which, in turn, is a major disadvantage of full-absorption costing systems.

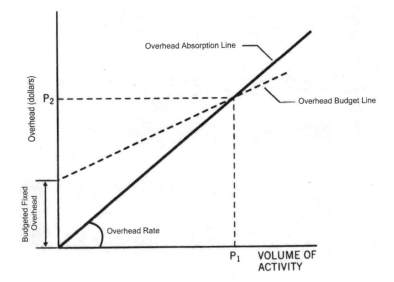

P_1 = Planned activity level
P_2 = Planned overhead at planned activity level

Figure 13A-2. Overhead accounting: full-absorption costing

Separating Overhead Volume and Spending Variances

Variance analysis in full-absorption costing is complex only because overhead is comprised of both fixed and variable cost elements. Thus, the discussion in this section applies to full-absorption costing but not to variable costing.

In full-absorption costing, the overhead rate absorbs both fixed and variable overhead elements. At the planned activity level, the amount of fixed overhead absorbed is just right—just enough and not too much. At any other activity level—above or below the plan—and the fixed overhead will be over- or underabsorbed relative to the budget.

Recall the statement earlier that overhead variance in full-absorption costing is driven by two effects: *spending* deviations from budget and *volume* of activity deviations from plan. To develop meaningful data for overhead cost control in a full-absorption environment, we must be able to isolate the spending effect from the volume effect.

Assume that department P at the Blume Manufacturing Company expects to use 4,500 hours of direct labor per month to accomplish its production plan for the forthcoming year. To support this activity level, the department estimates that total overhead expenditures will be $144,000 per month (including all overhead cost elements: variable, semivariable, and fixed). The resulting overhead rate for department P for the year is $32 per direct labor hour ($144,000/4,500 hours).

This $32 rate is used throughout the accounting period (the year). Since the overhead vehicle is direct labor hours, if job 246, for example, consumes four direct labor hours, it is charged with four times the overhead rate, or $128; in the jargon of cost accounting, job 246 "absorbs $128 of overhead". Throughout the accounting period, many similar entries are made, as direct labor is expended on department P's jobs. At the end of the period, the aggregate credit in the Overhead Variance account equals the total overhead absorbed by all jobs, and the aggregate debit is the total amount actually expended on the overhead during the same period.

Assume that, for department P, the Overhead Variance account shows the following at the end of October:

Overhead Variance

148,500	150,400

At first glance, we might conclude that the department P supervisor did a good job of controlling overhead expenses: the amount actually spent ($148,500) is less than the amount absorbed ($150,400). However, because Blume operates a full-absorption cost system, embedded in this single variance are two primary effects:*

1. *Volume:* If department P was more active this month than planned—that is, utilized more direct labor hours—more overhead was absorbed than planned.
2. *Spending or Efficiency:* In comparison with estimates (on which the overhead rate was based), Blume may have used more or less power, more or fewer indirect labor hours, or more or fewer production supplies, and so forth.

The task is to separate the single overhead variance into the two components just described: **overhead spending variance** and **overhead volume variance.** Put another way, the volume effect must be isolated from the spending effect.

Note that the truly useful management information is a comparison of actual overhead spending with a *revised* estimate of appropriate overhead expenditures, the revision to adjust from planned to actual production activity.

---

*Price variation is a third possible effect: Prices per kilowatt-hour for power, or wage rates for indirect labor, may have been more or less than the assumed prices and wage rates in the original overhead estimates. Separating price and usage is generally not warranted with respect to overhead (although it frequently is warranted for direct material and direct labor). For convenience, this appendix combines the price and usage effects.

As illustrated in Figure 13A-1, a volume-adjusted overhead budget is self-defined under variable costing. Not so in full-absorption costing, where the volume-adjusted overhead budget must be derived.

Overhead Budget Equation (Line). Overhead expenditures can be rebudgeted or adjusted based on the actual activity level in department P in October. The data compiled at the beginning of the year to establish the overhead rate indicates how total overhead expenditures are expected to vary with changes in volume. (Refer now to Figure 13A-3.)

At the beginning of the year, Blume's cost analyst estimated total overhead expenditures at several different activity levels; these data were needed for the company-wide budgeting exercise that led finally to the production activity plan calling for 4,500 direct labor hours per month. The analyst estimated that total overhead expenditures would be as follows:

Direct Labor Hours	Estimated Overhead Dollars
4,000	$138,000
4,500	144,000
5,000	150,000

These points are labeled *A, B,* and *C,* respectively, on Figure 13A-3. By fitting a line to these three points, the analyst developed an algebraic expression for the overhead budget line. The intersection with the vertical axis (at $90,000) signals the fixed-cost portion of overhead. The slope of the line, equivalent to $12.00 per direct labor hour, represents the variable cost portion of overhead. (Note that $12.00 would be the overhead rate in a variable cost system, since all fixed overhead costs would be charged as period expenses.)

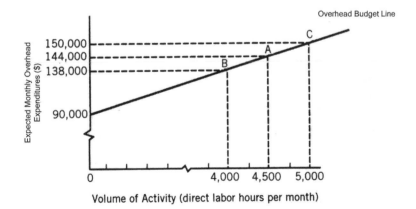

Figure 13A-3. Overhead budget estimates

In this example, the equation becomes

$$\text{Overhead} = \$90,000 + \$12.00 \times \text{direct labor hours.}$$

The slope and shape of this budget line is valid for only modest variations in production activity (for example, 20 percent above or below the planned level.) If actual production greatly exceeds or falls way short of the planned level (say, by 50 percent), the straight-line relationship between overhead and volume probably will not hold.*

Blume Manufacturing now has the information to separate the single overhead variance into its spending and volume variance components, because now a volume-adjusted overhead budget can be calculated based on the actual production volume in department P in October. A comparison of this volume-adjusted overhead budget to actual overhead expenditures reveals how well department P controlled spending on overhead.

During October, department P utilized 4,700 direct labor hours, verified as follows. The aggregate credits in the overhead variance account resulted from 4,700 direct labor hours: Each direct labor hour absorbed $32 of overhead, and the total overhead absorbed during the month was $150,400, or 4,700 hours $\times$ $32 per hour.

Activity at the rate of 4,700 direct labor hours is, obviously, somewhat above the original plan of 4,500. Accordingly, total overhead expenditures for October should logically be above the planned $144,000 per month, since certain overhead cost elements vary with volume. However, total overhead expenditures should not be as high as $150,400, the amount absorbed, since still other overhead cost elements are fixed and thus unaffected by this increased production volume.

Substituting 4,700 hours into the overhead budget equation yields a volume-adjusted overhead budget of

$$\text{Volume-adjusted overhead budget} = \$90,000 + \$12(4,700 \text{ hrs}) = \$146,400$$

Interpreting the Separated Variances. Now the single overhead variance shown above can be separated into these two variances, shown in T-account format:

---

*With more data points, a more complex curve (and resulting budget equation) could be developed to describe the variation in overhead with variations in volume. Typically, however, such added sophistication and complexity are not warranted.

Overhead Spending Variance		Overhead Volume Variance	
Actual overhead expenditures	Volume-adjusted overhead budget	Volume-adjusted overhead budget	Absorbed overhead (overhead rate × quantity of overhead vehicle)
148,500	146,400	146,400	150,400

Note several things about these two variances:

1. The accounting system is still in balance, debits equal credits; the credit balance in the Overhead Spending Variance account equals the debit balance in the Overhead Volume Variance account.
2. The algebraic sum of the two separate variances equals the total overhead variance:

$$
\begin{array}{lll}
\text{Overhead Spending Variance} & = (\$2,100) & \text{debit} \\
\text{Overhead Volume Variance} & = \underline{4,000} & \text{credit} \\
\text{Total overhead variance} & = \$1,900 & \text{credit}
\end{array}
$$

3. An overhead volume variance is always favorable when actual production activity exceeds plan, and always unfavorable when it falls short of plan. Here the activity was above plan—probably not news to Blume's managers and by itself not worth the analysis. However, only by isolating the overhead volume variance can Blume's analyst deduce the overhead spending variance.
4. The overhead spending variance is indeed useful. Blume's managers now know that overhead expenditures during October were $2,100 more than the adjusted spending plan—that is, the plan adjusted to actual activity volume.

Why Is All This Important?

Why does the interpretation of overhead variances in a full-absorption cost environment warrant such extensive analysis? The reason is straightforward: Surprisingly few managers know what over- or underabsorbed overhead means. As a result, misinterpretations of overhead variances are widespread. You need to be alert for the need to separate spending and volume effects whenever you are interpreting variances in an environment that contains both fixed and variable cost elements—and that includes almost all environments. Full-absorption costing predominates in manufacturing today, and analogous systems are used in many other kinds of operations, including most research, consulting, and professional service firms such as the following:

- Law firms
- Auto repair firms
- Architectural firms
- Contract engineering firms
- Research departments within fully integrated manufacturing firms
- Universities where research is government funded

Variable Costing

Recall that the complexity of separating volume and spending overhead variances disappears when we use a variable costing system. In variable costing, fixed manufacturing costs are treated as period expenses, and only variable costs—direct labor, direct material, and variable overhead—are treated as product costs, matched to revenue. A conventional gross margin, then, cannot be derived. Instead, a **contribution margin** is derived for a particular product or for an accounting period: it equals the difference between revenue and the variable costs that can be matched to that revenue.

The contribution margin comes much closer than does gross margin in suggesting the impact on profit of moderate savings in manufacturing activity. That is, producing a few more or a few less of a product is unlikely to cause changes in fixed costs, and, thus, the variable cost closely defines incremental costs—and it is incremental costs and revenues that are most relevant to operating decisions, as discussed in Chapter 14.

Advantages and Disadvantages of Variable Costing. While use of full-absorption costing has long been the tradition for industrial cost accounting systems, recently, variable costing has gained popularity as its advantages become more widely recognized. What are those advantages?

First, the clear and consistent separation of variable and fixed manufacturing costs readily yields data for those many analyses that require a focus on incremental or variable costs: make-or-buy decisions, capital investment decisions, and certain marketing decisions. Second, as we have just seen, overhead variance analysis is a great deal simpler under variable costing than under full-absorption costing.

Third, under full-absorption costing, if planned production volumes decrease over several years, the predetermined overhead rate spreads fixed production costs over fewer units of output, with a resulting increase in the overhead rate. As volume decreases, each unit appears to cost more. And, the reverse, as planned volume increases over several years, the overhead rate now spreads the same fixed production costs over more units of output; each unit bears a smaller share of overhead, and therefore appears to cost less.

This sounds sensible enough: the concept of economy of scale. But here the "scale" is unchanged. No change in the capacity or configuration of the factory or service operation occurs. The change is simply of output from an operation of certain scale. The change in apparent product costs, then, arises solely from the accounting treatment of fixed overhead. These changes in overhead per unit, illustrated in Table 13A-1, are easily misinterpreted and may lead to unfortunate decisions.

For example, suppose prices are set at a certain markup over cost—an illogical but frequent pricing rule. A company utilizing full-absorption costing will increase prices when planned production volumes decline, with the result that orders may decline further, and decrease prices when production volumes increase, just when strong demand suggests no need to lower prices.

In short, the contribution margin concept is more useful than gross margin. Of course, the fact that certain production costs are fixed makes them no less real. But if the level of fixed costs cannot be altered in the short run—if, in fact, they are fixed regardless of short-term swings in production output—it makes more sense to treat them as period costs, just as selling expenses and administrative expenses are treated.

A disadvantage of variable costing is that the accounting profession does not now sanction it as a generally accepted accounting principle to value inventory and cost of goods sold. Before you cast aside variable costing for that reason alone, however, recognize that it still can be used for internal management reporting; then, as financial statements are prepared for external audiences, the values of inventory and cost of goods sold can simply be restated to their equivalent full-absorption costs. Fortunately, these adjustments are easily made.

Effects on Net Profit

Are reported profits always the same under the two cost accounting methods? No, although they will generally be very close. Variable costing and full-

TABLE 13A-1. Overhead Rates at Alternative Planned Volumes: Full-Absorption and Variable Costing

	Planned Volume of Activity		
	Normal	Slow	High Output
Estimated production, in units per month	2,000	1,500	2,500
Estimated overhead:			
Variable ($1.50/unit)	$3,000	$2,250	$3,750
Fixed	4,000	4,000	4,000
Total	$7,000	$6,250	$7,750
Overhead rate, full absorption	$3.50 per unit	$4.17 per unit	$3.10 per unit
Overhead rate, variable costing	$1.50 per unit	$1.50 per unit	$1.50 per unit

absorption costing result in identical operating profits if (but only if) in-process and finished goods inventories remain the same during the period. When finished goods inventory increases during a month, full-absorption costing includes, in the valuation of that additional inventory, a portion of the month's fixed production costs; variable costing, by contract, treats all fixed production costs as period expenses of the month. Thus, in time of expanding inventories, profit is (slightly) higher if calculated under full-absorption costing; the reverse is true when inventories decline. Another advantage of variable costing, then, is that managers will not be tempted to overproduce for inventory solely to absorb more fixed overhead costs and thereby improve reported profits.

Summary

The sole difference between variable and full-absorption costing relates to the accounting treatment of the fixed cost elements of indirect production costs (overhead). When fixed overhead is absorbed into product costs—that is, when full-absorption costing is used—the Overhead Variance account needs to be decomposed into its spending and volume components. Variable cost accounting has significant advantages in providing useful data for operational decisions.

New Terms—Appendix 13A

Contribution margin. The difference between revenue and those variable costs that are matched to revenue. Contrasts with *gross margin.*

Direct costing. Another name for *variable costing.*

Full-absorption costing. A cost accounting system that defines product costs as direct material, direct labor, and all overhead, both fixed and variable. Contrasts with *variable costing.*

Overhead spending variance. A comparison of actual overhead expenditures with volume-adjusted (rebudgeted) overhead for the period.

Overhead volume variance. A comparison of volume-adjusted (rebudgeted) overhead with the overhead absorbed by actual production during the period.

Variable costing. A cost accounting system that defines product costs as direct material, direct labor, and variable overhead, while fixed overhead costs are treated as period expenses. An alternative name is *direct costing.* Contrasts with *full-absorption costing.*

APPENDIX 13B: STANDARD COST ACCOUNTING

This appendix considers the interesting opportunity to bring budgeting techniques (described later in Chapter 15) to bear on cost accounting. We shall see that so-called standard costing facilitates *management by exception,* by setting budgets (standards) and then tracking the difference between actual and budget (that is, developing variances) for product costs.

Standard costing sounds complex—and, in a sense, it is—but the benefits of standard cost accounting systems over actual cost systems have led most manufacturers of standard (or even semistandard) products to use them. While standard cost systems are a challenge to conceptualize, they are typically less expensive to operate day to day than are actual cost systems. Furthermore, while it may seem intuitively obvious that actual costs are in some sense better, more proper, or more accurate than standard costs, please withhold judgment until you know more about standard costing.

Incidentally, the choice between actual and standard cost accounting systems is independent of the choice between variable and full-absorption costing discussed in Appendix 13A.

What Are Standard Costs?

Our discussion of cost accounting thus far presumes that we track actual material costs as materials flow into and out of raw material inventory, then to in-process inventory, then to finished goods inventory, and finally to cost of goods sold. Similarly, as actual direct labor hours are spent, each hour is valued at the actual direct labor rate paid, and this direct labor cost increases the value of in-process inventory.

Alternatively, we could track standard costs of materials, labor, and overhead as these costs accumulate through in-process inventory, to finished goods inventory, and finally to cost of goods sold. To employ a standard cost system requires the following:

1. Standards must be set—that is, someone (for example, a cost accountant or an industrial engineer) must determine in advance—in effect, must budget—prior to actual production:
   a. The number of direct labor hours required per unit of output, and the hourly wage rate
   b. The number of pounds, kilos, meters, feet, or pieces of direct material required per unit of output, and the budgeted price per pound, kilo, meter, foot, or piece.

   Setting standards is simply detailed budgeting applied to product costs.
2. The differences between actual costs and standard costs must be identified in some appropriate, and hopefully useful, manner. That is, variances must be developed.

The cost model requires that a unit in inventory be valued at its *historical* cost—the cost of direct labor, direct material, and overhead incurred in making the unit. A standard cost system simply defines historical cost as what it *should* cost to produce the unit: the unit's predetermined or standard cost.

Advantages and Disadvantages of Standard Cost Systems

The primary disadvantage of standard cost systems is the expense associated with setting standards: predetermining, or budgeting, the standard costs for all the various products the company manufactures. In order to value inventories and cost of goods sold at what they *should* cost to produce, rather than what they actually did cost, someone must determine what they *should* cost. If the variety of products produced is wide and the manufacturing processes complex, setting product cost standards can be costly.

Mitigating this disadvantage is the usefulness of the cost standards themselves. They can assist managers in making a host of everyday operating decisions, such as these:

1. Pricing standard products, as well as products to be customized for a particular customer
2. Make-or-buy decisions (that is, whether to subcontract certain manufacturing or servicing activities)
3. Decisions to redesign products to reduce cost or to enhance performance
4. Production scheduling, for which the standard labor times and standard machine hours are necessary data inputs

Also, standard cost systems provide useful variance information for line operating managers. Differences between incurred (actual) production costs and budgeted (standard) production costs are highlighted to facilitate management-by-exception. If the direct labor standards accurately assess how much labor *should* be expended to produce a unit (or group of units), the difference between actual and standard direct labor costs is relevant to managers who are responsible for efficiency and cost control. A job-by-job analysis of this difference, or variance, helps managers pinpoint troublesome production jobs. Analysis of the sum of the variances on all jobs within a particular department helps managers to evaluate departmental efficiency. Similarly, variances can be developed and analyzed for direct material.

In addition, standard costing eliminates two shortcomings of actual costs systems. First, under an actual cost structure, physically identical units of finished goods inventory typically have different values if they were produced at different times. These differences are a nuisance when valuing inventories and the cost of goods sold. Recall the discussion of valuing merchandise purchased at different prices: the first-in, first-out (FIFO), and last-in, first-out (LIFO) methods. The same problem arises in valuing inventories of manufactured products. The use of standard costs eliminates the problem by assuring that physically identical units have identical product costs.

Standard cost systems are typically easier and less expensive to operate than actual cost systems because record keeping is less extensive, cumbersome, and costly. Differences between actual and standard costs are *drawn off* (isolated), at each manufacturing step. Once this difference, or variance, is removed, there is no longer a need to keep track of the actual costs of prior manufacturing steps as the product proceeds step by step from raw material to finished goods inventory.

In summary, standard cost systems are gaining in popularity over actual cost systems for three primary reasons:

1. Variance information points up exception areas—products, departments, or cost categories for which costs are not in line with plan.

2. Standard cost systems are less expensive to operate, both because cost flows are simpler and because identical inventory items are valued identically (at standard) even though manufactured at different times and therefore at different actual costs.

3. Standard costs provide handy reference data for marketing managers making pricing and product-line decisions, for purchasing agents making buying decisions, for production schedulers loading the shop, and for top management making investment decisions.

Interpreting Product Cost Variances

Standard cost variances are general ledger accounts that compare actual performance (actual costs) with budgeted or planned performance (standard costs) for selected segments of the operation. A wide variety of variances is possible. Variances can be developed, for example, for individual departments in a manufacturing or servicing company. They can be compiled across departments for specific products and individual cost elements (for example, labor or material). These variance data help managers at many levels—project leaders, department managers, plant managers, division general managers—focus on areas that are not operating on plan.

Table 13B-1 shows a variance report for the month of July for department M, one of several departments in the Exeter Manufacturing Company. To

TABLE 13B-1. The Exeter Manufacturing Co.: Variance Report, Manufacturing Department M for July

	Standard July	Actual July	Variance[a]
Direct labor wages ($000)	$18.0	$19.2	($1.2)
Direct materials used ($000)	$27.1	$29.6	($2.5)
Units of Malex produced	4,000	4,320	320

[a]Indicates effect on earnings. Amounts in parentheses reduce earnings.

simplify the example, assume department M produces only one product, Malex, and that Malex is produced entirely within this department.

You see immediately from Table 13B-1 that department M's total annual expenses exceeded the standard by about 6 percent. What do you conclude? That department M controlled expenses poorly during July, since the total expense variance is a negative (debit, or unfavorable) $3,900? Don't be too hasty. The causes of variation between actual and standard values need to be pinpointed before corrective action can be formulated, if indeed any action is warranted. Note the important nonfinancial information in the last line of the table: the number of Malex units produced, standard and actual.

In addition to volume of productive activity, the other key causes of variation between actual and plan (between actual and standard) are price changes and efficiency. Exeter may be paying more than standard prices for material; or perhaps changes in the work force mix have caused the average wage rate to vary from standard. Of course, the third cause—the one we tend to think of first—is efficiency.

To summarize, in addition to the effect of volume changes, two other factors influence how much is spent on direct labor wages: the efficiency of the labor force and the wage rates paid. Obviously, the same can be said about direct material expenditures: Exeter may be more efficient or less efficient in its use of materials, and, in addition, it may be paying more or less than standard prices for the materials used in July. We turn now to isolating these effects: volume; price, or wage rate; and efficiency, or usage.

Separating Price and Usage Variances

If standards are established with respect both to usage and to price, then separate variances can be generated. While such refinement is not warranted for all variances, it is particularly appropriate for direct labor and direct materials, both because they are large costs and because the necessary data to set the standards are generally at hand. With these standards, the cost accountant can develop separate **wage rate variances** from **labor efficiency (or usage) variances,** and also separate **purchase price variances** from **material usage variances.**

Table 13B-2 provides the information needed to develop material and labor variances, including separate usage and price variances, in Exeter's department M for July. Part A shows standard costs of material and labor for one unit of Malex. Note that separate standards have been established for the quantities and the prices of the inputs (material and direct labor). Part B of the table details the actual results for the month of July. Obviously, you need both sets of data—standard and actual—in order to develop variances. Table 13B-3 derives a single variance for each of direct material and direct labor. These variances combine the effects of variations in price (wage rate) and usage (efficiency). The $29,600 actual material cost for the month compares with the standard cost of $30,068 (4,320 units produced times the standard

TABLE 13B-2. Exeter Manufacturing Company: Background Data for Analysis of Material and Labor Variances for July

A. Standard Material and Labor Costs (per Unit of Malex)			
Material: Standard Quantity × Standard Price = Standard Material Cost			
Item A	24 grams	$0.20/gram	$4.80
Item B	2 units	$1.08 each	2.16
		Subtotal, material	$6.96
Labor: Standard Hours × Standard Wage Rate = Standard Labor Cost			
Skill P	0.21 hours	$14.00	$2.94
Skill Q	0.12 hours	12.00	1.44
		Subtotal, labor	$4.38
Total standard material and labor costs			$11.34

B. Actual Results, July	
Quantity produced from Table 13B-1)	4,320 units
Direct material used:	
Item A	103,000 grams
Item B	8,700 units
Total material cost (from Table 13B-1)	$29,600
Direct labor hours used:	
Skill P	895 hours
Skill Q	510 hours
Total labor cost (from Table 13B-1)	$19,200

material cost per unit of $6.96), resulting in a credit, or favorable, balance of $468. (Note that the standard cost has been adjusted for actual volume produced, 4,320 units of Malex. That is, the $30,068 standard is the volume-adjusted budget for material.) This single variance doesn't tell you whether the favorable balance arose from lower material prices, reduced material usage, or some combination of the two—or whether perhaps some inefficiency in the material use was more than offset by actual prices well below standard, or vice versa. Similarly, the total labor variance is a $278 debit, but the separate effects of wage rate variations and labor efficiency or inefficiency are not discernible.

Table 13B-4 separates the total material variance ($468 credit) into its price and usage variance components. (The total labor variance can be decomposed in an exactly analogous manner.) The price variance compares actual and standard prices for the actual volume of material used; thus, neither the total production volume nor the efficiency of material usage (that is, the extent of wastage or scrap) affects this variance. The credit entry to this purchase price variance (actual material quantity used times standard prices) is balanced by the debit entry to the usage variance. Thus, the usage variance compares

TABLE 13B-3. Exeter Manufacturing Company, Variances (Combined) for Material and Labor for July

Material Variances		Labor Variance	
Actual total material cost	Standard material cost of output	Actual total labor cost	Standard labor cost of output
$29,600	4,320 units × $6.96 = $30,068	$19,200	4,320 units × $4.38 = $18,922
	Balance = $468 credit	Balance = $278 debit	

TABLE 13B-4. Exeter Manufacturing Company, Separate Price and Usage Variances for Material for July

Material Purchase Price Variance		Material Usage Variance	
Actual total material cost[a]	Actual material quantity used × standard prices	Actual material quantity used × standard prices	Standard material cost of output
$29,600	A: 103,000 grams × $0.20 = $20,600	$29,996	4,320 units × $6.96 = $30,067.20
	B: 8,700 units × $1.08 = $9,396		
	Total A + B = $29,996		
	Balance: $396 credit		Balance: $71.20 credit

[a]Actual material quantity used × actual prices.

actual and standard usage, both valued at standard prices. Variations in material prices do not affect this variance, as standard prices determine both the debit and credit entries.

Now, how should we interpret these variance data? Note that the $468 credit in the total Material Variance account (Table 13B-3) shows the combined effect of actual prices below standard ($396 credit) and efficient material usage ($71 credit). That is, the arithmetic sum of Table 13B-4's separate price and usage variances equals the corresponding total variance in Table 13B-3 (with allowances for rounding):

Material purchase price variance	$396 credit
Material usage variance	71 credit
Total material variance	$467 credit

Thus, the single variance has been decomposed into its two component parts.

A strong argument for separating price and usage variances is that the two effects—price and efficiency—are the responsibility of different managers within the organization. The purchasing department negotiates material prices; the efficiency with which materials are used—that is, the extent of spoilage, waste, or rework—depends to a large degree on the care and competence of the direct labor force. Another argument for separation is that a favorable price variance might lead to an unfavorable usage variance if, for example, the company purchases lower-cost but inferior materials, resulting in high wastage or spoilage. And, at other times the reverse can be true.

Effects of Tight and Loose Standards

The discussion thus far of standard costing may imply that standards—that is, budgeted product costs—can be set with precision. In fact, they are no more than estimates, even if rigorously arrived at. Moreover, managers frequently adopt the philosophy, or policy, of setting standards on the tight side (low and therefore a challenge to achieve). The magnitude of variances is, of course, greatly influenced by the philosophy that managers adopt. Some managers set standards tight because they believe standards should provide an incentive, a goal, toward which the organization should strive. If standards are set tight, negative (debit) variances typically occur. Managers who believe in **tight standards** anticipate debit variances and plan accordingly. Managers who use **loose standards** anticipate credit variances.

Because tight standards lead to marginally lower product costs and debit variances, they result in marginally lower values of inventory and cost of goods sold. In the long run, these two effects—low product costs and high debit variances—exactly offset each other. But not necessarily so in the short run. In fact, a company's profits are unaffected by the tightness or looseness of standards only if production volume equals sales volume and, thus, inventories are unchanged over the period. If inventories grow when standards are tight, the value added to inventory is low (by comparison with the value added under looser standards) and the debit variances generated in the production of these inventoried products reduce profit in the current period.

Other Variances in Standard Costing

This discussion does not include all the possible variances used in practice. As we will discuss in the next chapter, managers often find it useful to develop variances on, for example, revenues or sales, as well as on costs and expenses. Recall that any comparison between actual results and budgeted or expected results is termed a variance. Deviations from standard or budgeted sales revenue can result from variations in numbers of units sold (volume effect) or from variations in sales price realized (price effect), or from some combination of the two. Using procedures just explained, these two effects can be isolated and reported.

Some Final Caveats

This discussion of standard cost systems needs to conclude with several caveats. First, as emphasized elsewhere, it costs time and money to develop, compile, and analyze accounting detail. These efforts must be balanced against the benefits derived from the additional information. Striking the balance between too little and too much detail is a challenge; too often, excessive detail is reported.

Second, don't be misled by the common jargon that refers to debit variances as unfavorable, or negative, and credit variances as favorable, or positive. Any deviation from standard—debit or credit—is worth investigating. The standard represents the plan, and managers need to understand the reasons why actual operations deviate from that plan, whether these deviations add to or reduce profits.

Third, it's important to consider the accuracy of the standards before hastily judging the meaning of variances. Variances can arise simply because standards were poorly set. Sometimes standards are little better than off-the-top-of-the-head guesses, particularly in companies operating with understaffed accounting departments or in volatile industries with rapidly changing product lines. In other companies with extensive cost histories and a staff of industrial engineers to establish and revise standards, cost standards may be very accurate indeed. In practice, the need for accuracy depends on the uses to which the standards are put. Where labor-hour standards are the basis for employee compensation—either as piecework (so much per unit produced) or as a group bonus—the standards must be set with precision, using detailed time studies or other sophisticated engineering techniques.

Regardless how the standards are set, they should be reviewed periodically for reasonableness, taking into account changes in the products, manufacturing methods, and staffing. Generally, standard product costs are reviewed when a major error is suspected; otherwise, not more frequently than once per quarter nor less frequently than once per year.

Finally, remember that employees are motivated—potentially constructively or detrimentally—by standard costs (and by budgets). Some managers argue that standards and budgets should be set low (tight) to spur employees to be more efficient; they feel that high (loose) standards condone a lax attitude toward cost control. Other managers argue that standards and budgets set too tight are likely to be ignored by employees as being unreasonable and therefore irrelevant.

Summary

Standard costing systems bring the benefits of budgeting and variance analysis to product costs. To implement these systems, product costs must be estimated (that is, predetermined or budgeted) in detail—an expensive activity and one prone to error under certain circumstances—including the amount of material to be used, prices of the material components, labor hours, wage rates, and machine hours. These data typically have important uses beyond the cost accounting department, for example by the scheduling, marketing, and design engineering departments.

Despite appearances, standard cost systems are typically less expensive to operate than actual cost systems because inventory valuation procedures are simplified.

Moreover, of course, the differences between standard and actual costs are useful to management in monitoring performance by department or by product line and deciding what, if any, corrective action needs to be taken. On the other hand, some standard cost accounting systems generate more detail than is really useful.

The decomposition of material and labor variances into their price (wage rate) and usage (efficiency) components is often warranted, since these decomposed variances have relevance to different parts of the organization. These same techniques are useful in analyzing revenue by decomposing sales variances into price, volume and mix components.

New Terms—Appendix 13B

Labor efficiency variance. A comparison of actual and standard direct labor hours, both valued at standard wage rates.

Loose standards. Contrasted with *tight standards,* loose standards are higher estimates of standard product costs.

Material usage variance. A comparison of actual and standard material quantities used, both valued at standard prices.

Purchase price variance. A comparison of actual and standard prices for the actual quantity of material acquired or used.

Tight standards. Contrasted with *loose standards,* tight standards are lower estimates of standard product costs.

Wage rate variance. A comparison of actual and standard wage rates for the actual number of hours of labor used.

EXERCISES

1. What are the primary purposes of cost accounting?

2. What costs are included in "overhead"? Which of these are variable costs, and which are fixed costs?

3. Why are overhead cost elements growing as a percentage of total manufacturing costs, while direct labor and direct material costs are shrinking?

4. What exactly does it mean to say that overhead has been "overabsorbed" during an accounting period? Is this a favorable or unfavorable condition?

5. Indicate whether each of the following statements is true or false.

   a. Product costs are independent of the particular overhead vehicle selected.

   b. Cost drivers in activity-based costing are similar to overhead vehicles.

   c. Manufacturing companies need cost accounting systems because they convert materials into other states.

   d. When we speak of products absorbing overhead, we mean that the manufacturing process has eliminated a portion of the overhead costs.

   e. By-products are a special category of joint products.

   f. Prime costs are defined as the sum of direct material and direct labor costs only.

   g. An auto repair shop generally uses labor hours as its overhead vehicle.

   h. Marketing managers are very dependent on cost accounting data in determining the price that should be charged for products.

   i. A make-or-buy decision involves the choice between producing a particular part or service within the company, or subcontracting the work to another firm.

   j. While some gray area exists between direct and indirect labor, the differentiation between direct and indirect material is unambiguous.

6. Indicate whether each of the following expenses should be considered wholly a manufacturing overhead expense (a product expense) or an op-

erating expense (such as selling, engineering, or administrative), or should be apportioned between the two categories on some rational basis:

a. Salaries in the accounting department

b. Salaries and wages in the shipping and receiving department

c. Rental of the company's facilities

d. Salaries of the manufacturing engineers

e. Salaries of the vice president of manufacturing and his immediate staff

f. Workmen's compensation insurance (insurance coverage for injured employees)

g. Utilities (heat, light, and power) expense

h. Depreciation on the company's delivery trucks

i. Operating costs for the company-wide communications network

j. The costs of employment advertising in connection with hiring additional assembly personnel

7. If you were in charge of a product engineering department in a mid-sized Japanese company producing small motorcycles, what use might you make of cost accounting information for the various motorcycles now in production?

8. If you were designing a cost accounting system (to be used primarily for billing purposes) for a large law firm, what would you choose as an overhead vehicle? Can you think of any alternative vehicles?

9. Why is cost accounting less constrained than financial accounting is by formal rules imposed from outside the firm?

10. What conditions within a particular company might lead you to conclude that machine hours would be preferable to labor hours as the overhead vehicle?

11. If a company's cost accounting system leans in the direction of including ambiguous costs as period costs rather than product costs, how will the company's reported profits be affected during periods when the firm's inventory of manufactured finished goods increases? Why?

12. Assume a conventional cost accounting system is in use and direct labor dollars is the overhead vehicle; is a product built in small lot sizes likely to have a higher or lower cost than its cost as revealed by an activity-based cost system? Why?

13. Since costs of joint products are quite arbitrarily arrived at, what possible use do they have in an accounting system?

14. A new corporation expects in its first year of operations to incur the following costs:

Materials for production	$290,000
Direct labor	126,000
Overhead	536,000
Engineering, selling & administrative	460,000
Total	$1,412,000

If this company produces 150,000 units, ships 130,000 of them (on which it expects to earn revenue of $10.80 each), what will be its reported first-year profits and value of year-end inventory given each of these scenarios?

a. It defines product costs as comprising solely direct material.

b. It defines product costs as comprising prime costs.

c. It defines product costs as comprising prime costs plus overhead.

d. It defines product costs as absorbing all costs, including engineering, selling and administrative expenses.

15. The Tanenbaum Company anticipates that in the coming year (a) it will produce 40,000 units of its sole product and require a total of two labor hours per unit; (b) its average labor wage rate will be $12.75 per hour; and (c) it expects to incur indirect production costs of $1,460,000 to produce these 40,000 units. What should its overhead rate be in the coming year if Tanenbaum uses an overhead vehicle of

a. Number of units?

b. Direct labor hours?

c. Direct labor dollars?

16. What is the key concept behind activity-based costing?

17. Yoshihara Corporation produces two products with the same total prime cost:

Product	Direct Labor	Direct Material
S	$146	$210
T	$200	$156

If Yoshihara uses units of production as its overhead vehicle, the two products also have the same total manufacturing costs. If the company uses direct material dollars as its overhead vehicle, which product do you think will end up costing more? Why?

EXERCISES—APPENDIX 13A

1. If company D operates a variable cost system and incurs a debit overhead variance of $2,800 for the year, what conclusion(s) can you draw?

2. What does the overhead budget line mean in full-absorption costing?

3. Indicate if each of the following statements is true or false:

a. The sole reason that an overhead volume variance exists in full-absorption costing is because fixed overhead costs are being absorbed.

b. Variable costing provides more useful information for make-or-buy decisions than does full-absorption costing.

c. Overhead volume variances occur frequently in manufacturing environments but seldom in service environments, where full-absorption costing is used.

d. The contribution margin is defined as the difference between revenue and variable manufacturing costs.

e. An argument against the use of variable costing is that it does not comply with generally accepted accounting methods.

f. When manufactured inventory grows during an accounting period, full-absorption costing reports slightly lower earnings than does variable costing.

4. Jansen Corporation reported a total overhead variance of debit $1,230 during an accounting period when 800 hours of direct labor were utilized (a total labor cost of $8,800, with direct labor dollars as the overhead vehicle). Jansen's overhead budget equation is $6,300 + 1.2 × direct labor dollars. Jansen's production plan assumed 750 direct labor hours.

a. What was Jansen's overhead volume variance for the period?

b. What was Jansen's overhead spending variance for the period?

5. The Lee Company reported the following overhead variances for an accounting period:

Overhead spending variance	$2,400 credit
Overhead volume variance	1,600 debit

a. What was Lee's total overhead variance for the period?

b. Was Lee's production above or below expected volume?

EXERCISES—APPENDIX 13B

1. In what ways does standard costing simplify inventory accounting?

2. Why does a standard cost accounting system better fit a relatively mature company producing stable product lines than a young company producing custom products?

3. Indicate whether each of the following statement is true or false:

a. Product standard cost data are often utilized by managers and analysts in departments other than the accounting department.

b. Standard cost accounting systems are expensive to operate.

c. Wage rate variance information is less useful to a production manager than labor efficiency variance information.

d. When standard costs are set tight and inventories are growing, a standard cost accounting system tends to report lower operating profits than does an actual cost accounting system.

e. A company could have a zero balance for total material variance for an accounting period while at the same time report credit variances for both the Material Price and Material Usage Variances accounts.

f. Material price variance information is more useful to the purchasing department than is material usage variance data.

4. The Carson Fabricating Company reported the following results for January 2006:

	Actual	Standard
Labor hours	28,410	29,060*
Wages paid	$336,000	—
Hourly wage rate	—	$12.10

Calculate the following for the month of January:

a. Total labor variance

b. Wage rate variance

c. Labor efficiency variance

5. Hecht Corporation budgeted sales, for the third quarter of 2004, of $3,840,000 and 60,000 units. If Hecht produces and markets only one product, had actual sales of 62,000 units for the quarter, and reported a total sales variance of $73,000 credit, what were Hecht's sales price variance and sales volume variance for the quarter?

---

*Units actually produced × standard labor hours per unit.

CHAPTER 14

ANALYZING OTHER OPERATING COSTS AND DECISIONS

This book focuses on management decisions, and this chapter focuses on analytical techniques (in addition to those already discussed) to assist managers in making a variety of common and recurring operating decisions.

Chapter 9 discussed how analysis of the three primary accounting system outputs, the balance sheet, the income statement and the cash flow statement, influence many decisions. Recall that analysis of liquidity and each flows may lead to more borrowing or to repaying a portion of present indebtedness. A review of working capital may reveal that inventory is too high or too low, that customers should be pressed for speedier payments, or that more or less trade credit should be used. Capital structure and cash flow analyses suggest when to raise more permanent capital, whether to use more or less borrowed funds, and what should be the dividend policy. And finally, a review of profitability—expenses relative to sales, and profit relative to investment—can influence a host of operating decisions.

Chapter 13, on cost accounting, stressed managers' use of product cost reports to evaluate operating efficiency, to decide when and where to take corrective action in manufacturing (and similar) operations, and to influence the marketing and pricing of individual products.

In addition to these important decisions, some having long-range implications for the company, managers—particularly mid-level managers—face numerous operating decisions that collectively determine the success and financial health of their companies. Accounting systems should be designed to yield data to inform these day-to-day decisions. Moreover, not only must the data exist in the accounting system, but they must also be disseminated in useful format to decision makers, who in turn must understand the relevance (and shortcomings) of the data.

EXAMPLES OF OPERATING DECISIONS

The following are examples of prevalent operating decisions discussed in this chapter:

1. The production engineer is considering whether to fabricate a particular part or subassembly within the company or subcontract it to a vendor—the classic make-or-buy decision. If the company has the internal capability and capacity, or could acquire it, the primary decision criterion is typically cost. The engineer can request from the cost accounting department the standard costs for internal manufacture and can obtain the vendors' price quotations, but is this the appropriate comparison?

2. The sales manager is considering whether to accept a particular large order from a nonregular customer. The price offered is low, but what is the minimum acceptable price? Is it ever wise to accept an order at a price below cost?

3. The financial analyst in the treasurer's department is analyzing the return on a proposed investment to automate a production step. What are the true costs of the new and old methods?

4. The design engineer and the material manager are trying to determine when, and if, a particular redesigned product should be introduced to the market. The decision is complicated by the existence of a good deal of raw and semifinished inventory that would be rendered obsolete by the new design.

5. The product line manager in marketing is evaluating the effect on sales and profit of varying the marketing inputs—price, promotional effort, product repackaging, change in incentive to the sales force, and so forth.

Outside of the manufacturing industries, managers face many similar, as well as some different, operating decisions. Service firms encounter pricing and product line decisions, as well as make-or-buy decisions, and so do many nonprofit operations. Chapter 12 emphasized that capital expenditure decisions are prevalent in virtually all organizations. Interrelationships between volumes, prices, and costs permeate public activities and private, highly competitive companies and regulated ones, service and manufacturing firms, large companies and small.

IMPORTANCE OF FRAMING ALTERNATIVES

All of the operating decisions enumerated earlier involve choices among alternatives: make-or-buy, accept the special order or decline it, introduce the product or don't, accept the capital equipment investment proposal or reject it, select among a spectrum of possible prices and promotional plans. Sometimes, the decision is of the "go, no-go" type, deciding either to take an action or not; at other times, the decision involves choosing from among a broad set of alternatives, but even here, one of the alternatives is typically to do nothing.

The optimum decision is possible only if all relevant alternatives are evaluated. If only suboptimal courses of action are considered, the decision will necessarily be suboptimal. I can hardly overemphasize the importance of properly framing the alternatives, considering not only the obvious courses of action, but also the less obvious, but potentially more beneficial, options. When a problem is reframed in a broad context, other options may become apparent.

This admonition—consider all relevant alternatives—seems so simple and obvious as to insult your intelligence. Poor problem definition is so prevalent, however, that a couple of examples help to demonstrate the point.

1. The make-or-buy decision is typically viewed in its narrowest context: Given the part's design, in-house capabilities, and vendor price quotations and specifications, should the work be subcontracted? Another alternative might be to improve, expand, or upgrade the company's in-house capabilities, particularly if similar parts are used in other products, or if other parts might be redesigned to use the new in-house capability. Still another alternative might be to redesign the part to take particular advantage of a known vendor's capability. Or perhaps the part can be combined with one or more other parts and subcontracted as a unit; or perhaps the search for vendors has been too restricted geographically, since the part's light weight makes air freight feasible.

2. The evaluation of investment in new automated equipment frequently narrows immediately to a choice between the present method, at zero incremental investment, and the proposed method with a high initial investment. It may be that this decision should first be framed as a make-or-buy decision; that is, perhaps, among the acceptable alternatives is outsourcing the operation so that no in-house method is used, present or proposed. Or maybe a larger capital investment, involving equipment with broader capabilities, more versatility, or higher operating speeds would be justified if the evaluation were more comprehensive.

3. The marketing manager, when considering the relative effects of price, promotion, and sales force incentive, might consider entirely different methods of marketing: for example, independent agents instead of a company sales force, direct-mail promotion instead of magazine space advertising, or cents-off coupons instead of price discounts to wholesales.

DIFFERENTIAL CASH FLOWS

Once all relevant alternatives have been framed, the analyst must then focus on their financial differences. Note that only the *differences* between *alternatives* are relevant to the decision. One need not define all of the economic consequences of each possible course of action but, rather, only that subset of consequences affected by the choice. Any conditions and financial flows unaffected by the particular decision may be ignored.

For example, if the selling price and unit volume are unaffected by the make-or-buy decision or the decision to redesign, you need not project total revenues, costs, and profits to be realized from the product but, rather, only those few cost elements that are affected by the decisions.

If the present method of manufacture and the proposed automated method (necessitating a capital expenditure) both utilize the same factory floor space and supervisory attention, these costs can be ignored. Attention is appropriately focused instead on the differences in economic consequences that arise because the two methods utilize different amounts of power, direct labor, supplies, and so forth.

Moreover, it is the *cash flow* differences among alternatives that are relevant. (This point was repeatedly emphasized in the Chapter 12 discussion of capital investment analyses.) Cash flow is the ultimate financial objective, for it is cash flow that is used to repay borrowings, to pay dividends to shareholders, to support community and social programs, and to pay salaries, wages, and bonuses to the staff.

Important among the cash flows of a for-profit company are income taxes. To the extent that an operating decision affects the company's taxable profits, its cash flow for taxes is also affected. Thus, the decision to invest in new capital equipment affects future depreciation expenses for the company; while depreciation expense itself is not a cash flow, the fact that depreciation expenses reduce income taxes must be included in the analysis. Scrapping (throwing away) inventory does not directly involve a cash flow, but because the book value of that scrapped inventory is deducted from taxable income, the resulting cash savings in income taxes should be included in the analysis. Tax increases may also result; if a fixed asset is sold for an amount in excess of its book value, the difference is taxable, and the cash outflow for this added tax must be considered in the decision.

This emphasis on cash flow may suggest that differences in reported profits are irrelevant to the kinds of operating decisions discussed here. That statement is a bit too global. Some management incentive systems concentrate attention on parameters other than cash flow, for example, on current reported earnings (accrual, not cash basis). The price paid by an acquiring company for the shares of an acquired company may be a multiple of current period profits, not cash flows. Companies that operate in a regulated environment (such as utilities and certain government defense contractors) often respond more to regulatory constraints or pressures than to the timing and amount of cash flow.

RECAP OF FIXED AND VARIABLE COSTS

We need to reemphasize the distinction between **fixed expenses** and **variable expenses** (or **costs**), that is, between those expenses that increase or decrease proportionately with increases or decreases in volume of activity, and those expenses that are independent of changes in volume. For example, in a manufacturing operation, as the volume of production (total output) is increased within relatively narrow limits, some costs necessarily increase, while other costs are uninfluenced by modest changes in manufacturing volume (that is, they remain fixed over the short run, measured typically in months, not years). The classic example of a variable cost is direct material: the amount of material required is a direct function of the volume of units manufactured. The classic example of a fixed cost is rent or depreciation.

Bear in mind that the terms *variable* and *fixed,* as used in this context, refer only to the behavior of costs with changes in *volume.* Costs vary for many other reasons, such as the passage of time, the rate of inflation or deflation, the number of persons employed, or the size of the production facility. However, as the terms are used in accounting, the variability is solely with respect to the volume of activity.

Why is this distinction between fixed and variable costs important? In this chapter, we will see that, sometimes, it is appropriate to omit the fixed costs when addressing the question, What does a particular product cost? For example, in weak economic times, if a factory is operating at well below capacity, the company may be wise to accept an order even if the price offered is below total (fixed plus variable) costs, as long as the price is above variable costs. Also, many financial analyses, including analyses of capital investment proposals (see Chapter 12), require information on differences in costs and revenue among the alternatives being studied; frequently, fixed costs are irrelevant to these analyses. Moreover, effective budgeting, discussed in Chapter 15, depends on a clear separation of variable and fixed expenses.

It was noted earlier (Appendix 13A) that indirect production costs (factory overhead) are composed of both variable and fixed elements. As just mentioned, rent or depreciation of the factory space is typically a fixed cost. So, too, is the production superintendent's salary, depreciation on the production equipment, and perhaps expenses of the production control department. At the other end of the spectrum, production supplies such as lubricants and coolants generally vary with the volume of production—the greater the output, the more supplies will be consumed. Often, power consumption, equipment maintenance, and costs of operating the stockroom are also directly variable with production volume.

It has already occurred to you, no doubt, that in all environments—manufacturing as well as selling—many expenses are really semivariable with volume: they are neither entirely fixed, nor do they vary in direct proportion to volume. Some expenses may change as in a step function. For example, the factory stockroom force may be able to handle a modest increase in volume, whereas a further increase would necessitate hiring another person. In the case of production equipment maintenance, a certain amount must be performed simply to take care of the ravages of time, but, typically, as volume increases, so too does the need for maintenance.

PERVASIVENESS OF FIXED COSTS

Virtually any set of costs or expenses you encounter in business, in government units, in nonprofit enterprises, or elsewhere is comprised of fixed, semivariable, and variable elements. Unfortunately, those charged with interpreting financial performance of these entities tend to forget or ignore that fact; they are disposed to make the implicit, but erroneous, assumption that all costs are variable.

We know from reading the public press that, when revenues fall by 20 percent in a particular company, profits typically decline by a much greater percentage. Why is this so? Because, while variable costs may decline by 20 percent, fixed costs are "sticky": they do not decline automatically, but only as a result of explicit management action. Even alert and aggressive managers, however, find it difficult to reduce fixed costs as fast as, and in proportion with, short-term declines in volume.

This fixed-cost stickiness is hardly surprising. Rent, occupancy costs, and depreciation of fixed assets are all fixed costs. Management simply cannot resize its operations—up or down—in response to short-term fluctuations in revenue. Another important element of fixed costs is salaries of managers, technical and sales personnel, and other individuals whose experience and personal relationships are critical to the long-term health of their companies. Such employees cannot be lightly dismissed in order to reduce expenses in periods of reduced sales volume, particularly if the revenue reduction is expected to be only temporary.

Conversely, a sudden surge in revenues will typically result in a substantially larger surge in profits, because, while variable costs will increase proportionate with the revenue increase, fixed costs will be sticky on the upside. Even if management scrambles to increase plant capacity and to augment the ranks of salespeople, development engineers, and managers, it takes time to build or acquire facilities and equipment as well as to hire and train personnel, and thus the corresponding fixed costs will be slow to build.

A school, church, or temple has a cost structure dominated by fixed costs: Few of its expenses are dependent on short-term swings in enrollment or membership. A public school that suffers a 10 percent decline in enrollment is expected by many taxpayers and legislators to reduce its expenses by a corresponding 10 percent. Have some sympathy for the school principal as he or she struggles to "resize" the institution. Ten percent fewer pupils spread among all classes may not permit the cancellation of a single class or the dismissal of a single teacher. The costs of operating the facility and paying the teachers are fixed. Usage of classroom supplies probably varies with enrollment (a measure of activity)—perhaps one of the only truly variable expenses—but classroom supplies are a trivial portion of the school's overall expenditures. In spite of the concentration of fixed costs in school operations, notice that schools are typically funded by state legislatures on the basis of average daily attendance; this funding basis impies that schools can adjust their expenditures directly and proportionately with changes in enrollment.

If book borrowing from the city library declines by 20 percent, the head librarian is going to have a tough time cutting expenses. The expenses associated with acquiring books and periodicals and staffing the reference and checkout desks are largely fixed.

At the other extreme is the flower stand in the local shopping mall. The primary expense item is the acquisition of flowers, a fully variable expense if we ignore spoilage and pilferage. The operator incurs essentially no heat, power, or other occupancy expenses, and even the rental paid to the mall may be a percentage of the stand's revenue and thus fully variable. If the stand owner hires the salesperson on a commission basis, even the stand's labor costs are variable with volume.

But it should be obvious that few businesses and almost no public agencies or nonprofit entities have cost structures like that of the flower stand. More typically, fixed costs are substantially greater than variable costs.

Moreover, consider the factors at work increasing the dominance of fixed costs (shifting variable costs to fixed costs):

- Automation—substituting capital for labor
- Increased skill levels required—reducing the ability or willingness of managers to alter employment levels in response to short-term swings in volume
- Modern human resources practices that argue against swings in employment levels
- Union contracts if they guarantee employment

Consider, at one extreme, a software company. Its variable manufacturing costs are trivial: the cost of reproducing the software on a medium suitable for delivery to the customer. The same can be said for many pharmaceutical companies.

On the other hand, several practices have become popular in recent years that push in the opposite direction:

- Outsourcing (for example, part or all of the information technology function; payroll preparation; custodial work)
- Subcontracting fabrication, which reduces heavy investment in capital equipment
- Use of contract personnel to whom the company makes no long-term commitment

CHALLENGES IN DETERMINING DIFFERENTIAL CASH FLOWS

Most accounting systems do not report costs and expenses in a way that facilitates determining differential cash consequences of decisions. The problems arise for one or more of the following reasons:

1. Accounting systems are based on the accrual concept of accounting, and yet many operating decisions would be better served by a cash-basis system. While accrual-basis reports are more valuable for many purposes, particularly for audiences external to the company, maintaining records and issuing reports that facilitate the separation of cash and noncash expenses and revenues are also important.
2. If the operating decision being analyzed affects volume to a modest extent, variable costs will be affected and fixed costs will not.
3. Many expenses are allocated among products or segments of the business. For example, the production overhead rate is used to prorate or allocate costs among products, and sometimes corporate marketing or development expenses are allocated—necessarily somewhat arbitrarily—across all product lines or divisions. These arbitrary allocations are seldom equivalent to incremental cash flows and, as a result, can lead to faulty financial analyses.

Thus, frequently one cannot simply accept data from existing accounting reports and plug them into the analysis. Return, now, to the five examples outlined at the beginning of this chapter, and consider the adjustments to accounting data needed to focus on differential cash flows.

Make-or-Buy Decision

Arata Corporation is considering whether to produce part number 4783 in its own shop or to subcontract the fabrication to Schreiber & Associates. Let's assume that other alternatives have been considered and eliminated, and that these two alternatives are equivalent in quality, delivery, and other nonprice considerations. Schreiber has quoted $473 per 100 parts, and the analyst determines (from cost accounting records) that the standard full-absorption cost of in-house manufacture is $557 per 100. The decision seems obvious: subcontract.

From Appendix 13A, however, we know that full-absorption costs include certain allocated indirect production costs. Will these overhead costs be avoided if the part is subcontracted? Most will not be. If the decision is to subcontract, part number 4783 will no longer carry these overhead allocations, but the costs will not evaporate. A reallocation of the overhead will occur, or negative overhead variances will result. What the decision maker needs to know is the amount by which Arata's total overhead will increase or decrease as a result of this decision.

Assume the detailed standard product costs for 100 of these parts are as follows:

Direct labor	$120
Direct material	173
Variable overhead	120 (100% of direct labor)
Fixed overhead	144 (120% of direct labor)
Total	$557

What cash flows will be saved if Arata subcontracts? Surely material. How about direct labor? If Arata shifts the displaced direct labor employees to other productive activities or dismisses them, then the direct-labor cost will also be avoided by subcontracting.

The analysis of overhead savings is more difficult. We are inclined to assume that the variable overhead will also be saved. However, note that variability is defined with respect to changes in total activity, not to changes in the mix of that activity. A decision to subcontract may increase the work of the purchasing and inventory control departments but decrease the work of the factory supervisors and production control. Thus, even variable overhead may not be entirely differential with respect to this make-or-buy decision.

The analysis of the fixed overhead product costs is still more complex. If the factory and its equipment are unaffected by this decision—a reasonable assumption—then the portion of rent and depreciation costs allocated to part number 4783 is irrelevant. It does not follow, however, that all fixed overhead costs are necessarily irrelevant. For example, if the shipping and receiving function is part of the fixed overhead, the decision to subcontract increases fixed costs to the extent that shipping and receiving takes on added tasks. Any such difference in fixed costs must be considered.

Make-or-buy decisions are aided by a variable cost system (see Appendix 13A). The assumption that variable manufacturing costs are differential to the make-or-buy decision, while fixed manufacturing costs are not, is typically a useful approximation, though (as just pointed out) not completely accurate. This approximation is almost surely preferable either to ignoring all overhead in the analysis or to assuming that all overhead costs are differential. Applying the approximation to this example, the "make" alternative at $413 (the total standard cost of $557 less the fixed overhead of $144) compares favorably with the "buy" alternative at $473. Part number 4783 probably should be made in-house.

When to Accept a Low-Priced Order

The sales manager at the Sedwick Company must decide whether to accept a particular order from a distant, foreign customer. The order is large, but the price is low. The customer is offering only $71,000 for goods having a normal list price of $97,000. Sedwick's cost accounting group estimates the standard manufacturing cost of these goods at $75,000.

The marketing ramifications here are numerous. Will Sedwick have trouble serving this foreign customer? If the order is not taken at this low price, might the customer return later and offer a higher price? Would Sedwick be setting a dangerous precedent if other customers learn about this price? In addition to all these important but noncash considerations (or, at least, nonimmediate cash), the sales manager must consider the present cash consequences before making a decision.

The decision framework seems simple: accept the order or decline it. This statement, however, is not a complete description of the possible alternatives. How busy is the company? If, by accepting this order, Sedwick commits its factory so that other, higher-priced orders would have to be declined, the consequences are different than if, by accepting this order, Sedwick merely puts to use capacity that would otherwise remain idle. Again, framing the alternatives is the important first step.

Let's assume the following with respect to prices and costs:

Normal price	$97,000
Price offered	71,000
Standard manufacturing costs:	
Direct labor	20,000
Direct material	25,000
Variable overhead	15,000
Fixed overhead	15,000
Total standard costs	$75,000

If Sedwick is operating below capacity, with no reasonable expectations of selling this added capacity, then accepting the order appears attractive. Direct labor, direct material, and variable overhead are probably incremental or differential to this decision. Since the sum of these three costs elements is $60,000 and the customer is offering $71,000, Sedwick is better off—by

$11,000—taking this order than not. If the labor for this job would, in the absence of the job, be idle, then even direct labor is not differential, and acceptance of the order becomes very compelling.

Let's assume, on the other hand, that, if this order is declined, Sedwick will have a high probability of selling all the goods it can produce at normal prices. Now the decision is between two order opportunities. The decision is simple: decline the order, since the normal-price order will cause Sedwick to be $37,000 better off (the $97,000 price less the $60,000 variable cost), while the foreign order will improve Sedwick's condition by only $11,000.

Note that the full-absorption product cost data were not used. Would these data be relevant in deciding whether to expand the factory's capacity so that, for example, both the normal-price business and the foreign order could be accepted? No, the full-absorption costs are not differential to that decision either, because (1) they contain noncash expenses (primarily depreciation), (2) they involve arbitrary allocations, and (3) they do not capture for the analysis the immediate cash outflows to acquire the expanded facilities. In fact, for almost no operating decisions are full-absorption product costs differential.

Capital Investment Decisions

Capital investment decisions occur throughout both the private and public sectors of the economy. Procedures, generally referred to as *capital budgeting,* aid such decisions. Recall from Chapter 12 that these procedures require that the analyst develop a complete schedule of the cash inflows and outflows occasioned by each investment alternative under study. Again, the emphasis is on the magnitude and timing of differential cash flows.

Table 14-1 presents the accounting data developed by Lipinsky Manufacturing Corporation relating to a proposal to automate one manufacturing step. Once again, to isolate cash flow consequences, these data must be recompiled, since they were developed on the accrual basis and include both noncash and allocated expenses. Note that the full cost of manufacture is $12.20 per part under the present method and $6.20 under the proposed method; the apparent savings is $6.00 per part. The investment to automate is $600,000, with an estimated life of six years and estimated additional annual maintenance charges of $10,000.

Table 14-2, part A, illustrates the type of analysis that results from the uncritical use of the accounting data. This analysis assumes that overhead will be saved at a rate of 200 percent of direct labor savings. Indeed, some of the variable overhead will probably be saved, if much of it is labor related, including fringe benefits. But correspondingly, much of the fixed overhead that has been allocated to this operation R—rent, building maintenance, heat, and manufacturing management—will continue to be incurred by Lipinsky even if the labor time required for operation R is reduced.

TABLE 14-1. Lipinsky Manufacturing Corporation: Proposal to Automate Operation R

Present costs of completing operation R ($ per part):	
Material	$ 2.00
Labor	3.40
Overhead (200% of labor)[a]	6.80
Total present costs	$12.20
Estimated costs of completing operation R after automation ($ per part):	
Material	$ 2.00
Labor	1.40
Overhead (200% of labor)[a]	2.80
Total estimated costs after automation	$ 6.20
Investment to automate:	
Equipment and installation	$600,000
Estimated life	6 years
Annual maintenance expense	$ 10,000
Estimated salvage value	0
Estimated annual volume	40,000 parts/year

[a]One-half of total overhead is variable.

So, the first problem with the part A analysis is that the operating savings of $240,000 is overstated. However, there are other problems. Deducting depreciation on the new equipment—$100,000 per year, assuming straight-line depreciation and zero salvage value—is incorrect, since depreciation is a part of overhead; there is some double counting here.

This misleading analysis indicates the ratio of annual pretax savings to initial investment is 21.7 percent—a quite handsome accounting return. Indeed, the investment is still more attractive (43.3 percent) if the estimated annual savings is compared with the investment's average book value—that is, the book value of the equipment at the end of three years when it is one-half depreciated.

The shortcomings of this analysis are that (1) the operating savings associated with fixed overhead are illusory, since the decision to invest will not materially alter the company's fixed overhead cash expenditures; (2) the savings are reduced by the depreciation expense, although depreciation is not a cash outflow to Lipinsky; and (3) no recognition is given to the timing of cash flows—the fact that the investment occurs at time zero while the returns are spread out over six years.

The second analysis of operation R in Figure 14-2 (part B) overcomes these deficiencies. Depreciation is ignored. The discounted cash flow technique (see Chapter 12) takes explicit account of the timing of cash flows. The assumption is made that variable overhead will be saved in proportion to labor savings. The apparent savings in allocated costs (fixed overhead) is omitted. The resulting pretax return on investment is a much less attractive 13.0 percent, a rate that will be acceptable in some circumstances but not others.

TABLE 14-2. Lipinsky Manufacturing Corporation: Analysis of Proposal to Automate Operation R

A. Misleading Analysis of Accounting Return

Operating costs:	
Present (40,000 parts × $12.20)	$488,000
Proposed (40,000 parts × $6.20)	248,000
Operating savings	$240,000
Less: Depreciation	(100,000)
Maintenance	(10,000)
	$130,000

$$\frac{\text{Savings}}{\text{Initial investment}} = \frac{130,000}{600,000} = 21.7\%$$

$$\frac{\text{Savings}}{\text{Average book value of investment}} = \frac{130,000}{300,000} = 43.3\%$$

B. Correct Analysis of Cash Flows—Required for ROI Calculation

	Present	Proposed	Difference
Material	$ 80,000	$ 80,000	0
Labor	136,000	56,000	80,000
Variable overhead	136,000	56,000	80,000
Annual maintenance	0	10,000	(10,000)
Net annual cash flow			$150,000
Initial (time-zero) cash flow		($600,000)	
Return on investment (ROI)[a]		13.0%	

[a]Using the internal rate of return method and interest tables in Appendix 2A.

Both of these analyses have still another shortcoming: they ignored income taxes. Income taxes are very real cash costs and are affected both by the operating savings and by the depreciation expense, which is deductible for tax purposes.

When to Introduce a Product Redesign

The design engineer and material manager at Carlos Manufacturing are considering when to introduce a product redesign. The marketing manager is pressuring for early introduction, but the production manager and controller are anxious to avoid having to scrap raw and semifinished goods inventories. Some facts relevant to this decision are outlined in Table 14-3. The new model, designated Mark II, is less expensive to manufacture but should enjoy the same reception and command the same selling price as Mark I. Note that Table 14-3 compares only the variable product costs, not the full-absorption costs.

TABLE 14-3. Carlos Manufacturing Corporation: Incorrect Analysis of Introduction of Model Mark II

	Mark I	Mark II
	Data Relevant to Analysis	
Variable cost of manufacture:		
Direct material	$ 1.00	$ 1.40
Direct labor	2.00	1.30
Variable overhead	2.00	1.30
Total variable cost	$ 5.00	$ 4.00
Expected sales per month (units)	10,000	10,000
Inventory:		
Raw: Dollars	$ 20,000	—
Units	20,000	—
In-process: Dollars	$120,000	—
Units	40,000	—

	Costs Through Next Six Months	
	Introduction in 6 months	Immediate introduction
Cost of goods sold: Mark I	$300,000	—
Mark II	—	$240,000
Write-off inventory: Raw	—	20,000
In-process	—	120,000
Total expenses	$300,000	$380,000

Because the Mark II design is radically new, any Mark I raw or in-process material in inventory at the time of introduction will be scrapped. All of Carlos's managers agree that the new design should be introduced not later than six months from now, when all existing Mark I inventory has been exhausted, but the marketing manager is arguing for immediate introduction. The controller is asked to study the economic consequences of the decision.

The economic comparison shown in Table 14-3 is inappropriate, even though it correctly identifies the consequences to the company's income statement. If the introduction is delayed six months, all inventory will be used up, and therefore no inventory write-off will occur, but the cost of goods sold will reflect the higher unit cost of Mark I. On the other hand, if the introduction is immediate, the lower unit cost of Mark II will be reflected in the cost of goods sold for the full six months, but the inventory of Mark I materials, now carried on the balance sheet at $140,000, must be written off.

The key point here is that existing inventory (carried at historical costs) cannot be affected by the decision to use or to scrap the material. These are sunk costs (see Chapter 12). The incremental cost of turning the inventory now on hand into finished units of Mark I is something less than the total variable cost, since the inventory is a sunk cost. Therefore, the write-off does not represent a cash flow, although it affects the company's profit-and-loss-statement. (But consider another step of complexity: These sunk costs are deductible for income taxes if the inventory is scrapped, resulting in a positive cash impact through a reduction in taxes.)

A more careful analysis of this decision recognizes a possible third alternative: introduce Mark II in four months, using up the in-process inventory and scrapping the raw material; that is, it may be economic to convert the in-process material into finished goods. Accepting the problem as first stated—introduce now or else six months from now—without questioning whether other alternatives may be viable, obscures the optimum decision.

A correct economic comparison requires more data than are available in Table 14-3. What we need is an analysis of the cash costs to the company of each of the three alternatives. The cost to complete the in-process inventory will be less than the cost to complete the raw material inventory, and both costs will be less than the $5.00 total variable cost of a Mark I unit.

Cash inflows can be ignored, as they are assumed to be the same under all three alternatives. The analysis can be further limited to just those out-of-pocket (that is, future) cash costs for Carlos over the next six months; the comparison of costs beyond that period is unnecessary, as it is obvious that the lower-cost Mark II unit should by then be in production.

INTERRELATIONSHIPS AMONG VOLUME, PRICE, COST, AND PROFIT

Business decisions impinge on each other, and opportunities for trade-offs are abundant. Changes in prices affect volume of sales and profits; changes in volume alone affect variable costs immediately and affect fixed costs over time; these changes in turn affect profit; changes in cost may necessitate changes in price, which in turn affect volume and profit; changes in cost (for example, increased marketing effort) may lead to increased volume along with or instead of higher prices. These interrelationships and trade-offs lead managers to ask a broad variety of what-if questions similar to these:

1. If prices are increased by 5 percent, causing a 3 percent decline in volume, will the company be better off? If commission rates paid to agents are increased by 2 percent (increased selling costs) and the agents' increased efforts result in a 5 percent increase in sales, will the company's profits be enhanced?
2. If higher-cost materials are used to produce a higher-quality product that commands a higher market price, will profits be improved?
3. If marketing expenses are increased by 7 percent in order to increase sales volume by 4 percent, and if resulting economies of scale cause a 5 percent reduction in variable production costs with no change in fixed costs, will profits improve?

4. If automation is enhanced, thus adding $40,000 to annual fixed expenses and reducing variable manufacturing costs by 5 percent, will profits suffer if volume drops by 3 percent? If volume increases by 3 percent?

5. If the revision of a rental agreement on retail facilities calls for a lower monthly fixed rent in exchange for a percentage of sales, thus turning a fixed expense into a variable expense, under what conditions would this be better for a retailer than continuing under the current fixed-rental plan?

The relationship between volume, price, cost, and profit for an engineering services firm, Chen Clinical Research Company, which provides services to the pharmaceutical industry, is pictured in Figure 14-1. This graph of operating results displays service volume along the horizontal axis and dollars (both revenue and cost) along the vertical axis. The straight "revenue" line begins at the origin; its slope equals the price. Costs are layered: the fixed costs—selling and administration, and then of operations—are represented by horizontal lines; then the variable costs are represented by layered upward-sloping lines. The difference between the "revenue" line and the "total expense" line is operating profit; at the lower left of the graph, the negative difference represents operating loss. This graph is highly simplified, useful primarily in thinking conceptually about the interrelationships of volumes, prices, costs, and profits. Expenses do not, in fact, follow straight lines as volume changes; fixed costs tend to change in step functions, and the slopes

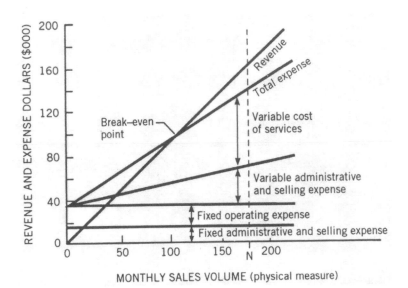

Figure 14-1. Chen Clinical Research Company: volume, price, cost, and profit relationships.

of the variable cost lines are different at different volumes. While fixed costs do not remain absolutely fixed as volume goes from, say, zero to 120 percent of capacity, we can reasonably assume that they are fixed for short-term swings of, say, 10 or 20 percent around normal volume. Similar assumptions apply to variable costs and revenue. What is of primary concern to managers, however, are changes induced by modest changes in costs, prices, and volumes around the company's present, or normal, operating point (N on the graph).

Figure 14-1 illustrates the four fundamental ways a company's profits can be changed: (1) prices can be increased or decreased, thus tilting the "revenue" line up or down; (2) activity (volume) can be increased or decreased, thus shifting to the right or left on the horizontal axis; (3) variable costs can be reduced or increased, thus tilting the "total cost" curve down or up (although its origin remains at the same point on the vertical axis); and (4) fixed costs can be reduced or increased, thus shifting on the horizontal axis the origin of the "total cost" curve down or up. Simple enough. However, each of these actions, as just pointed out, has second-order effects.

Table 14-4 is an example of a **contribution statement;** it displays financial data for Chen at normal volume; the cost-volume-profit relationships are those shown graphically in Figure 14-1. **Contribution** is defined as the difference between revenue and variable costs—all variable costs, not just those incurred in manufacturing. A contribution statement is designed to highlight the contribution by aggregating the variable costs above the "contribution margin" line and aggregating the fixed costs below that line. Note that *gross margin* is not shown in Figure 14-4, but *contribution*—revenue minus variable costs—is. The contribution at normal sales volume ($170,000 per month) is $68,000, 40 percent of revenue. You can see immediately that a $10,000 increase in monthly sales leads to a $4,000 (40 percent of $10,000) increase in contribution, and, if fixed costs truly remain fixed with this volume increase, operating profit also increases by $4,000.

TABLE 14-4. Chen Clinical Research Company: Monthly Contribution Statement—Normal Operations

		$000		Percent of Sales
Revenue		$170		$100.0%
Variable expenses:				
Cost of services	$80		47.1%	
Selling and administrative expenses	22		12.9	
Subtotal		102		60.0
Contribution margin		68		40.0%
Fixed expenses:				
Operating expenses	28			
Selling and administrative expenses	18			
Subtotal		46		
Operating profit		$ 22		

At normal volume N, as shown in Figure 14-1 and Table 14-4:

$$\text{Operating profit} = 0.4 \times (\text{revenue}) - \text{fixed expense}$$
$$= 0.4(\$170,000) - \$46,000 = \$22,000$$

Now, with these data, consider some simplified what-if scenarios for Chen:

1. Suppose Chen's managers believe that an increase of 1 percent in salespersons' commission (from 3.5 to 4.5 percent of sales) could substitute for $2,000 of advertising without affecting total revenue volume. The contribution margin would then be reduced to 39 percent, but fixed expenses would also be reduced by $2,000:

$$\text{Operating profit} = (0.39 \times \$170,000) - \$44,000 = \$22,300$$

   A modest increase in operating profit from $22,000 to $22,300 would result.

2. Because additional marketing effort seems productive, suppose the sales commissions are increased still more—now by 1.5 percent—and advertising is not cut (no change in fixed expenses). Management feels that sales will then increase by 5 percent. The contribution margin would be decreased, but the sales volume would increase:

$$\text{Operating profit} = (0.385 \times \$170,000 \times 1.05) - \$46,000 = \$22,700$$

   The result: a modest increase in operating profit. Note that, if a lesser increase in volume (say, 3.5 percent) is stimulated by increased commissions, the action would be unwise:

$$\text{Operating profit} = (0.385 \times \$170,000 \times 1.035) - \$46,000 = \$21,700$$

3. Suppose service prices are changed, resulting inevitably in volume changes. Managers forecast that a 5 percent price increase will bring about a 7 percent decrease in volume. Now the former contribution margin no longer applies to the new revenue volume figured at new prices; the changed pricing assumption requires a new contribution margin calculation. Service activity is expected to decrease by 7 percent to revenue of $158,100, stated in terms of the former prices; the old variable cost percentage, 60 percent, can be applied to this figure to determine the total variable cost. The new contribution margin is substantially higher than before (42.9 percent) because prices have been increased without any increase in the cost of services rendered. This price increase appears attractive: operating profit would increase from $22,000 to $25,100 even before Chen considers what fixed cost reductions might be possible given the lower volume.

4. A price decrease of 3 percent would stimulate revenues. If this decrease leads to a 5 percent increase in volume, the operating profit at Chen will suffer. Again, the former contribution margin cannot be used directly:

$$
\begin{aligned}
\text{Revenue} &= \$170,000 \times 1.05 \times 0.97 = 173,145 \\
\text{Variable costs} &= 0.6 \text{ (new volume at old prices)} \\
&= 0.6 \times 170,000 \times 1.05 = \$107,100 \\
\text{Operating profit} &= \$173,145 - 107,100 - 46,000 = \$20,045
\end{aligned}
$$

5. Chen now subcontracts to another firm certain laboratory services at $5,000 per month plus 5 percent of revenue. Alternatively, it could offer its subcontractor a fixed amount per month—say, $12,000—regardless of activity volume. (Chen would thereby turn a variable expense into a fixed expense; the implications of this conversion will be discussed in a moment). Chen's contribution margin will increase to 45 percent, more than enough to compensate for the increase in fixed expenses; therefore, the company's operating profit will improve by $1,500 per month:

$$
\begin{aligned}
\text{Operating profit} &= (0.40 + 0.05) \times \$170,000 - (\$46,000 + \$7,000) \\
&= \$23,500
\end{aligned}
$$

6. The last what-if scenario contemplates more far-ranging and ambitious changes. Suppose Chen feels that increasing volume is the key to improving profit; to get the added volume, management considers increasing sales commissions to 5 percent and simultaneously increasing advertising expenditures by $5,000 per month. To service the added volume, additional space and equipment will have to be rented at an incremental fixed rental cost of $4,000. Management believes that these changes can lead to a 20 percent increase in unit volume, with no changes in price, and that the higher volume, in turn, can permit Chen to achieve some economies of scale in providing services, with the result that variable operating costs will decrease from 47.1 percent of revenue to 46.0 percent. The result: the increased volume more than compensates for erosion in the contribution margin and for somewhat higher fixed expenses:

$$
\begin{aligned}
\text{Operating profit} &= (0.400 + 0.011 - 0.015)(170 \times 1.2) \\
&\quad - (46 + 5 + 4) = \$25.8
\end{aligned}
$$

Operating profit as a percentage of sales is reduced from 12.9 to 12.6 percent by this set of changes. And you can quickly calculate, if the volume expands by only 10 percent, rather than the forecasted 20 percent, operating profit will be only $19,100, a $2,900 reduction from present levels. Remember, too, that Chen may well have to invest somewhat more in accounts receivable and perhaps inventory (that is, invest more in working capital) to finance this expansion. When considering this final scenario, management will need to look at the return on assets.

Similar what-if questions could be asked with respect to adding sales personnel, changing the mix of services offered, extending more liberal credit terms to customers, and so forth.

Modeling an Operation

The graph in Figure 14-1 is a simple model of a company. It assumes a single product or service, sold at a single price, complete linearity in all costs and expenses, and a direct relationship between variable expenses and dollars of revenue. Of course, most businesses and other institutions are much more complex. Fortunately, it is increasingly possible to model these complexities, though the resulting model requires a computer, not a two-dimensional graph. A sophisticated computer-based model of an operation can accommodate multiple products at multiple prices, step functions in expenses, and variability of expenses in terms of parameters other than sales volume. Such a model facilitates analysis of what-if scenarios. Increasingly, managers use computer-based financial models of operations to test the financial consequences of alternative actions.

Forecasting the specific changes in sales or margins for use in the analyses above can be difficult. Some interactions among volume, price, cost, and profit are indisputable, but determining the extent of interdependency and the exact dollar effects is quite another matter. Single-point or discrete estimates can be avoided by performing analyses in terms of probability distributions.

Backing into the Answer

Another possibility is to rephrase the questions in terms of an **indifference point.** To illustrate, consider scenario 2 in the earlier section. The original question was phrased as follows: Should Chen increase sales commissions by 1.5 percent if such an increase would lead to 5 percent more sales? Because the precise increase in revenue—here estimated at 5 percent—is difficult to forecast, we might restate the question: by how much would sales have to be stimulated by a 1.5 percent commission increase in order for that increase to be warranted? The contribution dollars must be sufficient to cover fixed expenses and current profits:

$$\text{Revenue} = \frac{\text{fixed costs} + \text{current profit}}{\text{contribution manager}} = \frac{\$46,000 + \$22,000}{0.40 - 0.015} = \$176,600$$

Thus, revenue must increase by $6,600 from the normal rate of $170,000, or 3.9 percent. From a profit standpoint, and leaving aside other considerations, Chen is indifferent whether it continues the present commission rate and revenues of $170,000 per month, or increases the commission rate by 1.5 percent and achieves revenues of $176,600. Chen can now focus on the simple "go, no go" question: Are sales likely to be stimulated beyond this indifference point? In probability terms, the expected value of the distribution of possible revenue levels should exceed this indifference point.

Any of the other scenarios discussed above could be recast in terms of an indifference point. Scenario 4 might be recast to ask, By how much will revenues have to increase to offset the margin erosion from a 3 percent price decrease? Alternatively, How much of a price decrease could Chen withstand (that is, keeping operating profit at present level) in return for a 5 percent volume increase? Decision makers often find it easier to make a "go, no go" judgment with respect to a single indifference point than to make single-point estimates regarding sales, prices, margins, volumes, competitor reactions, and so forth.

Break-even Point

In Figure 14-1, the intersection of the "total revenue" line and the "total expense" line is labeled the **break-even point.** At that point, Chen's sales revenue is just sufficient to cover all expenses, with no profit or loss. Or, rephrased, at break-even revenue volume, Chen's contribution earned will be just sufficient to cover the fixed expenses with no contribution to profit. Since the company's contribution margin is 40 percent (see Table 14-4), the break-even revenue is:

$$\frac{\text{Fixed costs}}{\text{Contribution margin}} = \frac{\$46,000}{0.40} = \$115,000$$

In any month that the company's revenue drops below this point, the company incurs a loss. For example, at a revenue volume of $100,000:

$$\text{Operating profit} = \text{contribution margin} \times \text{revenue} - \text{fixed costs}$$
$$= 0.40 \times \$100,000 - 46,000 = -\$6,000 \text{ (loss)}$$

Stated another way, the loss equals the contribution margin times the amount by which revenue fell below the break-even volume:

$$(\$115,000 - \$100,000) \times 0.40 = \$6,000 \text{ (loss)}$$

New companies and inadequately financed companies need particularly to focus on break-even volumes. A company consistently operating below break-even will ultimately run out of financial resources and fail.

Two refinements of this break-even analysis are frequently useful. First, note that this break-even point is given in terms of reported profits measured on the accrual basis. Even more critical to a company is reaching the *cash* break-even. A company with large noncash expenses (primarily depreciation) can operate for a long time below the profit break-even as long as it is operating above the cash break-even. If $10,000 of depreciation and amortization is included among Chen's $46,000 fixed expenses, then the company's cash break-even is

$$\text{Revenues required to break even} = \frac{\$36,000}{0.40} = \$90,000.$$

A second refinement considers points other than the zero-profit point. Many companies have obligations beyond just meeting their expenses. For example, suppose Chen's five-year term loan agreement requires certain monthly principal payments. Other companies feel strongly about maintaining a certain dividend level, funding a certain employee bonus plan, or providing resources for a capital investment program; break-even analyses are useful in assessing the companies' risks in achieving these goals.

OPERATING LEVERAGE

This discussion of break-even makes clear that the higher a company's break-even point, the greater its risk of operating at a loss. A high break-even point typically implies high fixed costs, costs that do not automatically reduce if the expected revenue volume fails to materialize.

On the other hand, the high contribution percentage that generally accompanies high fixed expenses provides an opportunity for handsome profits if actual revenue exceeds the expected volume. That is, if most expenses are fixed, total expenses increase only modestly as revenues increase; much of the increased sales volume is then reflected in higher profits.

The mix of fixed and variable expenses defines a company's **operating leverage.** A company with high fixed costs—with their attendant high risk of loss and opportunity for profit—has high operating leverage. To the extent that most of the company's expenses are variable, it has a relatively low risk of loss (because expenses decline as sales decline) and a relatively low opportunity for extraordinary profits (because expenses increase as sales increase); such a company has low operating leverage. The conditions for high and low operating leverage are illustrated in Figure 14-2.

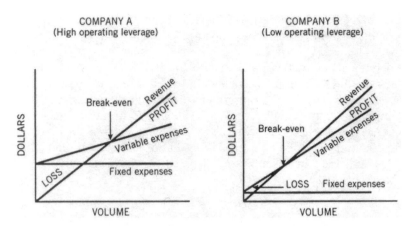

Figure 14-2. The effects of operating leverage.

Recall the discussion of debt leverage in Chapter 10. A company has high debt leverage if a large proportion of its total capital is obtained from borrowings on which it is obligated to pay interest. While debt leverage and operating leverage are different phenomena, their effects are parallel. Both high debt leverage and high operating leverage are risky, but both create an opportunity for improving profit.

Companies with high operating leverage tend to be quite capital (as opposed to labor) intensive. As companies become more automated, increasing investment in their plant, they become more highly leveraged. Electric power and telephone utilities have high operating leverage, since most of their expenses derive from fixed plant: the generating and distribution facilities of electric power utilities, and the telephone lines, switching facilities, and central offices of telephone utilities. The incremental expenses incurred by the utility—that is, variable costs—associated with a customer making another long-distance telephone call or leaving the porch light on overnight are very small. Companies that develop and sell software or manufacture high-price, low-cost pharmaceuticals also enjoy—and are exposed to the risks associated with—high operating leverage.

Labor- (as opposed to capital-) intensive operations have low operating leverage. Personal service firms, such as janitorial services or secretarial services, are examples. These companies require very little capital equipment; most of their expenses are people related. If business picks up, more janitors or stenographers are hired; and when business slackens, they are removed from the payroll again.

Managers make the trade-off between risk of loss and opportunity for profit in deciding whether to increase or decrease their operating leverage. In the example of Chen Clinical Research, renegotiating the service contract to a higher fixed monthly fee (from a lower fixed fee plus a percentage of revenue) increases Chen's operating leverage. New and smaller firms with limited financial resources are well advised to maintain variability in as many expenses as possible, in order to reduce risk. On the other hand, a stable, secure, well-financed larger company may choose to increase its operating leverage in order to provide opportunities for increased profits, accepting the attendant higher risk. Companies with high operating leverage frequently offset this risk by arranging to finance with little debt leverage.

SUMMARY

The primary users of accounting data—operating managers—have been the focus throughout this book, but particularly so in this chapter. As important as the traditional accounting statements are—operating statement, balance sheet, cash flow, and cost accounting reports—detailed financial data should also be available to managers to help analyze the economic consequences of their operating decisions.

Audiences of financial reports often make the implicit and erroneous assumption that expenses and costs are variable, when, in fact, the great proportion of them are fixed—or, more accurately, fixed in the short term, although by management action they can be made to change over the long term.

The key to operating decisions is considering the differences in cash flows (both amounts and timing) between relevant alternatives. The first challenge is to frame properly the alternatives, searching for less obvious options that may prove optimal. Deriving the differential cash flows from traditional accounting data is complicated because (1) accounting records are maintained on an accrual, not a cash basis; (2) variable and fixed expenses are likely to be affected quite differently by the decisions; and (3) many accounting reports, particularly cost accounting reports, include expenses that have been allocated in an arbitrary way, which can confuse decision makers.

A clear understanding of the relationships between volume of sales or revenue, selling price, operating costs (fixed and variable), and profits help managers to analyze many day-to-day operating decisions. Simple or sophisticated models that describe these interrelationships are now widely used to test alternative action plans before final decision—that is, to analyze what-if questions. Such models can also help to focus attention on operating leverage and break-even points.

NEW TERMS

Break-even. The condition of zero profit or loss. Break-even occurs when the total contribution equals the total fixed cost.

Contribution. The difference between total revenue (sales) and total variable costs.

Contribution statement. A form of income statement that clearly separates variable and fixed costs and thus derives the contribution margin instead of the gross margin.

Fixed expense (costs). Those costs that remain unchanged with modest changes in volume of activity.

Indifference point. The value of a parameter (typically profit, cash, or revenue) that is the same under alternative decisions being evaluated.

Operating leverage. The extent to which a company's total costs and expenses are fixed. A company with high fixed costs but low variable costs is said to have a high operating leverage. A company with a high operating leverage has a high risk of incurring operating losses but also opportunities for leveraging itself to higher profits.

Variable expense (cost). Those costs that vary directly and proportionately with modest changes in volume of activity.

EXERCISES

1. You are considering replacing your delivery trucks every three years instead of every four years because truck maintenance expenses have, in your opinion, been growing too fast. What other alternatives might you consider besides simply increasing the frequency of replacement?

2. The Pacheco Company is analyzing a possible investment in constructing a new facility in the suburbs, comparing its costs and efficiencies to Pacheco's current site. What other alternatives might it consider?

3. Cite three reasons why, in modern industry, the mix of variable and fixed costs keeps moving to a higher percentage of fixed costs.

4. Why might your company choose to take actions to lower its operating leverage?

5. Indicate whether each of the following statements is true or false:
   a. A company is always wise to take actions that will turn fixed costs into variable costs.
   b. A company's break-even point refers to the revenues it achieved in the previous fiscal year.
   c. When a retail store switches from a lease with a fixed dollar rental to one that substitutes a percentage of revenue for a portion of the fixed dollar rental, it is decreasing its operating leverage.
   d. A company should never accept an order at a price less than full-absorption cost of the products or services ordered.

e. Contribution is defined as the difference between revenue and variable manufacturing costs.

f. For any particular operation, the contribution margin is greater than the gross margin.

g. A company with high operating leverage should match that risk with high debt leverage.

h. Pharmaceutical manufacturers tend to have higher operating leverages than do electronic assembly operations.

i. When a company takes action to lower its break-even point, it also typically reduces its operating leverage.

j. Sales commissions set at a certain percentage of revenue are fixed costs for the company.

6. Fill in the word or phrase that best completes the following statements:

a. _____ operating leverage is more risky than _____ operating leverage.

b. The break-even point of a company for whom all expenses are variable is _____.

c. Electric power utilities can tolerate high operating leverage along with high debt leverage because their revenues are _____.

d. Former expenditures that cannot be recouped are considered _____ costs.

e. The key financial criterion for selecting between two alternatives is the difference in _____.

f. When a rental contract for office space has an inflation escalation clause specifying a 3 percent annual increase in rent, the rent expense is _____ (fixed/variable).

7. What are some reasons why many companies are outsourcing (subcontracting to other firms) an increasing portion of their activities?

8. Describe how you would go about analyzing a make-or-buy decision when:

a. Your in-house production capability is not being used to full capacity.

b. You do not have the in-house production capability but could acquire it.

c. You have the in-house capability, but demand for its use exceeds full capacity.

9. You manage a branch sales office. The company controller charges your branch with a percentage allocation of the home-office marketing expenses based on actual revenues of your branch and other branches.

a. Are these expenses fixed or variable for your branch?

b. For the company as a whole, are these expenses fixed or variable?

c. What can you do to control these expenses?

10. Suppose a company has a contribution margin of 40 percent and total fixed costs of $3 million per year:

a. What is its break-even point in revenues?

b. If its fixed costs increase by 10 percent, and its contribution margin remains unchanged, by what percentage of revenue does its break-even point increase?

c. By how much would its contribution margin increase if it could raise prices by 3 percent with no changes in variable or fixed costs?

11. Would you accept a $165,000 order for 3,000 units of product T46 from Sturgis Limited, a South African company? The list price of T46 is $93.00, and its manufacturing costs are as follows:

Full-absorption cost	$61.80
Variable cost	48.15
Prime cost	37.30

State any assumptions you make.

12. The Tan Toy Company is a medium-sized manufacturer of toys. Its engineering department aggressively seeks product cost reductions through redesign (often referred to as *value engineering*). One attractive opportunity is to substitute a molded plastic chassis (frame) for the present fabricated sheet-metal chassis on a particular toy. The engineering department has proven the technical feasibility of the substitution and, in order to make and test prototypes, earlier acquired for $10,000 the tooling required for the molded-plastic chassis. Here are comparative full-absorption manufacturing cost data (per unit):

	Plastic	Metal
Material	$0.88	$0.41
Labor	0.13	0.75
Overhead	0.18	1.07
Total	$1.19	$2.23
Tooling	$10,000	Fully depreciated

The plastic unit requires 0.02 hour of labor, while the metal unit requires 0.12 hour. Estimated annual usage is 12,500 units. The overhead rates are $8.92 per direct labor hour (full absorption) and $2.75 per direct labor hour (variable). A subcontractor will mold the plastic unit, and minor additional processing will be required at Tan Toy. The raw material for the metal chassis was sheared to size by a metal distributing company and then punched, formed, and finished at Tan. The 0.12 hour of processing labor is represented by these activities:

0.06 hour for punching
0.04 hour for forming
0.02 hour for finishing

a. Should the substitution be made?
b. Assume the following existing inventories (related only to the metal unit):

	Number of Units	Inventory Value
Raw material	3,000	$1,230
Work-in-process	4,200	$5,544

The work-in-process units have been punched but not formed or finished. The scrap value of the metal parts is about 5 cents per unit. When should the substitution be effected?

13. Tan Toy (see exercise 12) is approached by Breakfast Bounty, Inc. (BBI), with an offer to purchase up to 10,000 units of a small game introduced by Tan Toy two years ago. Units sold to BBI would be modified slightly from Tan's standard design and would carry the BBI label rather than the Tan label. BBI would offer these to its customers through a coupon on or in cereal boxes at a price that would appear to customers to represent a bargain. The primary terms of BBI's offer to Tan Toy are as follows:

Delivery   During April, May, and June – not less than 1,500 and
           not more than 3,500 in any single month; BBI to give
           30-day notice of the exact quantity required
Terms   Net 30 days
Price $5.35 per unit

Tan Toy's estimated standard full-absorption cost for the unit is as follows:

Direct material	$1.27
Direct labor (0.35 hour)	2.19
Overhead	3.12
	$6.58

(See exercise 12 for overhead rates.) Sales of the comparable standard game were 5,000 units in the first year, 12,000 units last year, and 15,000 (estimated) units this year and next, after which time volume is expected to fall off quickly. The list price to the customer is $28.50, and Tan Toy receives a price of $11.00 from its distributors. Tan Toy's sales manager is anxious that the company not jeopardize the near-term profit potential from this game by diverting either its own production capabilities or retail customer demand to BBI. The manufacturing manager is attracted by the timing of deliveries to BBI: coming right in the heart of Tan Toy's slow season. By early or mid-June, Tan Toy is typically operating at or near full capacity. Should Tan accept this offer from BBI? Explain your answer in detail.

CHAPTER 15

BUDGETING AND FORECASTING

Every person and every organization operates to a plan, whether explicitly or implicitly. Well-managed enterprises operate to a documented, explicit annual financial plan: an operating **budget.** The best-managed operations—businesses, schools, governmental units, and social organizations—also generate detailed reports to compare actual financial performance with their budgets. This chapter focuses first on the budgeting process and on the role budgets can play in analyzing performance. The chapter then turns to longer-term financial planning: developing multiyear forecasts of financial statements, referred to as *pro forma statements.* The chapter concludes with an illustration of short-term cash budgeting.

GUIDELINES FOR BUDGETING

It is tempting to dive right into the details of budgets and budgeting. How much should you spend on salaries in department X? How should salary levels be set? How much should you spend on advertising and promotion, and how should that relate to the company's present size or future growth? But the critical first step in budgeting is the same as for most complex tasks: establishing the objectives. If you don't know where you're trying to go, you can't plan how to get there!

Setting Objectives

In a profit-seeking enterprise, management typically has in mind some profit objective. The profit objective may be set in terms of return on equity or return on assets. Alternatively, it may be set in terms of profits of previous years—for example, increase profits 10 percent over last year, return to the record profitability of three years ago, or hold the profit decline to only 15 percent in the coming recession.

Managements typically have other financial objectives as well; profit may not be the overriding one. The company may currently be geared more to revenue growth or to gaining market share. For a fledgling company, the objective may be simply to break even this year, that is, avoid a loss. Or, financial objectives may be framed in light of commitments to repay $X to creditors or pay dividends to shareholders of $Y per share.

In a nonprofit organization, the objective on which the financial budget focuses is often simply to achieve an equilibrium of cash inflow and cash expenditures; to generate a specified surplus of inflow over outflow; to increase revenue by, for example, 10 percent to permit an expansion of services; or to reduce expenditures by 7 percent, say, in the face of declining membership. In recent years, the budgeting struggle for many national governments, including the United States and most member countries of the European Union, has been to limit government's operating deficit, since the prospect of a balanced budget or surplus is so remote as to provide no realistic budgeting target.

Whatever the financial objectives, they are a distillation in financial terms of the myriad qualitative and quantitative objectives to which the organization is committed.

Budgeting: One Element of Planning

Chapter 1 emphasized that financial statements cannot provide a full and complete history. Much goes on in any operation that cannot be immediately reduced to monetary terms: personnel are hired or leave, stubborn technical problems are encountered or solved, the firm enhances or erodes its reputation with customers.

Financial statements are to history what budgets are to the future. Operating budgets simply state explicitly management's estimate of future financial results, the monetary consequences of the plans for the future. Thus, just as financial statements do not completely record history, neither do budgets spell out complete plans for the future. Budgeting is just one aspect of planning.

Budgeting can neither precede detailed operational planning nor follow it, but must be an integral part. Financial budgets don't dictate plans—although operating plans are often tempered by financial realities revealed by the budgets—nor should plans be finalized without a careful review of their budgetary implications. Planning is iterative: Some planning must be completed to provide a basis for financial budgeting, but some replanning is typically required to adjust plans to budget constraints. Sometimes several rounds of replanning and rebudgeting are necessary before both operating plans and financial budgets become compatible with the organization's overall purposes, objectives, and available resources.

Bottom Up Versus Top Down Budgeting

Who sets the budget? How is an acceptable operating budget finally assembled? Just as one person cannot do all of the planning (except in the smallest of organizations), neither can one person establish the budget.

Some top-level managers are inclined to impose detailed budgets and budgetary constraints on their organizations. Once established, they expect their subordinates to accept responsibility for their respective operating units, to make and revise plans, to be decisive, to implement actions as required, and thus to operate within these budgets. But most mid- and lower-level managers will resent being held responsible for living within a budget they had no hand in setting. How can managers be asked to take operating but not financial responsibility for their business units? Therefore, an important budgeting guideline is that each manager play a strong role in setting the budget for his or her operating unit. The budget of the entire organization is then built up from unit budgets. The most useful budgets are developed from the *bottom up* rather than from the *top down*.

To illustrate, consider the process of developing the annual selling expense budget for the McCarthy Company, a manufacturer of specialty chemicals. The national sales manager, with offices at the Baltimore headquarters, may be tempted to dictate to the regional sales managers budgets for both sales (or incoming orders) and for selling expenses for the regions for the coming year. Suppose you are the regional sales manager in Tampa, Florida. You are more knowledgeable than the national sales manager about such matters as the relative positions of McCarthy and its competitors in the Florida market; the need for additional training for the Tampa sales force; and the amount of travel, entertainment, and telephone expenses to be incurred by the Tampa office during the coming year. Moreover, the national sales manager undoubtedly wants to hold you responsible for the Tampa region's sales and expenses for the coming year. If so, you want to have a say in establishing sales and expense targets for your region.

The nationwide sales and selling expense budgets are, of course, the responsibility of the Baltimore-based, national sales manager. These budgets will generally be both more realistic and more useful if built up from the budgets established in and by each of the regions. However, each regional manager needs some guidance from headquarters regarding planned new product introductions, national sales promotions in the coming year, expected actions by competitors, and, most importantly, the company's overall objectives and financial constraints. The process, then, of establishing the Tampa regional budget is a joint one.

Most large organizations have a budget department headed by a person with a title such as Manager of Budgeting. While too many such managers, unfortunately, attempt to impose budgets on their organizations, the budgeting department's appropriate role is that of facilitator and consolidator: articulating budgeting guidelines that are consistent with the organization's goals, assisting individual managers at all levels to develop realistic budgets consistent with those guidelines, and assembling the organization's total budget from the segment budgets.

Cornerstone of the Budget: The Sales Forecast

The cornerstone of the budget is the forecast or budget of revenue for the coming period. In a profit-seeking company, the forecast of sales becomes the critical first step in the budgeting process. A nonprofit organization estimates revenue from dues, gifts, services, and all other sources. A governmental unit estimates tax revenues, fee revenues, and revenues from the sale of services.

Since most operations must incur expenses to generate revenue, and revenue provides the wherewithal to pay the expenses, neither the sales budget (or forecast) nor the expense budget can be set independently of the other. This situation leads to "chicken first, egg first" arguments. As the Tampa regional sales manager, you are reluctant to commit to a certain incoming order level for the coming year without a reciprocal commitment regarding your expense budget. At the same time, the national sales manager is reluctant to approve your expense budget without a reliable estimate from you of the region's sales.

But sales revenue is the place to start. Necessarily, the first sales forecast must be considered tentative. Once the entire budget is compiled, McCarthy may discover that resources will be available to invest, for example, in opening several new sales offices or in launching a long-delayed new product; these investments may cause an upward revision in the tentative sales forecast. Or McCarthy may discover that a decision to reduce promotional expenses and increase spending on research and development will make the tentative sales forecast difficult to achieve. In either case, the tentative sales forecast needs to be modified, and in turn this revision may necessitate still another revision in the expense budgets. To repeat, budgeting is an iterative process.

Many companies develop both a sales target and a sales budget. The target, somewhat higher than the budget, represents a goal toward which the sales organization will strive, but which has a probability of only, say, 50 percent being achieved. Management is unwilling to commit itself to expense levels based on such a low-probability sales estimate. The lower sales budget—the sales forecast on which the expense budget is based—is at a level that management is perhaps 80 percent confident can be met or exceeded.

History as Prologue

The budgeting process relies heavily on historical financial data. Since an operating budget is simply an estimate of the income statement for a future period, past operating statements are obviously useful guides. Frequently, budgets for the coming year are established simply by incrementing last year's actual expenditures up or down. In any case, a manager needs a detailed understanding of current expenditure levels as he or she considers appropriate future budget levels.

If you know that telephone expenses in the Tampa sales office have been running about $700 per month this year, you will consider factors that could affect this expense item next year: for example, telephone rates are expected to increase by 5 percent, one more salesperson will join the office early in the new year, or e-mail can reduce the number of calls to the Baltimore headquarters. All factors considered, you may decide that an appropriate budget level for next year is about 7 percent above current levels, or about $750 per month.

A caution: the simple incrementing of historical expenses has the unfortunate tendency of confirming present expenditure levels as appropriate and necessary. For example, you may not question whether telephone expenses this year of $700 per month are appropriate. Perhaps the sales force would be just as efficient with less use of the telephone, or, on the other hand, perhaps you should encourage greater use of the telephone to increase sales or decrease travel expenses. The inevitable but unfortunate tendency is to rely on present expense levels as the best indicator of what will be, or must be, spent in the future period.

A budgeting technique referred to as **zero-base budgeting** deemphasizes the use of historical data as the basis for budgeting. The technique, originally introduced in certain agencies of the federal government but now popular in the private sector as well, requires that each manager justify every dollar spent, not solely the budget increment. That is, the manager must justify from a zero base the need for each staff member and each expenditures for supplies, telephone, travel, maintenance, computer time, and so on.

Under zero-base budgeting, you would need to build up the Tampa office telephone expense budget for the coming year by justifying the number of telephone lines coming into your office, the number of telephone sets and cell phones, the number and duration of long-distance telephone calls, and so forth. The advantage of such a procedure is that you may thereby determine that, for example, a special leased line connecting the Tampa and Baltimore offices would reduce long-distance telephone charges; or that the office could get along with fewer telephone sets since the salespersons rely primarily on their cell phones. Zero-base budgeting is time-consuming and often frustrating, but the benefits in terms of increasing the efficient use of resources—reducing expenditures in some areas and increasing them with good effect in others—can far outweigh the costs.

Commitment to Budgeting

The effectiveness of budgets and the budgeting process varies widely among organizations. Where budgeting is a meaningful planning step, budgets become working documents that are both useful and used. In others, budgeting is simply a required but perfunctory exercise that is accorded little time and less thought; the resulting budgets are filed away and seldom, if ever, referred to again. The difference is a function of the commitment made to the budgeting process, commitment not just by the manager of budgeting but also by managers at all levels and particularly by senior managers.

If the head of the organization—president, chair of the trustees, principal, or whoever—is serious about budgeting, everyone else will also take the process seriously. If the chief executive takes time to articulate meaningful objectives and budget guidelines, to review and analyze budget requests, to require rebudgeting until an acceptable overall budget is achieved, and to compare actual operating results throughout the year to the budget, then lower levels of management will also devote the time required to develop, and redevelop, meaningful budgets and to use them in operating their departments.

Bottom-up budgeting requires the deep involvement of all management levels. Each manager negotiates his or her department budget with the next manager up the hierarchy. Once agreed to, a departmental budget becomes a commitment, a kind of contract between the department manager and his or her supervisor.

Frequency of Revision

Most organizations undertake a major budgeting exercise just prior to the beginning of each fiscal year. However, many also require interim rebudgeting as well, perhaps semiannually or even quarterly. The more volatile and unpredictable the operation, the more frequent and extensive must be the budget revisions. Revisions should not be so frequent—for example, monthly—that the budget, always in a state of flux, cannot serve as a guide; moreover, managers should not devote unwarranted amounts of time to budgeting. But neither should rebudgeting be so infrequent that the operating budget becomes unrelated to changed operating conditions facing the company.

RESPONSIBILITY ACCOUNTING

Central to the concepts of both bottom-up budgeting and management commitment is **responsibility centers:** subdividing the organization into departments, segments, or units for each of which a single manager is responsible. The first-line supervisor is responsible for a relatively small segment of the business—perhaps only several employees and a few expenses—while managers farther up the management hierarchy have responsibility for whole functions, divisions, or groups that are in turn composed of a number of smaller segments. Finally, the organization's chief executive has the ultimate responsibility for all operations.

Again, assume you are manager of McCarthy's Tampa office; you head a responsibility center that includes all of the company's activities in Florida.

Reporting to you are several managers, each in turn heading a responsibility center—for example, the service manager responsible for the service force, their travel, and associated expenses; the office supervisor responsible for the support and clerical staff, janitorial services, and the cost of supplies; and three area sales managers, each responsible for the activities and expenses of several salespersons. You report to the national sales manager, who has responsibility for overseeing regional offices throughout the country.

Responsibility centers are categorized by the extent of control exercised by the responsible manager. A segment of the business for which the manager has responsibility only for expenses is referred to as a **cost center.** Good performance requires that the manager develop and obtain approval for a realistic budget, and then control expenditures to that budget. Where the manager is responsible for both expenses and revenue, but not for the investment of long-term resources, the center is referred to as a **profit center.** A profit center manager is expected to make operating decisions keeping in mind the profit of the business segment, not concentrating on just revenues or solely expenses. The Tampa regional office is a profit center. Your performance can be judged in terms of the profitability of the Tampa region. Finally, if the manager is accountable for the company's investment in the particular center, as well as for the near-term revenues and expenses, the center is referred to as an **investment center.** Autonomous divisions of large companies are investment centers; the division manager is held accountable for both the assets at his or her command and the profits generated by those assets. In Tampa, you are not operating an investment center nor, presumably, is your boss, the national sales manager; the division manager to whom the national sales manager reports, however, is in charge of an investment center, and its financial performance is evaluated in terms of the return on assets, or a similar measure.

RECAP OF THE REASONS TO BUDGET

Before turning to the use of budgets in analyzing performance, let's review the primary benefits of an effective budgeting process, well managed in terms both of setting budgets and of utilizing those budgets as working documents.

The single most important benefit of budgeting is that it causes explicit planning. Operating plans have a tendency to be vague until their financial implications are reduced to budgets. As the Tampa regional manager, you may be inclined to talk about increasing the company's market share in Florida, or utilizing more efficiently the sales force's travel time, or making initial contacts with 200 new customers. However, when you turn to setting next year's budget for incoming orders and selling expenses, you must become very explicit in planning: How many salespersons will be employed, how long will they be trained, who will make the contacts with new customers, how many cars will be leased and how many airplane trips taken, and how much will telephone expenses increase if the amount of travel is to be reduced? Since you need to do this planning to make sensible staffing and other operating decisions, the benefits of detailed and explicit planning extend well beyond simply setting budgets.

This explicit planning is neither easy nor comfortable. Most of us do not enjoy budgeting because we know that once we commit plans to monetary terms, we are less free to change plans and we are increasingly accountable for them.

A second important benefit of budgeting is communication. Once plans are translated, to the extent possible, into monetary terms, they are more easily communicated. As Tampa regional manager, you communicate in part by budget, as well as orally and in writing, with the national sales manager. You also communicate with others working in the Tampa office; they understand the commitment that you, their supervisor, has made to headquarters for orders to be generated during the coming year and expense levels to be met. Your Tampa regional budget also communicates to other groups within the company: the training department learns how much training support Tampa will need, and the local advertising plans for the Florida market are communicated to the advertising department. Similarly, you learn about plans in other company departments by referring to their budgets.

Finally, budgets serve as a basis for comparing actual results. It is this third benefit that is most widely recognized and discussed, and the one to which we now turn.

ANALYZING PERFORMANCE: BUDGET VERSUS ACTUAL

Good managers focus their attention not on departments or divisions that are operating in accordance with plans, but rather on those operating at odds with plans. Referred to as **management by exception,** this technique requires that the company's financial reporting system highlight those business segments that are not "on plan." The differences between actual and planned results are referred to as variances, here **budget variances:** the differences between actual and budgeted costs and revenues.

Table 15-1 shows an operating report for the Tampa regional office that you received early in August. This report, prepared by the home-office accounting staff, was forwarded to you and to others at McCarthy, including the national sales manager. It shows actual financial results for July, as well as for the first seven months of the company's fiscal year, and compares those results with the budget.

Table 15-1 and the discussion of variances in this chapter follow the convention of enclosing in parentheses variance amounts that reduce profit. Thus, when current actual expenditures are in excess of the plan, the amount is enclosed in parentheses. Parentheses around sales or revenue variances indicate revenues that are below budget or plan. Thus, variances in parentheses are equivalent to debit balances.

The report indicates that, for July, the Dallas region was over budget in shipments by $120,000—a favorable condition—but under budget for the

TABLE 15-1. The McCarthy Company: Operating Report—Actual Versus Budget, Tampa Region, July ($000)

	July			Seven Months Year to Date		
	Budget	Actual	Variance	Budget	Actual	Variance
Sales	$1,250	$1,370	$120	$8,600	$8,450	($150)
Regional expenses:						
Salaries	40.0	41.1	(1.1)	280.0	282.5	(2.5)
Sales commissions	25.0	27.4	(2.4)	172.0	169.0	3.0
Discounts and freight allowed	5.0	5.4	(0.4)	34.4	33.5	0.9
Travel and entertainment	16.0	19.0	(3.0)	112.0	110.5	1.5
Telephone	4.0	3.8	0.2	28.0	29.0	(1.0)
Advertising	8.5	8.0	0.5	59.5	57.0	2.5
Rent and other occupancy	3.5	4.6	(1.1)	24.5	24.3	0.2
Total regional expenses	$ 102.0	$ 109.3	($ 7.3)	$ 710.4	$ 705.8	$ 4.6
Allocated expenses:						
Headquarters sales expense	20.0	20.5	(0.5)	140.0	142.5	(2.5)
National advertising	17.5	16.5	1.0	105.0	101.0	4.0
Trade shows	11.0	9.0	2.0	35.0	34.0	1.0
Total allocated expenses	$ 48.5	$ 46.0	$ 2.5	$ 280.0	$ 277.5	$ 2.5

year to date by $150,000. (The region must therefore have been $270,000 under budget after six months, that is, at the beginning of the month.) July was a good month for sales.

In expenses, the region was over budget for the month—an unfavorable condition—but under budget for the seven-month period. The over-budget condition in expenses may, of course, have resulted in the strong July sales.

Note that the report contains two categories of expenses: regional and allocated. As the regional manager, you are in charge of a responsibility (profit) center and accountable for those expenses over which you have control. Presumably, you do control salaries, travel, telephone, and the other regional expenses. On the other hand, you don't decide the staffing at headquarters, the national advertising campaign plans, or the trade show schedule, and therefore you cannot be held responsible for these allocated expenses. Thus, when judging your expense control performance, you and your boss focus on the regional expenses alone.

A number of expense categories have unfavorable variances for July and favorable variances for the seven-month period. The reverse is true for telephone expenses. Since some randomness in monthly expenditures is inevitable, year-to-date data are useful, as are the monthly data. The under-budget condition for telephone expense in July might be explained simply by several salespersons being on vacation, or it might be the result of improved management of telephone expenses. The year-to-date data help in judging whether an over- or under-budget condition persists, or whether a particular month's variance is simply a random event.

INTERPRETING VARIANCE BALANCES

In the previous section, the terms *favorable* and *unfavorable* appear several times to describe variances that increase and decrease profit. Variance balances enclosed in parentheses are popularly referred to as **negative** (or **unfavorable**) **variances**—negative in the sense that they reduce profit. Similarly, variances that add to profit are popularly called **positive** (or **favorable**) **variances.** This nomenclature is both understandable and unfortunate. Any variance indicates actual performance different from the plan. It does not necessarily indicate good or poor performance.

For example, the Tampa region was under budget in advertising expense for July and for the seven-month period as well. Is this a favorable variance? Yes, in that it added to this year's profit, but it may not be favorable that the region is doing less advertising than originally planned. Is the reduced advertising level contributing to the lower-than-anticipated year-to-date sales? Advertising is a discretionary expenditure, requiring management judgment. The precisely correct amount of advertising for the Florida market is difficult, probably impossible, to ascertain. You, the Tampa regional manager, can be commended if you are carrying out the planned advertising campaigns at reduced cost; however, if you are achieving the under-budget condition by reducing total advertising exposure, this may or may not be a favorable situation, particularly over the intermediate and long term.

Consider your possible motivations. If you are under great pressure from (that is, if you feel threatened or coerced by) the national sales manager to live within your monthly expense budget, there is much you could have done to avoid July's $7,300 over-budget situation. You could have reduced advertising, dismissed one salesperson, or forgone the repair of the facility. Or, you could have cut back on janitorial or maintenance cost, saving money in the short run while running the risk of a deteriorating office and unhappy employees. Each of these actions would have yielded short-term financial benefits—and reduced the threat of your boss disciplining you—but the actions might not have been consistent with the company's long-term health.

As you interpret variance reports, be alert to the possibility that expenses that are heavily discretionary are being inappropriately cut in order to "make the budget." That's a great temptation! Ask the sales managers and R&D managers of the world; they will tell you that, when times are tight and budgets must be cut, the first budgets to be attacked are, all too frequently, the advertising budgets—particularly for so-called image advertising—and research budgets, particularly for work on programs or products that won't reach the market for years. Too much of this behavior is clearly inappropriate.

The concept of materiality certainly applies to variance analysis. Some variance is almost bound to occur for virtually all expenses in almost all accounting periods. For example, in Table 15-1, the favorable $200 variance in rent and other occupancy costs for the seven months is hardly material in relationship to the $24,500 budget for the same period; for this cost element, we can conclude that the Tampa region is about on plan for the year to date.

Finally, note that the usefulness of the Table 15-1 report depends very much on the timeliness with which you receive it. You are interested in seeing how your operation is performing relative to the plan to which you committed. If the operation is off plan, you are anxious to take corrective action. You may be startled to see that travel expenses were well over budget; not recalling any unusual circumstances occasioning extra travel in July, you may want to take immediate corrective action. The longer the delay in obtaining financial feedback, the longer an out-of-control condition is likely to persist. Accounting departments should attach high priority to the timely reporting to operating managers of budget-versus-actual information.

FLEXIBLE BUDGETING

Recall from Chapters 13 and 14 the distinction between variable and fixed expenses. How can the separation of variable and fixed expenses improve the quality of the budgeting process? Table 15-2 recasts the July operating expense budget for the Tampa region to make this separation. The exhibit also

TABLE 15-2. The McCarthy Company: Recast Expense Budget—Flexible Budget, Tampa Region, July

	July Budget	
	$000	Percentage
Sales	$1,250.0	100%
Regional expenses:		
Variable:		
Sales commissions	25.0	2.0
Discounts and freight allowed	5.0	0.4
Total regional expenses	$ 30.0	2.4%
Fixed:		
Salaries	$ 40.0	3.2%
Travel and entertainment	16.0	1.3
Telephone	4.0	0.3
Advertising	8.5	0.7
Rent and other occupancy	3.5	0.3
Total fixed expenses	$ 72.0	5.8%
Total regional expenses	$ 102.0	8.2%

shows the percentage that each expense element represents of the budgeted sales volume for the month. The key budget data for the variable expenses are the percentage figures; if sales for the month turns out to be other than $1.25 million, then the sales commissions budget should be fixed not at $25,000 but at 2 percent of *whatever* sales turn out to be. On the other hand, the $8,500 advertising budget for the month is independent of the sales amount realized for the month; there is no reason that advertising expenditures should be driven off plan by the fact that this month's sales revenue is different from the plan. The budget for the fixed expenses is best stated in absolute dollar amounts, not percentages.

A **flexible budget** stating variable expenses in percentages and fixed expenses in absolute dollars, would be particularly useful for the Tampa region:

$$\text{Expense budget} = 2.4\% \text{ of sales} + \$72,000$$

Volume-Adjusted Budgets

This flexible budget allows us to construct, for the Tampa region, a **volume-adjusted budget,** giving effect to the impact that changes in sales volume have on budgeted expenses. Table 15-3 is a reconstruction of Table 15-1, comparing actual results with volume-adjusted budgets. A comparison of these two tables reveals the following:

TABLE 15-3. The McCarthy Company: Operating Report—Actual Versus Volume-Adjusted Budget, Tampa Region, July ($000)

	July			Seven Months Year to Date		
	Volume-adjusted Budget	Actual	Variance	Volume-adjusted Budget	Actual	Variance
Sales	$1,370	$1,370		$8,450	$8,450	
Regional expenses:						
Variable:						
Sales commission (2.0%)	27.4	27.4	—	169.0	169.0	—
Discounts and freight allowed (0.4%)	5.5	5.4	0.1	33.8	33.5	0.3
Total variable expenses	32.9	32.8	0.1	202.8	202.5	0.3
Fixed:						
Salaries	40.0	41.1	(1.1)	280.0	282.5	(2.5)
Travel and entertainment	16.0	19.0	(3.0)	112.0	110.5	1.5
Telephone	4.0	3.8	0.2	28.0	29.0	(1.0)
Advertising	8.5	8.0	0.5	59.5	57.0	2.5
Rent and other occupancy	3.5	4.6	(1.1)	24.5	24.3	0.2
Total fixed expenses	72.0	76.5	(4.5)	504.0	503.3	0.7
Total regional expenses	$ 104.9	$ 109.3	($4.4)	$ 706.8	$ 705.8	$ 1.0

1. For the month, sales commissions are just equal to the volume-adjusted budget, although Table 15-1 indicated that they were $2,400 over budget. That over-budget condition was fully explainable by the fact that actual sales exceeded budgeted sales.
2. Discounts and freight allowed were closer to the plan, for both the month and the year to date, than suggested by Table 15-1, which shows a $400 unfavorable variance for July; on a volume-adjusted basis, this variance changes to a $100 favorable variance.
3. Overall for the month, regional expenses were $4,400 over the volume-adjusted budget—much closer to the plan than indicated by the $7,300 unfavorable variance shown in Table 15-1.
4. On the other hand, the $4,600 under-budget condition in total expenses for the year to date, as shown in Table 15-1, is misleading. Most of this underbudget condition was the result of lower-than-expected sales volumes, not of effective control of expenses. As compared with the volume-adjusted budget for the year to date, actual expenses have been only $1,000 under budget.

Recall my persistent urging that variable and fixed costs be separated and clearly identified. Such a separation is fundamental to an accurate assessment of performance for virtually any enterprise, profit seeking or nonprofit.

HUMAN BEHAVIOR CONSIDERATIONS

The budgeting process, and particularly variance interpretation, is ripe with human behavior complexities. Human motivations infuse both setting budgets and living with them. Many of the words used in this discussion elicit visceral reactions: *commitment, responsibility, control, unfavorable, discipline, positive* and *negative,* and *performance evaluation.* These strong words suggest personal interaction and occasional confrontation. While we frequently speak of an organization in impersonal terms—set standards, exercise control, and alter direction—in fact, of course, an organization is not a machine but a collection of individuals, some of whom are designated managers. Control is exercised by managers and so, if the organization is to change course or correct problems, individuals have to take action.

Goal Congruency

A truism of human behavior is that people take action, in their professional lives as in their personal lives, only because they are motivated to do so. That motivation derives from a desire to satisfy some need, and each of us has a set of needs we strive to satisfy. To satisfy these needs, employees are motivated to take actions, some of which may be beneficial to the organization, while others are detrimental.

If an individual manager's needs are closely aligned with those of the organization as a whole, then that manager's actions will benefit the organization's goals. Conversely, if the individual's needs deviate from those of the organization, the actions to which the manager is motivated may not be in the organization's best interest. Obviously **goal congruency**—that is, good alignment between the manager's goals and those of the organization—is desirable, although absolute and complete goal congruency is an ideal that is seldom achievable.

What action do budgets motivate a manager to take? The manager participates in setting the budgets, yet the manager's subsequent performance is judged, in part, by reference to the budgets: expense control for the cost center manager, profits for the profit center manager, and return on investment for the investment center manager. Motivations depend very much on top-level management's attitudes toward budgets and performance analysis.

Pessimistic View

Suppose top management decides to use budgets in a threatening manner, indicating implicitly or explicitly that if managers fail to meet the budget (whether expense, profit, or return on investment)—that is, if negative variances prevail—they will be disciplined and perhaps lose their jobs. This attitude implies a quite pessimistic and demeaning view of middle managers. In essence, top management is saying, "Middle managers are interested in just getting by, not in doing a good job. They are somewhat incompetent and generally lazy. They can't be trusted to make decisions in the best interest of the company. Therefore, we must impose tight budgets and coerce the managers into meeting these budgets." In such an environment, middle-level managers are motivated to avoid discipline and to retain their jobs, whatever actions that may require.

When budgets are being established, the middle-level manager is motivated to "sandbag" top management, to attempt to negotiate a budget that can be easily met. The profit center manager who thinks that a $1.5 million profit can be realized next year will negotiate for a budgeted profit of perhaps $1.3 million or $1.4 million. In turn, the manager's boss, suspecting sandbagging, reacts by pressing for a profit budget of greater than $1.5 million. While some tension in budget negotiations between two levels of management is both inevitable and healthy, carried to an extreme it leads to much loss of time, poor communication, mutual suspicion, and an unrealistic budget.

In this environment, financial reports, particularly those that compare actual results with the budget, are viewed as top management tools, not as information aids to help middle managers. The accounting department is viewed as a spying operation for top management, reporting financial data to permit top management to coerce, dominate, and discipline.

If top management views budgeting as a device to threaten or coerce mid-level managers, the latter will take actions to cause their operations to appear to fit the budget, to minimize negative variances and maximize positive variances. These actions may or may not be in the best long-term interests of the company. A middle manager may be tempted to charge certain expenses to the wrong account, to defer certain discretionary expenses such as maintenance or training, to shift expenses to another responsibility center, or to speed up the recognition of revenue. All these actions merely improve the *appearance* of the center's financial results; the company's fundamental economic position remains unchanged. These actions are less than honest and, moreover, wasted motion. If threatened sufficiently, most people will go to great lengths to doctor-up financial results, and a few will succumb to illegal actions. (Recall the "cooking the books" section in Chapter 11.)

Optimistic View

Alternatively, suppose top management views budgets and internal financial reports primarily as tools to aid managers at all levels in doing a better job of managing. This view derives from the more optimistic top management assumption that all employees are competent, are motivated to do a good job, and will respond constructively when they have information that suggests corrective action. Coercion is not required. Rather, the organization's goals and those of the individual are sufficiently congruent that mid-level managers will, by seeking to satisfy their personal needs, take actions that also satisfy the organization's needs.

As budgets are negotiated, mid-level managers in this more optimistic and supportive environment are willing to discuss openly the problems and opportunities facing the operation; they are confident that top management seeks realistic budgets. With the threat of discipline greatly lessened, managers are willing to take actions that are in the company's best long-run interests, even if these actions lead to negative expense or revenue variances in the short term. They welcome accurate financial reports from the accounting department, particularly those that compare budget and actual amounts, because such reports help them make better decisions. They are confident that thoughtful explanations of the variances (but not excuses) will be heard. The possibility is acknowledged that budgets may be unrealistic or accounting data inaccurate. All this communication takes place in an atmosphere free of threats or suspicion.

In summary, then, top managers can use budgets and variances analyses as a blunt club to force middle managers to take action, the motivation being to avoid discipline. On the other hand, they can be viewed as tools to facilitate communication and assist managers in running their operations, the manager's primary motivation being the desire to do a good job. Most mid-level managers prefer to operate in the second environment. In addition, there is good evidence that the budgeting process is much more efficient and useful when the management environment is supportive.

OTHER TYPES OF BUDGETS

Operating budgets are analogous to income statements: the budget is prospective, while the financial statement is historical. Just as there are other financial reports, so there are other types of budgets, some of which are analogous to the other financial reports.

Capital Budgets

The capital budget focuses on the longer-term need for the generation of investment capital. Capital budgets should be developed for a time frame of several years, typically five and in some cases 10, in order to provide plenty of advance warning of capital shortages or excesses. Raising additional capital and the judicious deployment of extra resources require considerable planning; once commitments are made, they cannot easily be changed. The capital budget takes into account not only the forecasted operating results for the company but also the many other decisions that do not immediately affect profit and loss; examples of these are the investment in or disposition of plant and equipment, the buildup or reduction in working capital as operations grow or shrink, the payment of dividends and the scheduled repayment of borrowings, and anticipated changes in the capital structure of the operation.

Long-Term Budgets

While discussion here has centered on budgets for one year and portions of a year, most companies also devote much attention to longer-term operating budgets. As part of the annual budgeting procedure, companies often develop budgets for each of the next three or five years, although typically in less detail than next year's budget. Focusing solely on a one-year budget, without considering the longer-term financial consequences of the year's operating decisions, can be dangerous. A five-year budget may reveal the need for additional facilities having a long construction lead time, or for additional equity or debt capital that may or may not be obtainable, or for a substantial increase in the sales force, or for the introduction of new products to replace those nearing the end of their life cycles.

Of course, a budget established now for a period beginning four years hence will be tentative at best. However, the budget will become firmer as it is revised in the course of each of the intervening annual budgeting cycles.

PRO FORMA FINANCIAL STATEMENTS

A natural extension of the detailed operational budgeting process is the development of **pro forma financial statements,** which are financial statements—income statements, balance sheets, and cash flow statements—projected for future periods as they will appear if the entity's plans are achieved. These statements are generally developed for a number of years into the future.

The several purposes of such statements include determining whether the company will need or accomplish the following:

- Meet its profit projections
- Generate cash flows sufficient to meet its investment needs, its debt re-payment requirements, and any planned dividend payments
- Evolve a capital structure consistent with the company's risk tolerances and its ability to borrow funds
- Need to raise additional long-term debt or equity capital
- Achieve returns on investment attractive to present and potential investors

Start-up companies—and start-up operations within a company—find pro forma statements essential elements of business plans that are developed to attract and convince investors.

The financial results for most enterprises—both for-profit and nonprofit—are driven most frequently by a relatively few key parameters. Making—and often remaking and testing—assumptions regarding those key parameters is fundamental to developing realistic pro forma statements. For example, the interaction of sales volumes and prices, and competitors' reactive price actions, are frequently key. Marketing decisions, such as selecting appropriate distribution channels and the type and extent of product promotion, in turn influence price and volume. Decisions regarding internal or subcontracted production influence the asset side of the balance sheet and, typically, margins as well. For financial institutions, prevailing interest rates in the capital markets may drive results. The mix of products sold and purchase payment plans utilized by customers (lease versus buy, for example) may drive results in other industries. Thus, forecasters should spend considerable time early in the process to define these key parameters and their interdependencies.

As with all budgeting processes, developing pro forma statements is typically an iterative process. Often, the forecaster has to alter underlying assumptions and then reproject the pro forma statements when the first iteration reveals problems in one of the five areas enumerated above. Moreover, planners often find it useful to produce multiple projections, perhaps "optimistic," "most likely," and "pessimistic." With the availability of computer-based spreadsheet programs, multiple iterations are relatively easy to produce.

Recall (from Chapter 9) that comparing financial results with those of similar companies in the same industry can be very revealing. The same can be said, of course, about pro forma statements. Industry benchmarks help financial forecasters assess the reasonableness of pro forma statements and may suggest sources of competitive advantage to differentiate the company from its competitors.

Getting Started

Where to begin? As with short-term budgeting, the key driver of long-term results and the future financial condition for the company is typically sales revenue. Sales projections deserve a great deal of attention, including convincing backup information. If the products or services produced are "me too" in nature—that is, they face considerable undifferentiated competition—the focus will be on the market share that can be captured by the company. When developing pro forma statements for a new entrant to the market, the forecaster needs to think hard about just how and to what extent the company can wrest market share away from entrenched competitors. If the product or service is thus far unknown to the market, the planner faces the greater difficulty of estimating the rate at which customers will accept it; the tendency is to overestimate the rapidity of acceptance. Customers can be slow to change their buying habits and are likely to be skeptical of the advantages claimed for the new product or service. But the reverse can also be true: occasionally demand for a new pharmaceutical drug, toy, or electronic gadget takes off in an unexpected rush, and an overly conservative producer of the new product, unable to meet the demand, gives up valuable market share to more nimble competitors. When a family of products or services is offered, the mix of business must be projected; product mix assumptions are particularly important when different products and services have substantially different margins or different marketing or engineering requirements.

A second key parameter is staffing. Salary costs (and costs related to salaries, referred to as *fringe benefits*) generally account for 50 to 80 percent of a company's total expenses. For small and start-up companies, the forecaster should identify, position by position, the timing and salary of each addition. For mature and relatively stable operations, the focus may be simply on incremental staffing changes, additions or layoffs. Because of the dominance of "people costs," forecasting attention focused on staffing pays off.

Finally, financial forecasting must be based on the enterprise's strategic plan, not the reverse. Of course, the strategy may have to be modified based on financial realities revealed by the pro forma statements, but it is a mistake to leave to the financial forecasters the task of defining the company's strategy.

A discussion of corporate strategy considerations is well beyond the scope of this book; nevertheless, I will outline just a couple of key strategic decisions that a new company might face, each having major financial implications:

- Is the company to be a technical leader or a technical follower? Surely, the answer will greatly affect expenditures on research and development. If a follower, will the company rely largely on licensed intellectual property?
- Does the company seek to be the low-cost producer in the field? If so, it will probably have to capture a major share of the market in order to

drive down its costs. Or, alternatively, will it seek to earn high margins by remaining on the cutting edge of product features offered? If the latter, the company is likely to have short product life cycles and must assemble a rapid-response product development team.

- Will the company develop its own field sales force, or sell through other channels of distribution such as agents, distributors, or other manufacturers?
- To what extent will the company look to vendors to manufacture or service its products?
- What are the risk tolerances of the owners and managers? Are they tolerant of—or perhaps even eager for—high debt leverage? Do they prefer a substantial safety stock of cash, or are they comfortable with a cash-lean operation? Is reliance on a single product acceptable, or are the owners and managers eager to spread their market risks across many products?
- Will the owners tolerate substantial equity dilution? If not, the company's strategy had better aim at moderate, not hyper, growth.

I could go on, but you probably get the idea!

Developing a Pro Forma Operating Statement

Now, to the details. One can talk endlessly about the general strategy of the firm, but financial forecasting requires the planning to become much more explicit. As with short-term budgeting, the tough part of forecasting is detailing operating plans sufficiently that they can be translated into dollar terms.

Let's work through an example for a start-up operation we'll call the Lehman Company. Lehman is founded on technology, now well protected by patents, for a new analytical instrument for use primarily in industrial laboratories. While Lehman plans to introduce only a single model to begin with, in time Lehman wants to offer both simpler and more sophisticated versions of the instrument. Given the high-tech nature of the product, Lehman is convinced it will have to deploy its own field sales force. Customer training will be important. Manufacturing will consist largely of assembly activities, because Lehman is located in a region with well-developed, low-cost, high-quality, reliable manufacturing subcontractors and because the instrument incorporates many standard subassemblies available from others.

The new instrument, incorporating new technology, will, Lehman believes, both directly compete with existing instruments and also open new opportunities not now addressed by others. The company founders have undertaken substantial, although informal, market research that serves as the basis of Lehman's sales forecasts. Lehman believes it can demand a price premium over the competition of about 20 percent, but that premium may erode as competitors react to Lehman's new offering. The timing of new model introductions is uncertain. Lehman's sales projections are shown in Table 15-4.

TABLE 15-4. Lehman Company: Sales Projections, 2004–2008

Product	2004	2005	2006	2007	2008
Model SX:					
Units	5	20	40	55	75
Price/unit ($000)	$160	$160	$155	$145	$135
Sales ($000)	$800	$3,200	$6,200	$7,975	$10,125
Model SY:					
Units			10	25	35
Price/unit ($000)			$120	$125	$130
Sales ($000)			$1,200	$3,125	$4,550
Model SZ					
Units				5	15
Price/unit ($000)				$210	$210
Sales ($000)				$1,050	$3,150
Total Sales	$800	$3,200	$7,400	$12,150	$17,825
% growth over previous year	—	300%	131%	64%	47%

Lehman's financial forecasters will want to test many assumptions regarding this sales projection: the rate of sales growth of the first instrument offered; the mix of "direct competition sales" and "new opportunity sales"; the 20 percent price premium and the rate of its erosion; the timing and selling prices of new instrument models to be introduced in future years.

Next, the cost of goods sold. Lehman may have the advantage of obtaining reliable price quotations from both producers of subassemblies incorporated in its new instrument and from manufacturing subcontractors in the area. Lehman still must estimate the costs associated with receiving, incoming quality inspection, scheduling, assembly setup, rental of the assembly space, employee training, maintenance and repair, power, assembly supervision—all manufacturing overhead costs—to say nothing of the direct labor time required to assemble, calibrate, test, and ship the instrument. Lehman may want to develop standard variable costs (or perhaps simply standard prime costs) for each model to be introduced; this information will help the forecasters do various what-if testing of the pro forma statements. Table 15-5 shows both the standard prime cost for each of the anticipated products and the fixed manufacturing overhead as it increases with increased manufacturing capacity.

Turning now to operating expenses, we need to consider the details of staffing:

- How many engineers and support personnel will the development function require? When will they come on board? and at what salary?
- How many field salespersons will Lehman need to cover the territory? "Subquestions" must be answered: How widely dispersed are the customers? How long is the "selling cycle"? Where will the field salespeople be located? When will they join the company (timing should allow

TABLE 15-5. Lehman Company: Projections to Determine Cost of Goods Sold

	Standard Product Costs (Prime Cost in $000)		
	Direct Labor	Direct Material	Total
Model SX	$15	$60	$75
Model SY	12	50	62
Model SZ	18	80	98

	Manufacturing Overhead ($000)				
	2004	2005	2006	2007	2008
Supervision	$ 90	$145	$205	$ 255	$ 275
Production control	35	50	120	205	215
Manufacturing engineering	55	125	185	215	215
Occupancy	45	50	135	180	205
Depreciation	10	30	80	140	180
Other	10	10	30	65	75
Total overhead	$245	$410	$755	$1,060	$1,165

them to be trained before they call on customers)? And what combination of salary and commission will constitute their compensation package?

• What set of functions are lumped under "general administrative?" Probably they should include executive management (chief executive officer, chief financial officer, chief scientific officer, chief information officer, and others), departments of accounting, human resources, and facilities.

Lehman should be able to count heads in each function and estimate both salaries and hire dates.

But, of course, there are nonsalary operating expenses as well. Here some rules of thumb may be useful. Pretty good data exist on the nonsalary costs of keeping a salesperson in the field, and the experience of other instrument development departments may help determine the nonsalary overhead in the development and engineering department (perhaps as a percentage of total salaries of the engineers). Within the administrative functions, Lehman might think about contracting out the payroll function, and perhaps administrative computing and human resources as well.

Lehman should also add a contingency allowance; despite best efforts, planners are almost bound to forget something. A contingency of 5 to 10 percent of expenses may be wise.

Wrapping all this together, Lehman might arrive at the set of assumptions shown in Table 15-6, assumptions that affect both the income and balance sheet. The pro forma income statements are shown in Table 15-7. Because at this point we don't have much basis for estimating expenses below the "operating profit" line—interest expense and income and income tax expense—

TABLE 15-6. The Lehman Company: Other Assumptions

1. *Production control and manufacturing engineering:* must grow substantially in advance of new product introductions.
2. *Facilities:* (leased) expansion in 2006.

3. *Staffing:*	2004	2005	2006	2007	2008
Development engineers	2	3	5	8	9
Manufacturing engineers	1	2	2	2	2
Field sales force	3	3	6	8	10
4. *Capital equipment ($000):*					
Leasehold improvements	$100	$200	$ 400	$100	$200
Manufacturing equipment	200	100	1,000	300	500
Office and sales equipment	200	100	500	150	150
	$500	$400	$1,900	$550	$850
5. *Prepaid expenses ($000):*	$100	$200	$ 400	$450	$450
6. *Ratios (year-end):*					
A/R collection (days)	63	60	57	53	50
Inventory flow (days)[a]	60	70	90	80	70
A/P payment (days)[a]	45	45	40	40	40
Other accruals:			equal to A/P		

7. *Other:*
  a. Cash reserve: 5% of next year's projected sales. Excess cash from financing held as cash equivalent.
  b. Short-term bank loans available from 2006 in amounts up to 70% of A/R.
  c. Interest income at 4%; interest expense at 9%.

[a]Days of standard production prime costs.

let's focus, for the moment, only on the numbers above that line. We'll need to work though the balance sheet projections and make some decisions about capital structure before we can get a good handle on these nonoperating items.

Developing a Pro Forma Balance Sheet

Now consider the investment in assets required to support the revenue and expense projections made so far. These assets have to be financed by a combination of current liabilities that arise from the operations, borrowing (perhaps both short-term and long-term), and owners' equity (stock purchased by founding and subsequent investors and, in time, retained earnings.)

Let's return now to the ratios discussed in Chapter 9. Again, benchmarks in the instrument industry are key: accounts receivable collection periods (any reason that Lehman's should be longer or shorter?); inventory turnovers (perhaps Lehman can accelerate its turnover because of its reliance on vendors); accounts payable turnover (perhaps as a function of the cost of goods sold or inventory); and accrued liabilities (perhaps as a function of total salary expense, as a substantial portion of accrued liabilities is salary related.)

TABLE 15-7. Lehman Company: Pro Forma Income Statement: 2004–2008 ($000)

	2004	2005	2006	2007	2008
Sales	$ 800	$3,200	$7,400	$12,150	$17,825
Cost of goods sold	375	1,500	3,620	6,165	9,265
Fixed manufacturing cost	245	410	755	1,060	1,165
Subtotal	$ 620	$1,910	$4,375	$ 7,225	$10,430
Gross margin	$ 180	$1,290	$3,025	$ 4,925	$ 7,395
Percent of Sales	23%	40%	41%	41%	41%
Contribution percent of sales[a]	53%	53%	51%	49%	48%
Operating expenses:					
Development and engineering	235	410	675	1,010	1,335
Sales and marketing	360	400	800	1,165	1,430
General administrative	480	530	610	815	1,020
Total operating expenses	1,075	1,340	2,085	2,990	3,785
Operating profit	($ 895)	($ 50)	$ 940	$ 1,935	$ 3,610
Percent of sales	neg	neg	12.7%	15.9%	20.3%
Nonoperating income and expense:[b]					
Interest income	40	56	26	—	—
Interest expense	—	(40)	(85)	(76)	(67)
Profit before taxes	(855)	(34)	881	1,859	3,543
Income taxes	—	—	—	743	1,417
Net income	($ 855)	($ 34)	$ 881	$ 1,116	$ 2,126

[a]Contribution percent of sales $= \dfrac{\text{Sales} - \text{Cost of goods sold}}{\text{Sales}}$

[b]These nonoperating incomes and expenses projections utilize the balance sheet projections contained in Table 15-8.

Just as hard and explicit thinking about staffing requirements is required, so too, regarding fixed-asset investments. Will Lehman purchase or lease its facilities? (Most new companies lease.) How much excess square footage should it acquire at the outset? And what will these facilities cost (depending, in part, on how long a lease Lehman is willing to sign)? What equipment will be required for manufacturing? for the sales force (vehicles, laptop computers, fax machines, and so on)? the engineering department?

As mentioned in Chapter 9, accounts receivables, inventories, and fixed assets comprise the lion's share of the total assets. Nevertheless, cash, prepaid expenses, and intangibles are also important. In the early years, Lehman will probably need to keep reasonable cash reserves (perhaps including some short-term cash-equivalent investments), since cash inflows from customers will be minimal and perhaps irregular. As the company matures, its need for cash reserves should diminish. If Lehman relies to any significant extent on acquired intellectual property, investments in these intangibles may be significant. As prepaid expenses are likely to be minor, we need not waste a lot of time forecasting them.

The lower portion of Table 15-6 lists some key assumptions used in forecasting the "assets" and "current liabilities" portions of the balance sheet (shown as Table 15-8). Before turning to consider the appropriate capitalization of Lehman, we should review for reasonableness the projections made to date. Noteworthy observations are:

- Lehman's projected contribution margin exceeds 50 percent, a very healthy but not extraordinary margin, if, in fact, Lehman can capture technological leadership of its industry segment.
- Experience tells us that technology-based businesses often spend 10 percent of their revenue on research and engineering, and in some cases more, particularly in the companies' early years, when sales are building. Given Lehman's technology strategy, we might expect development and engineering expenses to exceed 10 percent of revenue in the early years but then to settle down to about that level after four or five years. This pattern prevails in these projections.
- We can also size up selling expenses vis-à-vis the competition. The type of capital equipment that Lehman intends to market typically requires numerous customer calls to close a sale, and thus selling costs will be

TABLE 15-8. Lehman Company: Projections of Required Permanent Capital, December 31, 2004–08 ($000)

	2004	2005	2006	2007	2008
			Assets		
Current assets:					
Cash	$ 96	$ 222	$ 365	$ 535	$ 535
Accounts receivable	138	526	1,156	1,764	2,442
Inventory	62	288	893	1,351	1,777
Prepaid expenses	100	200	400	450	450
Subtotal	396	1,236	2,814	4,100	5,204
Fixed assets	480	840	2,600	2,900	3,500
Total	$876	$2,076	$5,414	$7,000	$8,704
		Liabilities & Owners' Equity			
Current liabilities:					
Accounts payable	$ 46	$ 185	$ 397	$ 676	$1,015
Other accrued liabilities	46	185	397	676	1,015
Bank loan payable	—	—	809	1,235	1,709
Subtotal	92	370	1,603	2,587	3,739
Permanent capital required	$784	$1,706	$3,811	$4,413	$4,965

relatively high. We might, in fact, question whether these projections shown an adequate allowance for selling expenses.

- On the balance sheet, the projected current ratio is very high in the early years, as one would expect due to extra cash reserves, but then settles down to reasonable levels (adequate liquidity.)

- Projected total asset turnover is high, once Lehman reaches reasonable sales levels. Given its reliance on subcontractors, this turnover rate may be reasonable, but the forecasters should double-check that the accounts receivable and inventory turnovers are reasonable, and that adequate allowance has been made for fixed-asset investments.

Note the bottom line of Table 15-8: "permanent capital required." These amounts equal total assets minus current liabilities. Permanent capital will come from three sources:

Long-term debt
Invested equity capital
Retained earnings

By reviewing the operating profit projections, assuming an income tax rate of 40 percent, and ignoring interest income and interest expense, we can see in Table 15-7 that the change in retained earnings will be negative for the first two years but will be about $1 million in 2007 and $2 million in 2008. Since the incremental increases in permanent capital in 2007 and 2008 (Table 15-8) are less than these incremental increases in required retained earnings, a reasonable assumption is that Lehman will be self-financing after three years. Accordingly, we can focus on how Lehman can raise $3.8 million of permanent financing, the amount shown on Table 15-8 at the end of 2006.

Now, let's try some financing scenarios and see how they work out in these pro forma projections. Assume that the Lehman Company will sell $1.5 million of common (or a combination of common and preferred) stock in late 2003, coincident with the start of operations, and another $1.0 million a year later. Further, assume that Lehman plans to borrow on a long-term basis, $250,000 in 2004, another $250,000 in 2005, and $500,000 more in 2006; the collateral for this long-term borrowing will probably be fixed assets and the borrowing may well be in the form of "financing leases" of equipment.

Now we have the necessary assumptions to complete the pro forma income statements (the portion of Table 15-7 below the "operating profit" line). Interest expense in 2004 is zero (since there won't be any borrowing), but interest revenue will be at the rate of 4 percent of the cash raised from the sale of stock and not yet invested in other assets or spent on start-up expenses; this excess cash should average about $1.0 million ($1.5 million on January 1, 2004, and about $500,000 at year-end.) Income tax expense is, of course, zero as Lehman will record a loss in 2004. Note that this loss can be carried forward, thus reducing income taxes in 2006, the first year Lehman is expected to be profitable; accordingly, the income tax expense for 2006 will be eliminated by this carry-forward, and the tax rate of 40 percent of profit before tax first applies beginning in 2007.

The bottom line of Table 15-7 shows net income. Since Lehman plans to pay no dividends during this time period, these amounts add directly to retained earnings. We are now in a position to develop a capital structure for Lehman (Table 15-9). Since, in 2007 and 2008, the permanent capital available is greater than the requirements shown on Table 15-8, the difference is shown as "excess cash available" at the bottom of Table 15-9. (For simplicity sake, and because the amounts are small, interest earnings on this excess cash are ignored in the pro forma income statements.)

The projections in Table 15-9 indicate that the financial structure suggested earlier doesn't quite work. In 2006, Lehman will have a deficiency of $369,000 in "excess" cash, an amount approximately equal to its projected "working" cash at that time, namely $365,000. Most would agree that this scenario is too close for comfort. The Lehman Company should raise additional permanent capital at the company's founding or soon thereafter.

Deriving the Cash Flow Statement

With the five-year projections of the income statement and the balance sheet completed, we can readily derive the cash flow statement in its conventional format (Table 15-10).

Modeling and Sensitivity Analysis

The projections illustrated in Tables 15-7 through 15-10 were arrived at by a rather brute-force method. To facilitate the analysis of the financial dynamics

TABLE 15-9. The Lehman Company: Capital Structure, 2004–2008 ($000)

	2004	2005	2006	2007	2008
Permanent Capital Required (from Table 15-8)	$ 784	$1,706	$3,811	$4,413	$4,965
Capital structure:					
Long-term debt	$ 250	$ 475	$ 950	$ 850	$ 750
Shareholders' equity:					
Capital stock	$1,500	$2,500	$2,500	$2,500	$2,500
Retained earnings	(855)	(889)	(8)	1,108	3,242
Subtotal	645	1,611	2,492	3,608	5,742
Total	$ 895	$2,086	$3,442	$4,458	$6,492
Cash excess (deficiency)	$ 111	$ 380	($369)	$ 45	$1,527

TABLE 15-10. The Lehman Company: Pro Forma Cash Flow Statements, 2004–2008 ($000)

	2004	2005	2006	2007	2008
Cash flow from operations:					
Net income	($ 855)	($ 34)	$ 881	$1,116	$2,126
Depreciation	20	140	240	300	350
Change in working capital	(304)	(562)	($345)	(302)	48
Subtotal	($1,139)	($ 456)	$ 776	$1,114	$2,524
Cash flow for investing:					
Fixed assets	($ 500)	($ 500)	($2,000)	($ 600)	($ 950)
Cash flow from financing:					
Long-term borrowing	$ 250	$ 250	$ 500	—	—
Debt repayment	0	(25)	(25)	(100)	(100)
Sale of equity	1,500	1,000			
Dividends payments	—	—	—	—	—
Subtotal	$1,750	$1,225	$ 475	($100)	($100)
Excess Cash	$ 111	$ 269	$ 749	$ 414	$1,474
Cumulative cash excess[a]	$ 111	$ 380	(369)	$ 45	$1,519

[a] Checks with the bottom line of Table 15-9.

of Lehman's business, a relatively simple computer-based model may be worth building. Such a model can build in the ratios that we used and also codify some of the key relationships between assumptions. For example, non-operating interest income and expense on the income statement will be functions of cash reserves and borrowing on the balance sheet. Similarly, sales expenses (compensation) will be directly affected by sales volume.

With a computer-based model, sensitivity analysis becomes reasonably straightforward. Earlier we reviewed a number of assumptions built into the sales projections that should be tested. For example, what is the effect on net income of a faster erosion in the price premium than Lehman is now counting on? Other sensitivities—or what-ifs—that might be tested include the following:

- If the complexity of model SX leads to more time spent on calibration and quality control, what will be the effect on gross margin and net income?
- If the competition drops prices when Lehman introduces model SX, how much reduction in either sales volume or unit price can Lehman tolerate and still break even (that is, have zero profit) in 2006?
- If product development delays postpone the introduction of model SX by one year, and other expenses are not reduced, how much additional capital will Lehman require?

- If the ramp-up in sales of model SX is either faster or slower than projected, what will be the effect on Lehman's net income and on its capital requirements?
- If Lehman is accused of infringing on another company's patent and decides to settle with the accuser by paying a royalty at the rate of 3 percent of sales, what is the effect on net income and on borrowing requirements?

And so forth, and so on!

CASH BUDGETING

The pro forma cash flow statements in Table 15-10 confirm Lehman's cash positions, but only at each year-end. A more fine-grained projection is often required to reveal cash positions at various points during the year. This need is particularly critical in seasonal businesses, where the seasonal generation of cash is not coincident with the seasonal pattern of cash usage.

Here's a simple example: Suppose the Phelps Company is a wholesaler that experiences its busiest season in the summer, June through August. Phelps must build up its inventory in advance of the summer season and double its staff during the three-month busy season. The company's accounts receivable collection period is 60 days (for example, March sales are collected in May), and its accounts payable payment period is 30 days (that is, in any particular month, payments to vendors equal merchandise received during the previous month). Salaries and all other expenses must be paid in cash during the month in which they are incurred. To simplify matters, assume the company is not growing from year to year.

Table 15-11 shows the month-by-month income statement for Phelps and a notation of the monthly merchandise received. The company expects an operating profit of $106,000 and (ignoring income taxes, new investments in long-term assets, and long-term financing transactions) cash generation of the same amount—see the "cumulative cash inflow" for December. Nevertheless, during five of the 12 months (March through July), the company has a negative cash flow, and a peak cumulative cash need of $104 at the end of July—right in the middle of its peak season when its monthly profitability is at its highest.

Thus, Phelps must either begin the year with working cash reserves of at least $104,000 or arrange for seasonal borrowing during its peak season—despite the fact that it expects to be very profitable during this upcoming year and to generate $106,000 of cash from operations.

SUMMARY

Budgets, an integral part of the planning process, provide a road map in financial terms. They facilitate the process of managing by exception.

TABLE 15-11. The Phelps Company: Short-Term Cash Budgeting ($000)

	Jan.	Feb.	Mar.	Apr.	May	June	July	Aug.	Sept.	Oct.	Nov.	Dec.	Total for Year
Sales	$50	$50	$50	$50	$70	$100	$100	$100	$80	$50	$50	$50	$800
Cost of goods sold	25	25	25	25	35	50	50	50	40	25	25	25	400
Salary expense	10	10	10	10	10	20	20	20	10	10	10	10	150
All other expenses	12	12	12	12	12	12	12	12	12	12	12	12	144
Operating profit	$3	$3	$3	$3	$13	$18	$18	$18	$18	$3	$3	$3	$106
Purchases ($000)[a]	25	50	50	70	40	40	20	20	20	20	20	25	400
						Cash Budget							
Cash receipts	50	50	50	50	50	50	70	100	100	100	80	50	—
Cash outflows													
Payment for purchases[a]	25	25	50	50	70	40	40	20	20	20	20	20	—
Salaries	10	10	10	10	10	20	20	20	10	10	10	10	—
All other expenses	12	12	12	12	12	12	12	12	12	12	12	12	—
Total outflows	47	47	72	72	92	72	72	52	42	42	42	42	—
Net cash inflow (outflow)	3	3	(22)	(22)	(42)	(22)	(2)	48	58	58	38	8	—
Cumulative cash inflow (outflow)	$3	$6	($16)	($38)	($80)	($102)	($104)	($56)	$2	$60	$98	$106	—

[a]Purchases received during month and paid for during the following month.

Key guidelines for establishing and using budgets are the following:

1. Objectives, generally both quantitative and qualitative, must be set before budgets are established.
2. Budgeting is only one element of planning. Much important planning cannot be translated into monetary terms.
3. Planning, including budgeting, is typically an iterative process; successive revisions occur until the plans and budgets are mutually consistent and are compatible with both the objectives and resources of the organization.
4. Budgets for an operation should be built from the bottom up—that is, from segment budgets—and not imposed from top management down.
5. While the setting of revenue and expense budgets must go hand in hand, typically the cornerstone of the process is the sales or revenue forecast.
6. Historical accounting data provide a basis for budgets. Nevertheless, past expenditure levels should not simply be incremented to arrive at budgets. Zero-base budgeting is a useful technique.
7. The budgeting process involves negotiation between levels of management. The important end result of these negotiations should be commitment on the part of each manager.

Budgets are organized by responsibility center. Typically, as a manager ascends the management hierarchy, he or she takes on broader responsibility centers—from cost centers to profit centers, and, finally, at the division manager level, to investment centers.

The three primary reasons to budget are (1) to cause explicit planning, (2) to communicate plans, and (3) to provide a basis for comparing actual and planned results.

Developing and analyzing variances facilitate management by exception—that is, focusing attention on areas of the business requiring corrective action. A variance provides a systematic comparison between budgeted (planned) and actual results. Beware the pitfalls of automatically treating debit variances as unfavorable and credit variances as favorable.

The establishment and use of budgets is replete with human behavior considerations. Enlightened and supportive managements view budgets, and subsequent performance reporting, as tools to permit managers at all levels to do a better job, rather than as devices to coerce, control, and discipline lower-level managers.

In addition to near-term operating budgets, several other types of budgets are useful planning documents: (1) capital budgets; (2) operating budgets covering longer periods, say, three- to five-year budgets; (3) pro forma (estimated for the future) financial statements; and (4) short-term (often month-by-month) cash budgets.

NEW TERMS

Budget. A description in monetary terms of the organization's plan.

Budget variance. The difference between an actual financial result (revenue or expense) and the planned, or budgeted, outcome.

Cost center. A responsibility center whose manager is accountable for the control of expenses but not revenues or investments.

Favorable variance. Credit balances in variance accounts. Contrasts with *unfavorable variance.* An alternative name is *positive variance.* The value judgments implied by these terms can be misleading.

Flexible budget. A form of budget wherein variable expenses are budgeted in relation to activity volumes (often stated as percentage of revenue) and fixed expenses are stated in absolute dollar values.

Goal congruency. The alignment of the goals (objectives) of the individual manager with those of the overall organization.

Investment center. A responsibility center whose manager is accountable for investment in assets as well as for profit (revenues and expenses) and for which performance may be evaluated in terms of the return on investment.

Management by exception. A practice of focusing management attention primarily on those segments of an operation that are not proceeding according to plan.

Negative variance. A debit balance in a variance account. An alternative name is *unfavorable variance.* Contrasts with *positive variance.* The value judgments implied by these terms can be misleading.

Positive variance. An alternative name for *favorable* (credit) *variance.* Contrasts with *negative variance.*

Pro forma financial statements. Financial statements that are projected, or estimated in advance.

Profit center. A responsibility center whose manager is accountable for both revenues and expenses, and therefore for profit, but not for investments.

Responsibility center. A segment or unit of a business or other organization for which a single manager can be held accountable.

Unfavorable variance. A debit balance in a variance account. Contrasts with *favorable variance.* An alternative name is *negative variance.*

Volume-adjusted budget. A budget that has been revised to reflect the actual volume of activity during the accounting period. Volume-adjusted budgets are particularly useful when both fixed and variable cost elements are present.

Zero-base budgeting. A budgeting technique that requires justification for each dollar to be spent, rather than simply justification for expenditure increases or decreases relative to the previous period.

EXERCISES

1. What are the primary reasons for budgeting?

2. What is meant by "top-down" versus "bottom-up" budgeting?

3. What does "commitment" refer to in the context of budgeting?

4. Explain why budgeting is generally an iterative process.

5. If you believe your job will be threatened if your profit center fails to achieve your budget for three successive quarters, what actions might you take to help assure that you meet budget? Are these actions ethical? legal?

6. Describe a company situation for which a month-to-month cash budget (or forecast) would seem necessary, in addition to an operating budget and pro forma financial statements.

7. Indicate whether each of the following statements is true of false.

   a. Debit (so-called negative) variances always indicate the need for corrective action.

   b. Flexible budgeting permits calculation of volume-adjusted budgets to which actual results can be compared.

   c. The manager of an expense center is responsible for controlling expenses but not revenues or investments.

   d. The primary purpose of budgeting is to supply information to top managers so that they can take corrective action swiftly.

   e. Pro forma statements refer to financial statements that are adjusted following the end of an accounting period.

   f. Managers should focus their attention solely on debit (so-called unfavorable) variances.

   g. The director of budgeting has the sole responsibility for developing appropriate budgets for the company.

   h. Goal congruency refers to the alignment between an individual's goals and those of his or her company.

   i. Flexible budgeting permits responsibility center managers to adjust their budgets as market conditions change.

   j. Budgets are valuable communication tools within an organization.

   k. Just as financial statements provide a complete historical record for an operation, so budgets articulate detailed plans for that operation.

8. Insert the word or phrase that best completes the following statements:

   a. The three types of responsibility centers are expense, _____, and investment centers.

   b. Variance analysis facilitates management by _____.

c. Investment centers can be evaluated by reference to _____.

d. Cash budgeting is particularly important in _____ businesses.

e. A budgeting technique that requires justification for each dollar to be spent, rather than simply justification for increments in expenses, is known as _____ budgeting.

f. In a relatively mature and stable business, budgets should probably be revised only _____.

g. In a start-up operation, pro forma statements are particularly helpful in forecasting the need for _____.

h. Flexible budgeting permits one to compare actual results to _____ budgets.

9. If you were designing the budgeting and financial reporting system for a large university, indicate which of the following units you would classify as cost, profit, and investment centers:

a. The mathematics department

b. The student union

c. The university-owned bookstore

d. The copy center, which charges other units for reproduction work that it performs

e. The campus safety force (campus police)

f. The ticket (box) office within the student union

g. The intercollegiate athletics department

10. Henry Corporation expects its revenues to grow 20 percent to $2.4 million next year, with a gross margin of 36 percent. Its working cash balance and prepaid expenses will remain at $150,000 and $30,000, respectively. The accounts receivable collection period should remain at 60 days, and inventory turnover at six times per year.

a. Develop a pro forma projection for Henry's current assets at next year-end.

b. If total-asset turnover is expected to be 1.1, and the current ratio 1.6, how much permanent capital (long-term debt plus owners' equity) will the company need at the end of the year?

c. If Henry wants to maintain a long-term debt to owners' equity ratio of not more than 1, what is the maximum long-term borrowing it can accommodate?

11. If the Sunwoo Company projects the need to grow its assets by $400,000 next year, no change in its liabilities (both short-term and long-term), net income of $600,000 next year, and a 25 percent dividend payout ratio, describe its financial position as of the end of next year.

12. What is the break-even revenue for an operation that has the following flexible budget?

$$\text{Expenses} = 0.72 \times (\text{revenue}) + \$48,000$$

13. If a company's projected sales for next year are $420,000 with no seasonality in sales and it's year-end A/R collection period is expected to remain steady at 45 days, what should the pro forma year-end balance sheet show for the Accounts Receivable balance?

14. The Branner Coffee Shop budget-versus-actual performance report for a particular fiscal quarter is as follows:

	Actual	Budget	Variance
Revenue	$142,500	$150,000	($7,500)
Cost of goods sold	103,000	110,000	7,000
Salaries and wages	26,000	28,500	1,500
Utility costs	5,500	5,300	(200)
Rent	3,000	3,000	—
Promotion costs	4,000	6,000	2,000
Profit	$1,000	($2,800)	$2,800

   a. If cost of goods sold is a fully variable cost, salaries and wages are described by the formula $6,000 + 0.15 (revenue), and all other costs are fixed, develop an actual-versus-*flexible* budget for Branner for the period.

   b. What advice would you give the manager about the coffee shop's performance over the period?

   c. According to its budget, what is Branner's break-even revenue?

   d. If Branner's building rental agreement were changed from a fixed $3,000 per month to 2 percent of revenue, what would be Branner's new break-even volume?

15. You are asked to prepare a pro forma income statement and balance sheet for 2006 for Omar Corporation in as much detail as possible, given the following information:

Projected sales	$780,000
Gross margin	40%
Operating expenses	$260,000
Non-operating income/expense	—
Tax rate	30%
Accounts receivable collection period, year-end	40 days
Inventory flow period, year-end	80 days

Cash	10% of revenue
Prepaid expenses at year-end, 2006	$30,000
Current ratio at year-end, 2006	2.0
Total asset turnover at year-end	1.3
Retained earnings (beginning of 2006)	$214,000
Dividend payout ratio	50%
Long-term debt to owners' equity ratio	0.5

16. Prepare a cash budget for Syed Corporation's busy retail season, October through January, given the following information:

Cash at September 30	$60,000
Monthly sales, October through January (up from $50,000 per month for the last three months)	$100,000
Salaries, payable in the current month	$30,000
Merchandise receipts (credit purchases) (down from $60,000 per month for the last three months)	$40,000
Other expenses, payable in the current month	$10,000
Accounts receivable collection period	45 days
Accounts payable payment period (on merchandise receipts)	30 days

a. What will Syed's cash balance be at January 31?

b. Does the company have sufficient cash to get through the busy season?

INDEX